Ceramic Conductors

Ceramic Conductors

Special Issue Editors

Maria Gazda
Aleksandra Mielewczyk-Gryń

MDPI • Basel • Beijing • Wuhan • Barcelona • Belgrade

MDPI

Special Issue Editors

Maria Gazda
Gdansk University of Technology
Poland

Aleksandra Mielewczyk-Gryń
Gdansk University of Technology
Poland

Editorial Office
MDPI
St. Alban-Anlage 66
4052 Basel, Switzerland

This is a reprint of articles from the Special Issue published online in the open access journal *Crystals* (ISSN 2073-4352) from 2018 to 2019 (available at: https://www.mdpi.com/journal/crystals/special_issues/Ceramic_Conductors).

For citation purposes, cite each article independently as indicated on the article page online and as indicated below:

LastName, A.A.; LastName, B.B.; LastName, C.C. Article Title. *Journal Name* **Year**, *Article Number*, Page Range.

ISBN 978-3-03897-956-2 (Pbk)
ISBN 978-3-03897-957-9 (PDF)

Contents

About the Special Issue Editors

Maria Gazda is a Professor at the Department of Applied Physics and Mathematics, Gdańsk University of Technology (GUT), Poland. She has been working with various conducting oxide glasses and ceramics since 1980, while preparing her MSc thesis. Later, she worked with ceramic high-temperature superconductors, glass–ceramic granular conductors and superconductors, semiconducting oxides, and ceramic ion conductors. Currently, her main research interests are focused on proton-conducting, oxygen ion-conducting, and mixed electronic ion-conducting oxides.

Aleksandra Mielewczyk-Gryń is an Assistant Professor at the Department of Applied Physics and Mathematics, Gdańsk University of Technology (GUT), Poland. She earned her doctoral degree in Physics in 2013 from the same University and conducted postdoctoral research at the University of California, Davis (2013–14). Her main research interests are functional materials for energy technologies, especially ion-conducting materials for solid oxide fuel cells and solid oxide electrolyzers.

Preface to "Ceramic Conductors"

Ceramic conductors are fabricated throughout the world by both industry and researchers. They are one of the most important types of materials for various technologies, especially those related to energy harvesting. The research on their properties is therefore crucial for the development of modern society. This Special Issue covers research on various ceramic materials from both the experimental and theoretical points of view.

Maria Gazda, Aleksandra Mielewczyk-Gryń
Special Issue Editors

crystals

MDPI

Editorial
Ceramic Conductors

Maria Gazda * and **Aleksandra Mielewczyk-Gryń**

Faculty of Applied Physics and Mathematics, Gdansk University of Technology, 80-233 Gdansk, Poland; alegryn@pg.gda.pl
* Correspondence: maria.gazda@pg.edu.pl

Received: 21 March 2019; Accepted: 21 March 2019; Published: 25 March 2019

For more than 4000 years, mankind has used and developed ceramics. Starting from basic sun-dried pots in Neolithic times, ceramics have evolved through Medieval clay sculptures to high-temperature superconductors in modern times. Nowadays, it is simply impossible to overestimate the importance of ceramic materials. Ceramics have been traditionally considered to be electrically insulating. Within this issue, only $NiCr_xFe_{2-x}O_4$ studied by Lin et al. [1] may be considered as an insulator, or rather n-type semiconductor; however, this material exhibiting interesting magnetic properties is far from being a traditional ceramic. Nevertheless, several groups of modern advanced ceramics are electrically conducting. Among them, electronic-, ionic-, as well as mixed electronic-ionic-conducting ceramics are very important groups of materials. Proton conductivity may be observed in acceptor-doped perovskite oxides such as $Ba_{0.9}La_{0.1}Zr_{0.25}Sn_{0.25}In_{0.5}O_{3-a}$, as studied by Skubida et al. [2], and terbium-substituted lanthanum orthoniobate [3]. Oxygen ions are mobile charge carriers in materials such as substituted bismuth vanadate [4] and doped cerium oxide [5,6], as studied by Ring and Fuierer, and to a minor extent and at high temperature, they are also mobile charge carriers in ceramic proton conductors. Fluorine ions, the first mobile ions studied in solid-state ionic conductors, are present in SrF_2-YF_3 solid solutions, as reported by Breuer et al. [7]. Finally, electronic-type charge carriers dominate in such ceramics as donor-doped and reduced strontium titanate, as reported by Presto and collaborators [8].

The electric and electrochemical properties of conducting ceramics, apart from their chemical composition, strongly depend on the material morphology, micro- or nanostructure, and porosity. This means that the properties may be modified and optimised by the proper choice of fabrication method. The most often used method of ceramic materials preparation, which is usually used as a first-trial method, is the solid-state reaction method. This method was used for the preparation of doped barium indate $Ba_{0.9}La_{0.1}Zr_{0.25}Sn_{0.25}In_{0.5}O_{3-a}$ [2], terbium-doped lanthanum orthoniobate [3], and doped strontium titanate [8]. On the other hand, the solid-state reaction route often does not allow the achievement of single-phase ceramics with expected properties. One of the interesting synthesis methods which may be used for manufacturing either fine ceramic powders [4] or even single crystals is molten-salt synthesis. Also, sol–gel self-combustion synthesis [5] and spray-pyrolysis [8] lead to the formation of nanosized ceramic powders. On the other hand, the application of mechanosynthesis not only produces nanoceramic powder but also facilitates the formation of phases which do not form with the use of other methods. By employing mechanosynthesis, a single-phase solid solution of YF_3 and SrF_2 was obtained before the $Sr_{0.7}Y_{0.3}F_{2.3}$ composition [7].

Moreover, the variety of phenomena related to ion and electronic transport in ceramics render them very interesting for applications. Indeed, these materials have been applied in gas sensors, solid oxide fuel cells, electrolysers, batteries, memory cells, and other devices. Most of the materials reported in this Special Issue are developed with the long-term aim of usage in high-temperature electrochemical devices. For example, samarium-doped ceria thin films deposited using e-beam evaporation [6] or other method are used for various purposes in solid oxide fuel cells (SOFCs). La-doped strontium titanate was studied as a ceramic component of a composite current collector in a

Crystals **2019**, *9*, 173

metal-supported SOFC [8]. Similarly, the development of proton-conducting ceramics [2,3] is aimed at future applications in SOFCs with proton-conducting electrolytes. SOFCs both with oxygen ion- and proton-conducting electrolytes operate at high temperatures, i.e., between 600 °C and 800 °C. Even the YF_3–SrF_2 solid solutions are considered as electrolytes for high-temperature all-solid-state fluorine batteries [7]. Electro-ceramic devices like solid oxide fuel cells, electrolysers, and batteries are multi-layered systems consisting of different materials. This feature brings about several important technological issues. One of them is the optimization of the layer deposition method. An example of this is presented by Sriubas et al., who studied the influence of the powder characteristics on the properties of Sm-doped ceria thin films [6]. Another important issue is the requirement of structural and chemical compatibility of materials in a wide temperature range. Moreover, the relevant data are usually difficult to obtain. The thermomechanical properties of a wide group of proton-conducting ceramics were reviewed by Løken et al., giving substantial data on numerous systems within this group of materials [9].

The variety of properties and applications of conducting ceramics makes them very important for the future of societies worldwide. Thus, the papers included in this Special Issue should not only be viewed as presenting scientific data but also as giving information in a much broader context. We believe that understanding the importance of both basic and applied research in the field of conducting ceramics is key for the future development of many industrial areas. The broad spectrum of materials presented in this issue reflects the variety of applications and possible modifications of modern ceramics.

References

1. Yang, H.; Lin, J.; Du, X.; Shen, H.; He, Y.; Lin, Q. Structural and Magnetic Studies of Cr^{3+} Substituted Nickel Ferrite Nanomaterials Prepared by Sol-Gel Auto-Combustion. *Crystals* **2018**, *8*, 384. [CrossRef]
2. Zheng, K.; Niemczyk, A.; Skubida, W.; Liu, X.; Świerczek, K. Crystal Structure, Hydration, and Two-Fold/Single-Fold Diffusion Kinetics in Proton-Conducting $Ba_{0.9}La_{0.1}Zr_{0.25}Sn_{0.25}In_{0.5}O_{3-\alpha}$ Oxide. *Crystals* **2018**, *8*, 136. [CrossRef]
3. Dzierzgowski, K.; Wachowski, S.; Gazda, M.; Mielewczyk-Gryń, A. Terbium Substituted Lanthanum Orthoniobate: Electrical and Structural Properties. *Crystals* **2019**, *9*, 91. [CrossRef]
4. Ring, K.; Fuierer, P. Quasi-Equilibrium, Multifoil Platelets of Copper- and Titanium-Substituted Bismuth Vanadate, $Bi_2V_{0.9}(Cu_{0.1-x}Ti_x)O_{5.5-\delta}$, by Molten Salt Synthesis. *Crystals* **2018**, *8*, 170. [CrossRef]
5. Neuhaus, K.; Baumann, S.; Dolle, R.; Wiemhöfer, H.D. Effect of MnO_2 Concentration on the Conductivity of $Ce_{0.9}Gd_{0.1}Mn_xO_{2-\delta}$. *Crystals* **2018**, *8*, 40. [CrossRef]
6. Sriubas, M.; Bockute, K.; Kainbayev, N.; Laukaitis, G. Influence of the Initial Powder's Specific Surface Area on the Properties of Sm-Doped Ceria Thin Films. *Crystals* **2018**, *8*, 443. [CrossRef]
7. Lunghammer, S.; Stanje, B.; Breuer, S.; Pregartner, V.; Wilkening, M.; Hanzu, I. Fluorine Translational Anion Dynamics in Nanocrystalline Ceramics: SrF_2-YF_3 Solid Solutions. *Crystals* **2018**, *8*, 122. [CrossRef]
8. Barbucci, A.; Viviani, M.; Presto, S.; Carpanese, M.; Costa, R.; Han, F. Application of La-Doped $SrTiO_3$ in Advanced Metal-Supported Solid Oxide Fuel Cells. *Crystals* **2018**, *8*, 134. [CrossRef]
9. Løken, A.; Ricote, S.; Wachowski, S. Thermal and Chemical Expansion in Proton Ceramic Electrolytes and Compatible Electrodes. *Crystals* **2018**, *8*, 365. [CrossRef]

crystals

MDPI

Article

Terbium Substituted Lanthanum Orthoniobate: Electrical and Structural Properties

Kacper Dzierzgowski[ID], Sebastian Wachowski[ID], Maria Gazda and Aleksandra Mielewczyk-Gryń *

Department of Solid State Physics, Faculty of Applied Physics and Mathematics, Gdańsk University of Technology, Narutowicza 11/12, 80-233 Gdańsk, Poland; kacper.dzierzgowski@pg.edu.pl (K.D.); sebastian.wachowski@pg.edu.pl (S.W.); maria.gazda@pg.edu.pl (M.G.)
* Correspondence: alegryn@pg.edu.pl; Tel.: +48-58-348-66-19

Received: 2 January 2019; Accepted: 6 February 2019; Published: 11 February 2019

Abstract: The results of electrical conductivity studies, structural measurements and thermogravimetric analysis of $La_{1-x}Tb_xNbO_{4+\delta}$ (x = 0.00, 0.05, 0.1, 0.15, 0.2, 0.3) are presented and discussed. The phase transition temperatures, measured by high-temperature x-ray diffraction, were 480 °C, 500 °C, and 530 °C for $La_{0.9}Tb_{0.1}NbO_{4+\delta}$, $La_{0.8}Tb_{0.2}NbO_{4+\delta}$, and $La_{0.7}Tb_{0.3}NbO_{4+\delta}$, respectively. The impedance spectroscopy results suggest mixed conductivity of oxygen ions and electron holes in dry conditions and protons in wet. The water uptake has been analyzed by the means of thermogravimetry revealing a small mass increase in the order of 0.002% upon hydration, which is similar to the one achieved for undoped lanthanum orthoniobate.

Keywords: lanthanum orthoniobate; terbium orthoniobate; protonic conductivity; impedance spectroscopy; thermogravimetric analysis; water uptake

1. Introduction

Proton conducting ceramics have attracted much interest due to their possible applications in energy conversion, as protonic ceramic fuel cells (PCFC), proton ceramic electrolyzer cells (PCEC), hydrogen sensors, and chemical synthesis [1]. Among them, several distinctive groups can be listed: materials based on barium cerate–zirconate solid solutions [2], rare earth niobates [3,4], as well as other materials, e.g., calcium zirconate, rare earth tungstates, and lanthanum ytterbium oxide [5–8].

For the last decade, multiple rare earth orthoniobates systems have been investigated, with the interest being put on their electronic [9–11] along with ionic conduction with oxygen ions [9,10] and protons [3] as mobile charge carriers. Also, other interesting properties were observed in these materials, e.g., paramagnetism, ferroelectricity, and luminescent emission [1,12].

Lanthanum orthoniobate as a proton conductor has been widely investigated since 2006, when Hausgrud and Norby introduced 1 mol% calcium into the lanthanum sublattice as a way of enhancing protonic conductivity [13]. In the following years, multiple dopants have been introduced both into the lanthanum (e.g., by calcium [14], magnesium [15], or strontium [16–19]) and niobium (e.g., by vanadium [20–22], antimony [23–25], arsenic [26], or cobalt [27]) sublattices. Recently, we also reported the influence of co-doping, using praseodymium as a rare earth dopant and calcium as an acceptor dopant in the lanthanum sublattice, on these system properties [28]. The presence of a mixed 3+/4+ cation in the lanthanum sublattice leads to enhanced electronic conductivity and yielding a mixed proton-electron conductor. From the other point of view, such a substitution can affect not only conductivity but also the structure of the material. For example, as it has been reported for $Ce_{1-x}La_xNbO_{4+\delta}$, the increase of lanthanum content in the cerium sublattice caused a decrease in the phase transition temperature [29]. Following the cerium and praseodymium substitutions, a natural next step was to introduce the third lanthanide with mixed valence—terbium. In this work, we present the structural and transport properties of terbium doped lanthanum orthoniobates. The influence of a dopant on the conductivity of the system has been determined by the means of electrochemical impedance spectroscopy and thermogravimetric analysis.

2. Experimental

Powders of $La_{1-x}Tb_xNbO_{4+\delta}$ (x = 0.00, 0.05, 0.1, 0.15, 0.2, 0.3) were prepared via the solid-state reaction route. La_2O_3 (99.99% Aldrich, preheated at 900 °C for 4 h), Tb_4O_7 (99.99% Aldrich), and Nb_2O_5 (99.99% Alfa Aesar, Haverhill, MA, USA) were used as starting materials. The stoichiometric amounts of the reagents were milled in an agate mortar in isopropanol. The obtained powders were uniaxially pressed at 400 MPa into 12 mm diameter pellets. The green bodies were calcined at 1000 °C for 12 h. After the first step of the synthesis, the specimens were ground into powders. In the second step, the powders were pressed again at 400 MPa and resintered at 1400 °C for 12 h.

The Powder X-Ray Diffraction (XRD) patterns were collected using Philips X'Pert Pro MPD with Cu Kα radiation. High-temperature XRD (HTXRD) analyses were carried out with an Anton Paar HTK-1200 high-temperature unit. XRD data were analyzed with the FullProf suite [30]. The density of the samples was determined by a vacuum-assisted Archimedes method. The liquid medium used for the measurements was kerosene. The samples were dried, soaked, and suspended in the medium prior to weighing. To enhance soaking, the samples were immersed in kerosene and placed under a vacuum in order to remove air from open pores.

The microstructure was characterized using FEI (Waltham, MA, USA) Quanta FEG 250 scanning electron microscope (SEM) equipped with EDAX Apollo-SD energy-dispersive X-ray spectroscopy (EDS) detector. The microstructure imaging was performed in High Vacuum mode with Everhart-Thornley detector working either in Secondary Electrons (SE) or Back-scattered Electrons (BSE) mode.

Thermogravimetric analysis (TGA) was performed using a Netzsch (Burlington, MA, USA) Jupiter® 449 F1. The as-prepared powders were heated to 1000 °C and held at this temperature for 0.5 h under dry air to remove water and possible surface carbon dioxide. The samples after dehydration were cooled to 300 °C in dry gas. After 2 h of stabilization, the dry purge gas was switched to the humidified gas (P_{H2O} = 0.023 atm), then after an additional 2 h, the purge gas was switched back to the dry gas.

Impedance spectroscopy measurements were performed to determine the electrical properties of the investigated materials. Impedance spectroscopy measurements were performed in the frequency range 1 Hz–1 MHz and 1 V amplitude on samples with ink painted platinum electrodes (ESL 5542). The measurements were performed in wet (2.4% H_2O) and dry technical air (20% O_2, 80% N_2) using Gamry Reference 3000 at the temperature range from 350 °C to 750 °C with 50 °C steps. Obtained data were analyzed with ZView software. Studies of conductivity as a function of pO_2 were performed within the range of 2×10^{-6} to 1 atm at 700 °C. Different values of pressure were obtained through mixing nitrogen (<2 ppm O_2) with oxygen (purity 99.999%) gas. Both dry and wet gases (2.4% H_2O) were used in the pO_2-dependency study.

3. Results and Discussion

Figure 1 shows the X-ray diffractograms of the synthesized sample powders. All observed reflections were indexed within the monoclinic $LaNbO_4$ (ICSD 01-071-1405), therefore the samples may be regarded as single-phase orthoniobates. The Rietveld profile of the pattern and the difference plots for $La_{0.85}Tb_{0.15}NbO_{4+\delta}$ is presented in Figure 2b. The unit cell parameters of the compounds, refined with the Rietveld method, are listed in Table 1. Terbium substituting lanthanum (La^{3+} ionic radius for CN = 8 is 1.16 Å) can be present in the lattice as Tb^{3+} or Tb^{4+}. Ionic radii of Tb^{3+} or Tb^{4+} for CN = 8 are 1.04 Å and 0.88 Å, respectively [31]. The general trend of a decrease of the unit cell volume with increasing terbium content (Figure 2a) can be attributed to the lower ionic radius of terbium.

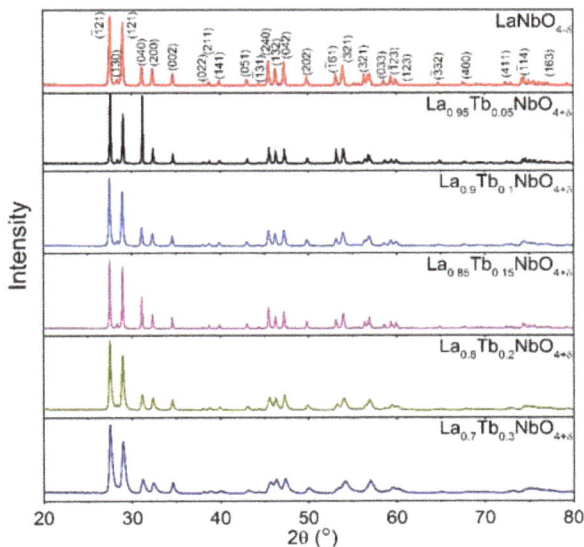

Figure 1. X-ray diffractograms of $La_{1-x}Tb_xNbO_{4+\delta}$ (x = 0.00, 0.05, 0.1, 0.15, 0.2, 0.3).

Table 1. Unit cell parameters and densities of $La_{1-x}Tb_xNbO_{4+\delta}$ (x = 0.00, 0.05, 0.1, 0.15, 0.2, 0.3); ϱ_t, ϱ_m, and ϱ_{rel} signify density calculated on the basis of unit cell parameters, density determined on the basis of Archimedes method, and relative density, respectively.

Sample	a (Å)	b (Å)	c (Å)	β (°)	V (Å³)	ϱ_t (g/cm³)	ϱ_m (g/cm³)	ϱ_{rel} (%)
$LaNbO_{4-\delta}$	5.5659	11.5245	5.2031	94.082	332.90	5.906	5.903	99.9
$La_{0.95}Tb_{0.05}NbO_{4+\delta}$	5.5509	11.4902	5.1954	94.091	330.52	5.963	5.214	87.5
$La_{0.9}Tb_{0.1}NbO_{4+\delta}$	5.5499	11.4934	5.1961	94.087	330.61	5.981	5.959	99.6
$La_{0.85}Tb_{0.15}NbO_{4+\delta}$	5.5532	11.4979	5.1969	94.092	330.97	5.995	5.259	87.7
$La_{0.8}Tb_{0.2}NbO_{4+\delta}$	5.5367	11.4648	5.1905	94.103	328.63	6.058	5.536	91.4
$La_{0.7}Tb_{0.3}NbO_{4+\delta}$	5.5202	11.4311	5.1812	94.103	326.11	6.145	5.525	89.9

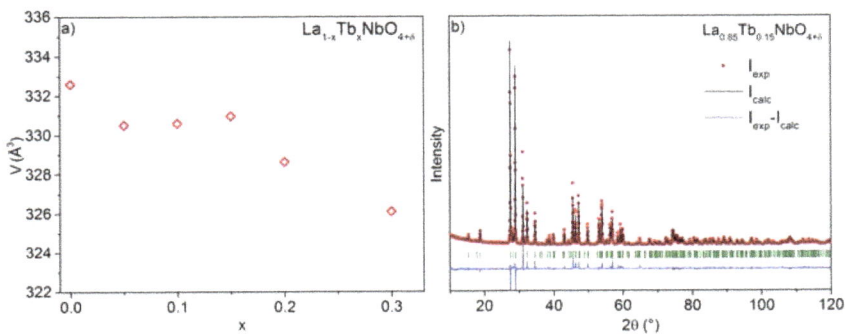

Figure 2. (a) Unit cell volume as a function of Tb content in $La_{1-x}Tb_xNbO_{4+\delta}$. (b) The Rietveld profile of the pattern and the difference plots for $La_{0.85}Tb_{0.15}NbO_{4+\delta}$.

The XRD patterns of $La_{0.9}Tb_{0.1}NbO_{4+\delta}$ obtained between 460 °C and 490 °C are presented in Figure 3. It can be seen that the reflections corresponding to the ($\bar{1}21$) and (121) planes of the monoclinic structure shift towards one another with increasing temperature and at 480 °C, they merge into one. This indicates the transition into the tetragonal phase. Similar behavior was observed in all studied samples. The phase transition temperatures were approximately 480 °C, 500 °C, and 530 °C for $La_{0.9}Tb_{0.1}NbO_{4+\delta}$, $La_{0.8}Tb_{0.2}NbO_{4+\delta}$, and $La_{0.7}Tb_{0.3}NbO_{4+\delta}$, respectively. Taking into consideration that terbium has a lower ionic radius than lanthanum, an increase of the transition temperature in $La_{1-x}Tb_{x}NbO_{4+\delta}$ with increasing terbium content is consistent with the other experimental results. For $RENbO_4$ with a decrease of rare earth metal ionic radius, an increase of phase transition temperature was observed [32].

Figure 4 shows exemplary SEM images taken of the sintered $La_{0.95}Tb_{0.05}NbO_{4+\delta}$, $La_{0.9}Tb_{0.1}NbO_{4+\delta}$, $La_{0.85}Tb_{0.15}NbO_{4+\delta}$, $La_{0.7}Tb_{0.3}NbO_{4+\delta}$ samples. In all samples, the observed fractures were apparently dense, without visible grain boundaries and with closed pores. Observed differences in porosity of the samples are consistent with the results of density measurements presented in Table 1. The relative densities vary from 87.5 ± 0.5 % ($La_{0.95}Tb_{0.05}NbO_{4+\delta}$) to 99.9 ± 0.3 % ($LaNbO_{4-\delta}$). The analysis performed with EDS and BSE detectors did not reveal secondary phases or impurities, which confirms the XRD results showing no phase separation.

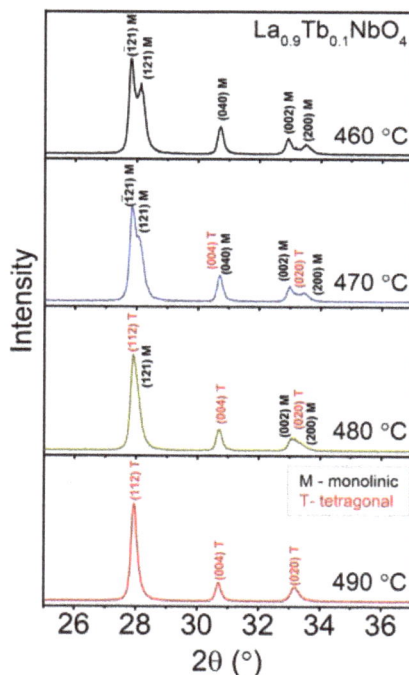

Figure 3. HTXRD patterns of $La_{0.9}Tb_{0.1}NbO_{4+\delta}$. Reflections of monoclinic and tetragonal phases were indexed with "M" and "T" letters, respectively.

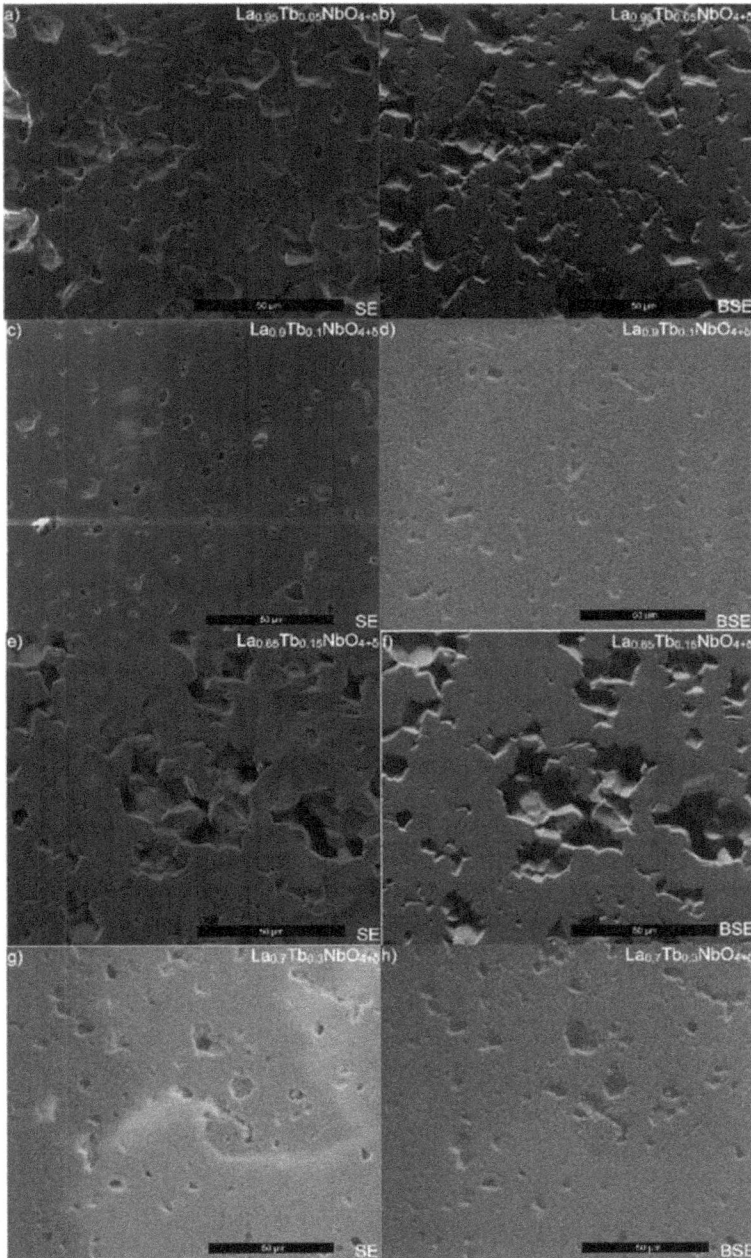

Figure 4. SEM image of polished fractures of $La_{0.95}Tb_{0.05}NbO_{4+\delta}$, $La_{0.9}Tb_{0.1}NbO_{4+\delta}$, $La_{0.85}Tb_{0.15}NbO_{4+\delta}$, $La_{0.7}Tb_{0.3}NbO_{4+\delta}$ taken in SE and BSE mode.

Thermogravimetric analysis was performed in order to determine the water uptake of the investigated samples. The results obtained for $La_{0.9}Tb_{0.1}NbO_{4+\delta}$, $La_{0.8}Tb_{0.2}NbO_{4+\delta}$, and $La_{0.7}Tb_{0.3}NbO_{4+\delta}$ are presented in Figure 5. One can see that the weight change during gas switch is of the order of 0.002%.

This is much lower than the results reported by Yamazaki et al. for barium zirconate system, who at 300 °C obtained 0.5% [33]. However, this is in accordance with the results achieved for undoped lanthanum niobate [34]. The relatively small water uptake suggests a rather small concentration of the protons within the samples structures. In the view of a small proton concentration in lanthanum orthoniobates, it may be considered strange why the conductivity differences between wet and dry air reach 50–400% at 700 °C (Table 2). The most probable reason for that is relatively high mobility of protons as well as the absence of defect association phenomena. The proton mobility and trapping effects were discussed extensively by Huse et al. showing their importance in the process of conductivity in this system [35].

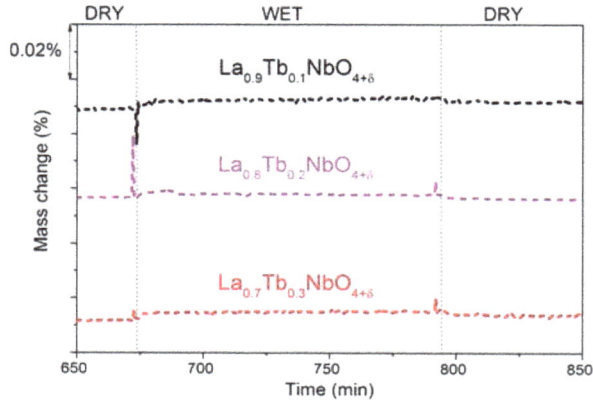

Figure 5. Results of TGA measurements performed for $La_{0.9}Tb_{0.1}NbO_{4+\delta}$, $La_{0.8}Tb_{0.2}NbO_{4+\delta}$, $La_{0.7}Tb_{0.3}NbO_{4+\delta}$.

Figure 6 presents an example of the acquired Nyquist plot for $La_{0.9}Tb_{0.1}NbO_{4+\delta}$. The clear curve separation into two semicircles can be seen. The brick layer model was used in the analysis of the impedance data [36]. The highest frequency semicircle is usually attributed to grain interior and the next semicircle is attributed to grain boundary conductivity [37]. Electrode responses, observed in the form of a low-frequency semicircle, as not in the scope of this study, were not analyzed. All impedance spectra were fitted with an equivalent circuit: $(CPE_g R_g)(CPE_{gb} R_{gb})$, where R is resistance and CPE is a constant phase element. For each semicircle, capacitances were calculated using the Formula (1).

$$C = Q_0^{\frac{1}{n}} R^{\frac{1}{n}-1} \tag{1}$$

where Q_0, the admittance and n, the angle of misalignment were calculated for each CPE. Typical values of capacitance obtained for high- and mid-frequency were 1×10^{-11} F/cm and 4×10^{-10} F/cm, respectively. The values are close to the ones typically observed in the literature [38] and confirm the semicircles were correctly attributed to grain, grain boundaries, and electrode processes. The conductivities of grain interior σ_g, specific grain boundary σ_{gb}, and total conductivity σ_{tot} were calculated with the use of Equations (2)–(4).

$$\sigma_g = S\frac{1}{R_g} \tag{2}$$

$$\sigma_{gb} = S\frac{1}{R_{gb}}\frac{C_g}{C_{gb}} \tag{3}$$

$$\sigma_{tot} = S\frac{1}{R_g + R_{gb}} \tag{4}$$

The *S* coefficient is a geometrical factor including sample porosity, thickness, and electrode area. Activation energies of total conductivity were calculated by fitting obtained values to the Equation (5), where σ is the conductivity, T is temperature, σ_0 is a pre-exponential factor, E_a is the activation energy, and k is the Boltzmann constant. Calculated values of activation energies are presented in Table 2.

Total conductivities of obtained samples in dry and wet air as a function of reciprocal temperature are presented in Figure 7. For all of the samples, the conductivity in wet air was higher than in dry air. The conductivity at 700 °C both in dry and wet air in the $La_{0.9}Tb_{0.1}NbO_{4+\delta}$ and $La_{0.85}Tb_{0.15}NbO_{4+\delta}$ samples reached 10^{-4} S/cm. Maximum conductivity at 700 °C was observed for the samples doped with 15% of terbium, while a further increase of Tb content led to the decline of total conductivity (Figure 8).

$$\sigma T = \sigma_0 e^{-\frac{E_a}{kT}} \tag{5}$$

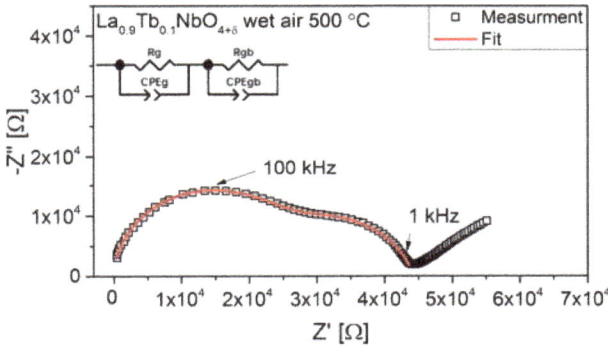

Figure 6. Nyquist plot of $La_{0.9}Tb_{0.1}NbO_{4+\delta}$ measured in wet air at 500 °C.

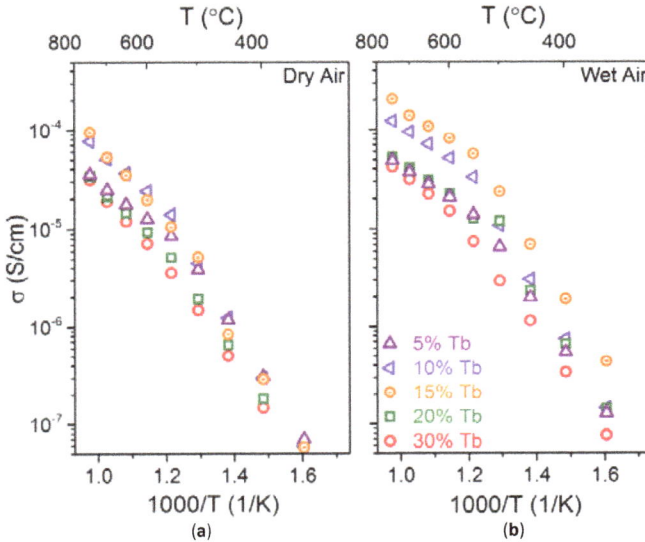

Figure 7. Total conductivity as a function of the reciprocal temperature of $La_{1-x}Tb_xNbO_{4+\delta}$ in (**a**) dry air and (**b**) wet air.

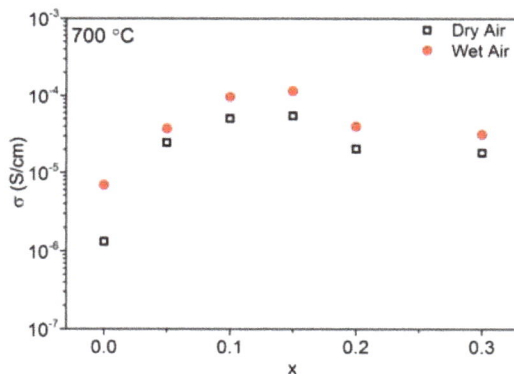

Figure 8. Total conductivities at 700 °C of $La_{1-x}Tb_xNbO_{4+\delta}$.

The change of slope of the plots observed at temperatures about 500 °C is related to the phase transition between the scheelite and fergusonite structures. Such behavior was observed before in many $LaNbO_{4-\delta}$-based materials [13,15,28,35]. The influence of the terbium content on the apparent activation energies of total conductivities (Table 2) is complex. In the low-temperature range, the values of activation energy in dry air decreases with the increase of Tb content. In the higher temperature range, the apparent activation energies are lower than these below 500 °C, however, no clear tendency of E_A dependence on Tb content can be observed. It should be noted that the $La_{0.8}Tb_{0.15}NbO_{4+\delta}$ sample, with the highest conductivities at 700 °C, shows also the lowest activation energies of conductivity in wet air, both for tetragonal (0.53 ± 0.02 eV) and monoclinic structure (1.16 ± 0.02 eV). Moreover, the strongest influence of humid atmosphere on the conductivity of this sample may be observed. The relative change of conductivity at 700 °C of $La_{0.8}Tb_{0.15}NbO_{4+\delta}$ reaches 107%, whereas in other doped materials, it is between 50 and 92%.

Table 2. Summary of electrical properties of $La_{1-x}Tb_xNbO_{4+\delta}$: total conductivities at 700 °C, a relative difference of total conductivities in wet and dry atmospheres at 700 °C, and apparent activation energies of conductivity in temperature ranges below and above 500 °C.

	σ_{tot}(700 °C)		$\frac{\sigma_{tot}^{wet} - \sigma_{tot}^{dry}}{\sigma_{tot}^{dry}}$	E_A Below 500 °C		E_A Above 500 °C	
	(10^{-5} S/cm)		(%)	(eV)		(eV)	
	Wet Air	Dry Air	at 700 °C	Wet Air	Dry Air	Wet Air	Dry Air
$LaNbO_{4-\delta}$	0.68	0.13	419	n/d	n/d	n/d	n/d
$La_{0.95}Tb_{0.05}NbO_{4+\delta}$	3.68	2.46	49	1.02 ± 0.02	1.29 ± 0.02	0.56 ± 0.02	0.68 ± 0.02
$La_{0.9}Tb_{0.1}NbO_{4+\delta}$	9.51	5.04	89	1.25 ± 0.02	1.25 ± 0.02	0.55 ± 0.02	0.68 ± 0.02
$La_{0.85}Tb_{0.15}NbO_{4+\delta}$	11.3	5.45	107	1.16 ± 0.02	1.24 ± 0.02	0.53 ± 0.02	0.86 ± 0.02
$La_{0.8}Tb_{0.2}NbO_{4+\delta}$	3.95	2.05	93	1.26 ± 0.02	1.11 ± 0.02	0.59 ± 0.02	0.74 ± 0.02
$La_{0.7}Tb_{0.3}NbO_{4+\delta}$	3.11	1.83	70	1.07 ± 0.03	1.15 ± 0.02	0.69 ± 0.05	0.85 ± 0.02

Figure 9 presents conductivity measured as a function of pO_2 in dry and wet atmospheres for $LaNbO_{4-\delta}$, $La_{0.05}Tb_{0.05}NbO_{4+\delta}$, and $La_{0.15}Tb_{0.85}NbO_{4+\delta}$. Under wet conditions, conductivity does not depend on oxygen partial pressure. Similarly, under dry conditions, the conductivity is almost independent for pO_2 below 10^{-4} atm, whilst above 10^{-3} atm it increases with increasing oxygen partial pressure. In terbium doped samples a stronger change of conductivity is observed, whereas in the $LaNbO_4$, change of conductivity is much smaller than in samples containing terbium. The observed slope of the conductivity at high pO_2 is close to $\frac{1}{4}$. Obtained results indicate that in wet conditions, materials are protonic conductors. In dry gases, the results show that at lower p_{O2}, the materials are oxygen ion conductors, whilst at high p_{O2}, the material is a mixed oxygen ion and electron holes conductor. The slope of the dependence observed in Figure 9 is characteristic of the situation where ionic charge carriers have dominating concentration, whereas holes are a minority defect, but due to much higher mobility, contribute to total conductivity.

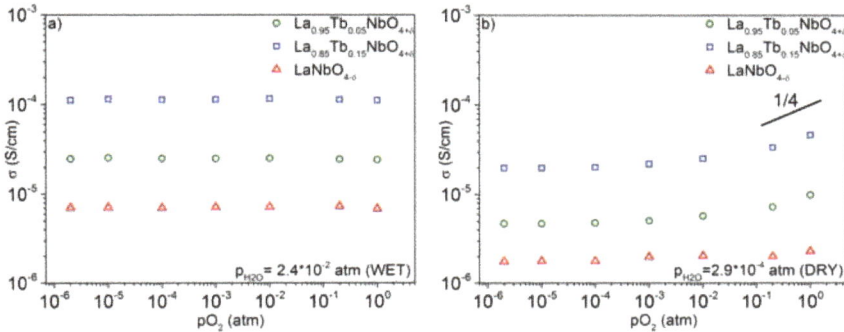

Figure 9. Total conductivity plotted as a function of pO_2 at 700 °C of $LaNbO_{4-\delta}$, $La_{0.05}Tb_{0.05}NbO_{4+\delta}$, and $La_{0.15}Tb_{0.85}NbO_{4+\delta}$ under (**a**) wet and (**b**) dry conditions.

The influence of Tb on the electrical properties of the material should be analyzed in view of two possible terbium oxidation states—terbium can be either isovalent (Tb^{3+}) with lanthanum (La^{3+}) or have higher oxidation state (Tb^{4+}) and serve as a donor. The source of terbium during the synthesis was Tb_4O_7, which contains a mixture of Tb^{3+} and Tb^{4+}. For simplicity in defect chemistry analysis, we will treat that oxide as a mixture of Tb_2O_3 and TbO_2. In the case of Tb_2O_3, in which terbium is isovalent with lanthanum, the incorporation of Tb follows the reaction (6).

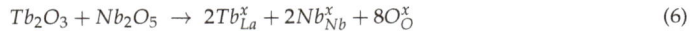

$$Tb_2O_3 + Nb_2O_5 \rightarrow 2Tb_{La}^x + 2Nb_{Nb}^x + 8O_O^x \tag{6}$$

In the case of TbO_2, the excess positive charge of Tb^{4+} can be compensated by the formation of either electron or oxygen interstitials, which is described by reactions (7) and (8).

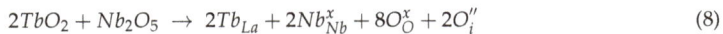

$$2TbO_2 + Nb_2O_5 \rightarrow 2Tb_{La} + 2Nb_{Nb}^x + 8O_O^x + 2e' + O_{2(g)} \tag{7}$$

$$2TbO_2 + Nb_2O_5 \rightarrow 2Tb_{La} + 2Nb_{Nb}^x + 8O_O^x + 2O_i'' \tag{8}$$

Another point to analyze is the preferred oxidation state of terbium in the crystal lattice of lanthanum orthoniobate. One can assume that the more preferred oxidation state of terbium is 3+, since the ionic radii difference between Tb^{3+} and La^{3+} is much smaller (0.12 Å) than that between Tb^{4+} and La^{3+} (0.28 Å). The structure with Tb^{3+} would be much less strained than with Tb^{4+}. Therefore, it may be expected that Tb^{4+} will be at least partially reduced into Tb^{3+}, which is accompanied by the generation of holes:

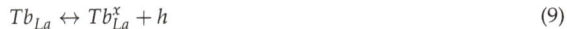

$$Tb_{La} \leftrightarrow Tb_{La}^x + h \tag{9}$$

Further analysis of the possible scenarios must be performed in relation to the knowledge of properties of undoped lanthanum orthoniobate. First of all, the most energetically favourable intrinsic defect in $LaNbO_{4-\delta}$ is an anion Frenkel pair, which is a pair of oxygen vacancy and oxygen interstitial [39]:

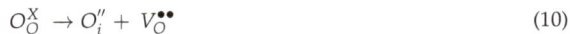

$$O_O^X \rightarrow O_i'' + V_O^{\bullet\bullet} \tag{10}$$

This indicates that the formation of interstitial oxygen ions' defects could be energetically preferred in the case of doped lanthanum niobate. Secondly, studies of $LaNbO_{4-\delta}$ [13] showed that in oxidizing conditions at 1200 °C, the material is a mixed electron hole and oxygen ion conductor. The electron holes are a minority defect (concentration-wise), however, due to the higher mobility of holes in the valence band in comparison to the mobility of ionic defects in the lattice, the overall effect is a mixed type of conductivity. The results for lower temperatures are not presented in the study by Haugsrud et al. [13], however by the studies of similar systems [38] it can be expected that with lowering temperature the contribution from holes diminishes and for temperatures of interest in our study, doped lanthanum orthoniobate should be an oxygen ion conductor in dry conditions. Indeed, our measurement of conductivity of lanthanum orthoniobate as a function of oxygen partial pressure shows that at 700 °C at lower p_{O2} it is an ion conductor, with increasing hole contribution at higher p_{O2}. The situation is the same for Tb doped lanthanum orthoniobates. This shows similarity of properties of Tb doped samples to the pure $LaNbO_4$.

The effect of Tb substitution given by reactions (7–9) could be twofold. In the first scenario it can be either that the incorporation of Tb^{4+} produces electrons (7), and then latter partial reduction to Tb^{3+} reduces that effect by either producing holes in the reaction (9), which can further recombine with electrons. The second scenario would be the creation of oxygen interstitials (8) accompanied by the formation of holes during a partial reduction of Tb to fit into the $LaNbO_{4-\delta}$ lattice (9). The latter hypothesis of the preferred formation of oxygen interstitials instead of electrons is supported by the results of the measurement of conductivity as a function of oxygen partial pressure in dry conditions. Moreover, a similar compound, $CeNbO_{4+\delta}$, is an oxygen ion conductor, in which oxygen interstitials dominate [40]. $CeNbO_{4+\delta}$ is a compound similar to $LaNbO_{4-\delta}$. It is isostructural with $LaNbO_{4-\delta}$, but cerium occupying the A-site may be present either as Ce^{3+} or Ce^{4+}. Therefore, $CeNbO_{4+\delta}$ is expected to be somewhat similar to terbium-doped lanthanum orthoniobate because terbium doping also induces a mixed 3+/4+ oxidation state in the A-site.

All considered, an overall effect with dominating reactions (8) and (9) should be that the oxygen ionic conductivity should increase due to the formation of ionic defects and the hole contribution to total conductivity should also increase with an increase of Tb content. Indeed, the conductivity of samples with x < 0.2 increases with increasing Tb content, which means that increasing Tb content leads to the increase of the concentration of mobile charged species. This is observed both in the wet and dry atmospheres, which supports the hypothesis of dominating ionic conductivity. Moreover, conductivity measured as a function of oxygen partial pressure in wet atmospheres shows a typical feature of a proton conductor, i.e., the conductivity does not depend on the pressure. Lower conductivity observed in samples with high Tb content (x $\geq$ 0.2) (Figure 8) can be explained by potential trapping of ionic charged species around the dopants. This is a common feature observed in heavily doped ionic conductors [41–43].

The observed increase of the conductivity in wet air compared to dry one may be considered as a typical feature of a high-temperature proton conductor related to the hydration reaction of an oxide. Typically, the hydration reaction is given by the following:

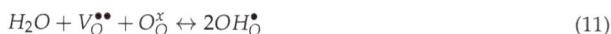

$$H_2O + V_O^{\bullet\bullet} + O_O^x \leftrightarrow 2OH_O^{\bullet} \tag{11}$$

In $La_{1-x}Tb_xNbO_{4+\delta}$, however, not many vacancies are expected to be present in the system, therefore a possible alternative reaction could be given:

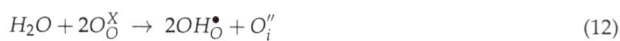

$$H_2O + 2O_O^X \rightarrow 2OH_O^{\bullet} + O_i'' \tag{12}$$

This reaction was previously proposed by Norby and discussed later by Islam for brownmillerite oxides [44,45]. Moreover, recent calculations showed oxygen ionic conductivity at high temperatures in undoped $LaNbO_4$, which may occur via oxygen interstitial positions [46]. Hydration of the doped materials may lead to the formation of both proton defects and interstitial oxygen ions. Thermogravimetric analysis showed that the mass change caused by the water uptake is rather small. The TGA results did not allow us to determine whether this depends on the terbium content. On the other hand, the analysis of the influence of the atmosphere on electrical properties shows that the relative increase of conductivity in wet atmospheres depends on the terbium content, in a similar way as the conductivity values (Table 2, Figure 7). It reaches the maximum in the $La_{0.85}Tb_{15}NbO_{4+\delta}$

sample, however, it is much lower than that in the undoped lanthanum orthoniobate—the relative change of conductivity for $LaNbO_{4-\delta}$ is 4 times higher than that for $La_{0.85}Tb_{15}NbO_{4+\delta}$. The proposed hypothesis, in which Tb presence increases the concentrations of oxygen interstitials and electron holes, could explain this observation. The hole conductivity occurs as a result of the partial reduction of Tb^{4+} in the lattice (reaction (9)), and this should not be affected by the hydration reactions (11) and (12). Therefore, with increasing terbium content, the hole conductivity also increases and the relative contribution of protons into the total conductivity should decrease.

Figure 10 presents the temperature dependence of the two components of the total conductivity, specific grain boundary, and grain interior (bulk) conductivities for $La_{0.9}Tb_{0.1}NbO_{4+\delta}$ sample, which is representative of the whole system. In the temperature range above 500 °C, the specific grain boundary conductivity is higher by over an order of magnitude than the bulk one. This is not typical behavior of solid electrolytes, however, similar phenomena were observed for $LaNbO_{4-\delta}$ co-doped with 10% praseodymium and 2% calcium [28].

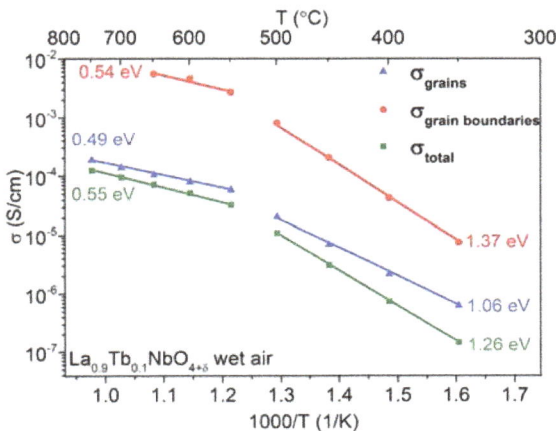

Figure 10. Temperature dependence of total conductivity, grain conductivity, and specific grain boundary conductivity of $La_{0.9}Tb_{0.1}NbO_{4+\delta}$ measured in wet air.

4. Conclusions

Single-phase samples of terbium doped lanthanum orthoniobates were prepared and studied. Unit cell parameters at room temperature of the studied compounds were determined. A decrease of the unit cell volume with increasing terbium content was observed. The results of high-temperature XRD showed that increasing terbium content in the lanthanum sublattice causes an increase of the phase transition temperature. The phase transition temperatures were approximately 480 °C, 500 °C, and 530 °C for $La_{0.9}Tb_{0.1}NbO_{4+\delta}$, $La_{0.8}Tb_{0.2}NbO_{4+\delta}$, and $La_{0.7}Tb_{0.3}NbO_{4+\delta}$, respectively. This behavior, as well as the observed decrease of the unit cell volume with increasing terbium content, was interpreted as related to the lower ionic radius of terbium.

Temperature dependence of electrical conductivity in different atmospheres was investigated and discussed. Both terbium content and the atmosphere influence the total conductivity of $La_{1-x}Tb_xNbO_{4+\delta}$. The highest conductivity was observed for the $La_{0.85}Tb_{0.15}NbO_{4+\delta}$ sample both in dry and wet atmospheres. For all samples, the total conductivity in the wet was higher than that in the dry air. On the basis of defect formation analysis supported by the comparison with the $La_{1-x}Ce_xNbO_{4+\delta}$, the hypothesis explaining the electrical properties of $La_{1-x}Tb_xNbO_{4+\delta}$ was proposed. In particular, it was proposed that Tb presence increases the concentrations of oxygen interstitials and electron holes, whereas the hydration of the doped materials leads to the formation of both proton defects and interstitial oxygen ions, which has been confirmed by pO_2 dependency measurements. This means that

Tb increasing content and humid atmosphere lead to an increase of the concentrations of three types of charge carriers. The non-monotonic dependence of the total conductivity on terbium content was interpreted as a result of a possible defect association phenomena for high Tb contents (above 0.15).

Author Contributions: Conceptualization, K.D., M.G. and A.M.-G.; methodology, A.M.-G., K.D.,S.W., M.G.; formal analysis, M.G.; investigation, K.D., M.G.,A.M.-G.; resources, A.M.-G.; data curation, K.D., S.W., A.M.-G.; writing—original draft preparation, K.D.; writing—review and editing, M.G., S.W., A.M-G.; supervision, M.G., A.M.-G.; project administration, A.M.-G.; funding acquisition, A.M.-G.

Funding: The research was financially supported by the Ministry of Science and Higher Education, Poland by Iuventus Scheme Grant No IP2015 051374.

Conflicts of Interest: The authors declare no conflict of interest.

References

1. Molenda, J.; Kupecki, J.; Baron, R.; Blesznowski, M.; Brus, G.; Brylewski, T.; Bucko, M.; Chmielowiec, J.; Cwieka, K.; Gazda, M.; et al. Status report on high temperature fuel cells in Poland—Recent advances and achievements. *Int. J. Hydrogen Energy* **2017**, *42*, 4366–4403. [CrossRef]
2. Gdula-Kasica, K.; Mielewczyk-Gryn, A.; Molin, S.; Jasinski, P.; Krupa, A.; Kusz, B.; Gazda, M. Optimization of microstructure and properties of acceptor-doped barium cerate. *Solid State Ionics* **2012**, *225*, 245–249. [CrossRef]
3. Haugsrud, R.; Norby, T. Proton conduction in rare-earth ortho-niobates and ortho-tantalates. *Nat. Mater.* **2006**, *5*, 193–196. [CrossRef]
4. Animitsa, I.; Iakovleva, A.; Belova, K. Electrical properties and water incorporation in A-site deficient perovskite $La_{1-x}Ba_xNb_3O_{9-0.5x}$. *J. Solid State Chem.* **2016**, *238*, 156–161. [CrossRef]
5. Hibino, T.; Mizutani, K.; Yajima, T.; Iwahara, H. Evaluation of proton conductivity in $SrCeO_3$, $BaCeO_3$, $CaZrO_3$ and $SrZrO_3$ by temperature programmed desorption method. *Solid State Ionics* **1992**, *57*, 303–306. [CrossRef]
6. Escolástico, S.; Vert, V.B.; Serra, J.M. Preparation and characterization of nanocrystalline mixed proton-electronic conducting materials based on the system Ln_6WO_{12}. *Chem. Mater.* **2009**, *21*, 3079–3089. [CrossRef]
7. Yajima, T.; Kazeoka, H.; Yogo, T.; Iwahara, H. Proton conduction in sintered oxides based on $CaZrO_3$. *Solid State Ionics* **1991**, *47*, 271–275. [CrossRef]
8. Sakai, T.; Isa, K.; Matsuka, M.; Kozai, T.; Okuyama, Y.; Ishihara, T.; Matsumoto, H. Electrochemical hydrogen pumps using Ba doped $LaYbO_3$ type proton conducting electrolyte. *Int. J. Hydrogen Energy* **2013**, *38*, 6842–6847. [CrossRef]
9. Haugsrud, R.; Ballesteros, B.; Lira-Cantu, M.; Norby, T. Ionic and electronic conductivity of 5% Ca-doped $GdNbO_4$. *J. Electrochem. Soc.* **2006**, *153*, J87–J90. [CrossRef]
10. Bayliss, R.D.; Pramana, S.S.; An, T.; Wei, F.; Kloc, C.L.; White, A.J.P.; Skinner, S.J.; White, T.J.; Baikie, T. Ferguson ite-type $CeNbO_{4+\delta}$: Single crystal growth, symmetry revision and conductivity. *J. Solid State Chem.* **2013**, *204*, 291–297. [CrossRef]
11. Li, C.; Bayliss, R.D.; Skinner, S.J. Crystal structure and potential interstitial oxide ion conductivity of $LnNbO_4$ and $LnNb_{0.92}W_{0.08}O_{4.04}$ (Ln = La, Pr, Nd). *Solid State Ionics* **2014**, *262*, 530–535. [CrossRef]
12. Huang, H.; Wang, T.; Zhou, H.; Huang, D.; Wu, Y.; Zhou, G.; Hu, J.; Zhan, J. Luminescence, energy transfer, and up-conversion mechanisms of Yb3+and Tb3+co-doped $LaNbO_4$. *J. Alloys Compd.* **2017**, *702*, 209–215. [CrossRef]
13. Haugsrud, R.; Norby, T. High-temperature proton conductivity in acceptor-doped $LaNbO_4$. *Solid State Ionics* **2006**, *177*, 1129–1135. [CrossRef]
14. Hakimova, L.; Kasyanova, A.; Farlenkov, A.; Lyagaeva, J.; Medvedev, D.; Demin, A.; Tsiakaras, P. Effect of isovalent substitution of La^{3+} in Ca-doped $LaNbO_4$ on the thermal and electrical properties. *Ceram. Int.* **2019**, *45*, 209–215. [CrossRef]
15. Mielewczyk-Gryn, A.; Wachowski, S.; Zagórski, K.; Jasiński, P.; Gazda, M. Characterization of magnesium doped lanthanum orthoniobate synthesized by molten salt route. *Ceram. Int.* **2015**, *41*, 7847–7852. [CrossRef]
16. Mielewczyk-Gryn, A.; Gdula, K.; Lendze, T.; Kusz, B.; Gazda, M. Nano- and microcrystals of doped niobates. *Cryst. Res. Technol.* **2010**, *45*, 1225–1228. [CrossRef]

17. Fjeld, H.; Kepaptsoglou, D.M.; Haugsrud, R.; Norby, T. Charge carriers in grain boundaries of 0.5% Sr-doped LaNbO$_4$. *Solid State Ionics* **2010**, *181*, 104–109. [CrossRef]

18. Mokkelbost, T.; Lein, H.L.; Vullum, P.E.; Holmestad, R.; Grande, T.; Einarsrud, M.-A. Thermal and mechanical properties of LaNbO$_4$-based ceramics. *Ceram. Int.* **2009**, *35*, 2877–2883. [CrossRef]

19. Nguyen, D.; Kim, Y.H.; Lee, J.S.; Fisher, J.G. Structure, morphology, and electrical properties of proton conducting La$_{0.99}$Sr$_{0.01}$NbO$_{4-\delta}$ synthesized by a modified solid state reaction method. *Mater. Chem. Phys.* **2017**, *202*, 320–328. [CrossRef]

20. Brandão, A.D.; Antunes, I.; Frade, J.R.; Torre, J.; Kharton, V.V.; Fagg, D.P. Enhanced Low-Temperature Proton Conduction in Sr$_{0.02}$La$_{0.98}$NbO$_{4-\delta}$ by Scheelite Phase Retention. *Chem. Mater.* **2010**, *22*, 6673–6683. [CrossRef]

21. Wachowski, S.; Mielewczyk-Gryn, A.; Gazda, M. Effect of isovalent substitution on microstructure and phase transition of LaNb$_{1-x}$M$_x$O$_4$ (M = Sb, V or ta; x = 0.05 to 0.3). *J. Solid State Chem.* **2014**, *219*, 201–209. [CrossRef]

22. Brandão, A.D.; Nasani, N.; Yaremchenko, A.A.; Kovalevsky, A.V.; Fagg, D.P. Solid solution limits and electrical properties of scheelite Sr$_y$La$_{1-y}$Nb$_{1-x}$V$_x$O$_{4-\delta}$ materials for x = 0.25 and 0.30 as potential proton conducting ceramic electrolytes. *Int. J. Hydrogen Energy* **2018**, *43*, 18682–18690. [CrossRef]

23. Wachowski, S.; Mielewczyk-Gryn, A.; Zagorski, K.; Li, C.; Jasinski, P.; Skinner, S.J.; Haugsrud, R.; Gazda, M. Influence of Sb-substitution on ionic transport in lanthanum orthoniobates. *J. Mater. Chem. A* **2016**, *4*, 11696–11707. [CrossRef]

24. Mielewczyk-Gryn, A.; Wachowski, S.; Strychalska, J.; Zagórski, K.; Klimczuk, T.; Navrotsky, A.; Gazda, M. Heat capacities and thermodynamic properties of antimony substituted lanthanum orthoniobates. *Ceram. Int.* **2016**, *42*, 7054–7059. [CrossRef]

25. Mielewczyk-Gryn, A.; Wachowski, S.; Lilova, K.I.; Guo, X.; Gazda, M.; Navrotsky, A. Influence of antimony substitution on spontaneous strain and thermodynamic stability of lanthanum orthoniobate. *Ceram. Int.* **2015**, *41*, 2128–2133. [CrossRef]

26. Wachowski, S.; Kamecki, B.; Winiarz, P.; Dzierzgowski, K.; Mielewczyk-Gryń, A.; Gazda, M. Tailoring structural properties of lanthanum orthoniobates through an isovalent substitution on the Nb-site. *Inorg. Chem. Front.* **2018**, *5*, 2157–2166. [CrossRef]

27. Li, M.; Wu, R.; Zhu, L.; Cheng, J.; Hong, T.; Xu, C. Enhanced sinterability and conductivity of cobalt doped lanthanum niobate as electrolyte for proton-conducting solid oxide fuel cell. *Ceram. Int.* **2019**, *45*, 573–578. [CrossRef]

28. Dzierzgowski, K.; Wachowski, S.; Gojtowska, W.; Lewandowska, I.; Jasiński, P.; Gazda, M.; Mielewczyk-Gryń, A. Praseodymium substituted lanthanum orthoniobate: Electrical and structural properties. *Ceram. Int.* **2018**, *44*, 8210–8215. [CrossRef]

29. Packer, R.J.; Skinner, S.J.; Yaremchenko, A.A.; Tsipis, E.V.; Kharton, V.V.; Patrakeev, M.V.; Bakhteeva, Y.A. Lanthanum substituted CeNbO$_{4+\delta}$ scheelites: Mixed conductivity and structure at elevated temperatures. *J. Mater. Chem.* **2006**, *16*, 3503. [CrossRef]

30. Rodríguez-Carvajal, J. *Recent Developments for the Program FULLPROF*; Commission on Powder Diffraction: Perth, Australia, 2001; Volume 26, ISBN 4971168915.

31. Shannon, R.D. Revised effective ionic radii and systematic studies of interatomic distances in halides and chalcogenides. *Acta Crystallogr. Sect. A* **1976**, *32*, 751–767. [CrossRef]

32. Stubičan, V.S. High-Temperature Transitions in Rare Earth Niobates and Tantalates. *J. Am. Ceram. Soc.* **1964**, *47*, 55–58. [CrossRef]

33. Yamazaki, Y.; Babilo, P.; Haile, S.M. Defect chemistry of yttrium-doped barium zirconate: A thermodynamic analysis of water uptake. *Chem. Mater.* **2008**, *20*, 6352–6357. [CrossRef]

34. Mielewczyk-Gryń, A. Water uptake analysis of the acceptor-doped lanthanum orthoniobates. *J. Therm. Anal. Calorim.* **2019**, submitted.

35. Huse, M.; Norby, T.; Haugsrud, R. Effects of A and B site acceptor doping on hydration and proton mobility of LaNbO$_4$. *Int. J. Hydrogen Energy* **2012**, *37*, 8004–8016. [CrossRef]

36. Abrantes, J.C.C.; Labrincha, J.A.; Frade, J.R. Applicability of the brick layer model to describe the grain boundary properties of strontium titanate ceramics. *J. Eur. Ceram. Soc.* **2000**, *20*, 1603–1609. [CrossRef]

37. Haile, S.M.; West, D.L.; Campbell, J. The role of microstructure and processing on the proton conducting properties of gadolinium-doped barium cerate. *J. Mater. Res.* **1998**, *13*, 1576–1595. [CrossRef]

38. Berger, P.; Mauvy, F.; Grenier, J.-C.; Sata, N.; Magrasó, A.; Haugsrud, R.; Slater, P.R. *Proton-Conducting Ceramics: From Fundamentals to Applied Research*; Marrony, M., Ed.; Pan Stanford Publishing: Singapore, 2016; Chapter 1; pp. 1–72.
39. Mather, G.C.; Fisher, C.A.J.; Islam, M.S. Defects, dopants, and protons in LaNbO$_4$. *Chem. Mater.* **2010**, *22*, 5912–5917. [CrossRef]
40. Packer, R.J.; Tsipis, E.V.; Munnings, C.N.; Kharton, V.V.; Skinner, S.J.; Frade, J.R. Diffusion and conductivity properties of cerium niobate. *Solid State Ionics* **2006**, *177*, 2059–2064. [CrossRef]
41. Wang, D.Y.; Park, D.S.; Griffith, J.; Nowick, A.S. Oxygen-ion conductivity and defect interactions in yttria-doped ceria. *Solid State Ionics* **1981**, *2*, 95–105. [CrossRef]
42. Guo, X.; Waser, R. Electrical properties of the grain boundaries of oxygen ion conductors: Acceptor-doped zirconia and ceria. *Prog. Mater. Sci.* **2006**, *51*, 151–210. [CrossRef]
43. Kilner, J.A.; Brook, R.J. A study of oxygen ion conductivity in doped non-stoichiometric oxides. *Solid State Ionics* **1982**, *6*, 237–252. [CrossRef]
44. Norby, T.; Larring, Y. Concentration and transport of protons in oxides. *Curr. Opin. Solid State Mater. Sci.* **1997**, *2*, 593–599. [CrossRef]
45. Islam, M.S.; Davies, R.A.; Fisher, C.A.J.; Chadwick, A.V. Defects and protons in the CaZrO$_3$ perovskite and Ba$_2$In$_2$O$_5$ brownmillerite: Computer modelling and EXAFS studies. *Solid State Ionics* **2001**, *145*, 333–338. [CrossRef]
46. Toyoura, K.; Sakakibara, Y.; Yokoi, T.; Nakamura, A.; Matsunaga, K. Oxide-ion conduction: Via interstitials in scheelite-type LaNbO$_4$: A first-principles study. *J. Mater. Chem. A* **2018**, *6*, 12004–12011. [CrossRef]

Article

Influence of the Initial Powder's Specific Surface Area on the Properties of Sm-Doped Ceria Thin Films

Mantas Sriubas [1], Kristina Bockute [1,*], Nursultan Kainbayev [1,2] and Giedrius Laukaitis [1]

[1]   Department of Physics, Kaunas University of Technology, Studentu street 50, LT-51368 Kaunas, Lithuania;
      mantas.sriubas@ktu.lt (M.S.); nursultan.kainbayev@ktu.edu (N.K.); giedrius.laukaitis@ktu.lt (G.L.)
[2]   Department of Thermal Physics and Technical Physics, Al-Farabi Kazakh National University,
      71 al-Farabi Avenue, 050040 Almaty, Kazakhstan
*   Correspondence: kristina.bockute@ktu.lt; Tel.: +370-37-300-340

Received: 30 October 2018; Accepted: 23 November 2018; Published: 27 November 2018

Abstract: The influence of a specific surface area of evaporating powder on the properties of thin Sm-doped cerium (SDC) oxide films has not yet been sufficiently investigated. Therefore, SDC films were deposited by e-beam evaporation using $Sm_{0.2}Ce_{0.8}O_{2-\delta}$ powders of 6.2 m^2/g, 11.3 m^2/g, and 201.3 m^2/g specific surface area on SiO_2, and Al_2O_3 substrates. X-Ray Diffraction (XRD) analysis showed that SDC thin films deposited on 600 °C SiO_2 substrates changed their preferred orientation from (111) to (311), (200), and (220) when evaporating 6.2 m^2/g and 11.3 m^2/g powders and using 0.2 nm/s, 1.2 nm/s, and 1.6 nm/s deposition rates. However, thin films deposited by evaporating powder of 201.3 m^2/g specific surface area do not change their preferred orientation. The crystallite size of the SDC thin films depends on the substrate temperature and specific surface area of the evaporating powder. It increases from 6.40 nm to 89.1 nm with increasing substrate temperature (50–600 °C). Moreover, crystallites formed by evaporating a powder of 201.3 m^2/g specific surface area are 1.4 times larger than crystallites formed by evaporating a powder of 6.2 m^2/g specific surface area. An impedance analysis revealed that the normalized resistance of "grains" is higher than the normalized resistance of grain boundaries. Moreover, a total conductivity depends on crystallite size. It changes from 4.4×10^{-7} S/cm to 1.1×10^{-2} S/cm (600 °C) when the crystallite sizes vary from 6.40 nm to 89.10 nm. In addition, the optical band gap becomes wider with increasing crystallite size proving that the Ce^{3+} concentration decreases with an increasing crystallite size.

Keywords: samarium-doped ceria (SDC); e-beam physical vapor deposition; solid oxide fuel cells (SOFC); thin films; ionic conductivity; specific surface area of powders

1. Introduction

The electrical properties of samarium-doped ceria (SDC) depend on the Sm dopant concentration, working temperature, oxygen pressure, migration path of oxygen ions, and microstructure. It has been found that oxygen diffuses to the surface of the electrolyte when the pressure of oxygen decreases and the temperature rises [1]. Ce^{4+} reduction to Ce^{3+} occurs at the same time and SDC becomes an ionic-electronic conductor [2,3]. Furthermore, CeO_2 doped with 15 mol% Sm has the highest ionic conductivity due to having the lowest vacancy activation energy at this dopant concentration [4]. On the other hand, it is possible to adjust the electrical properties of ceramics by changing the crystallite size. Concentrations of vacancies are different in grains and in grain boundaries. The concentration of oxygen vacancies is lower and the concentration of electrons is higher in the space charge zone near the grain interior [5,6]. According to the Brick layer model [7], oxygen ion conductivity is higher in grains than in grain boundaries and electron conductivity is higher in grain boundaries than in grains. Therefore, ionic conductivity decreases with decreasing grain size and electronic conductivity increases with decreasing grain size because the ratio between the grain size (L) and the grain boundary size ($2b$) increases ($2b/L$) [8].

Grain size, surface morphology and other properties of thin films depend on the deposition method and formation parameters. Thin SDC films can be formed using various formation techniques, e.g., magnetron sputtering, chemical vapor deposition, sol-gel method, spray pyrolysis, pulsed laser deposition, ion beam assisted deposition, thermal evaporation, metal-organic vapor deposition, electrostatic spray assisted vapor deposition, and e-beam evaporation [9,10]. The latter has some advantages over the other methods. During an e-beam evaporation process, evaporating materials can be heated up to 3000–4000 °C temperatures. Therefore, it is possible to deposit ceramics with high melting temperature, for example, CeO_2. In addition, it is possible to adjust the deposition rate between 0.02 nm/s and 10 nm/s and control the grain size using this method. Moreover, thin films are clean and uniform. In comparison, it is hard to achieve high deposition rates and to control them using the magnetron sputtering technique because oxides have a low sputtering yield. Another example is the sol-gel method. The minimum thickness of coatings is 10 μm using the sol-gel method due to the thermal mismatch between the coating and substrate, and cracking at a very low thickness (0.5 μm) resulting from trapped organics within the coating [11].

The properties of thin films deposited using the e-beam evaporation method depend on the deposition rate, substrate temperature, pressure, and specific surface area of the powders. The influence of the substrate temperature and the deposition rate has been investigated already. However, the influence of a specific surface area of powders on the properties of thin films has been studied less. There are few papers written by D. Virbukas et al., where the influence of the specific surface area of powders is discussed. Authors studied $Gd_{0.1}Ce_{0.9}O_{2-\delta}$ (S_{BET}: 6.44; 36.2; 201 m^2/g), and $Sm_{0.15}Ce_{0.85}O_{2-\delta}$ (S_{BET}: 39.3 m^2/g) thin films deposited on room temperature substrates and did not find a significant influence of the specific surface area of ceria powders on the microstructure of thin films [12,13]. These investigations were carried out with the thin films deposited at room temperature without investigating the thin films deposited on higher temperature substrates.

Therefore, the influence of the specific surface area of powders on the properties of $Sm_{0.2}Ce_{0.8}O_{2-\delta}$ thin films deposited on higher than room temperature (50 °C, 150 °C, 300 °C, 450 °C, and 600 °C) substrates was investigated in this work. For this reason, the powders of different specific surface areas (6.2 m^2/g, 11.3 m^2/g, and 201.3 m^2/g) were deposited on SiO_2 and Al_2O_3 substrates. In addition, thin films were deposited using 0.2 nm/s, 0.4 nm/s, 0.8 nm/s, 1.2 nm/s and 1.6 nm/s deposition rates. Microstructure, surface morphology, and electrical properties of thin films were also studied.

2. Materials and Methods

Sm-doped cerium oxide thin films were deposited on SiO_2 and Al_2O_3 substrates. The substrates were ultrasonically cleaned in pure acetone for 10 min. Thin films were formed with an e-beam physical vapor deposition system "Kurt J. Lesker EB-PVD 75, Kurt J. Lesker Company Ltd., Hastings, England", at a 0.2 nm/s ÷ 1.6 nm/s deposition rate and substrate temperatures from 50 °C to 600 °C. The $Sm_{0.2}Ce_{0.8}O_{2-\delta}$ powders (Nexceris, LLC, Fuelcellmaterials, Lewis Center, OH, USA) of 6.2 m^2/g, 11.3 m^2/g, and 201.3 m^2/g surface area were used as the evaporating material. They were pressed into pellets using a mechanical press (Carver Model 3851 C, Carver Inc., Wabash, IN 46992-0554, USA). The pellets were placed into a crucible and vacuum chamber, which was depressurized up to 2.0×10^{-4} Pa. After that, the substrates were treated with Ar$^+$ ion plasma (10 min) and preheated up to the working temperature. The pressure during deposition was approximately 2.6×10^{-2} Pa. The thickness (1500 nm–2000 nm) and deposition rate were controlled with a INFICON (Inficon, Bad Ragaz, Switzerland) crystal sensor. The structure of the deposited thin films was studied using an X-ray diffractometer (XRD) "Bruker D8 Discover, Bruker AXS GmbH, Karlsruhe, Germany" at a 2Θ angle in a 20°–70° range using Cu Kα ($\lambda = 0.154059$ nm) radiation, a 0.01° step, and the Lynx eye PSD detector. EVA Search–Match software and the PDF-2 database were used to identify diffraction peaks.

Measured patterns were fitted using Pawley's method. The crystallite size was estimated according to the Scherrer's equation [14]:

$$\langle d \rangle = \frac{K \lambda_R}{\beta \cos\theta} \qquad (1)$$

where $\langle d \rangle$ is the crystallite size of the material, λ_R is the wavelength of the X-ray radiation, β is the full width at half the maximum, K is the correction factor, and θ is the angle of diffraction.

The texture coefficient $T_{(hkl)}$ was determined using the formula [15]:

$$T_{(hkl)} = \frac{I(hkl)}{I_0(hkl)} \left[\frac{1}{n} \sum_1^n \frac{I(hkl)}{I_0(hkl)} \right]^{-1} \qquad (2)$$

where $I(hkl)$ is the intensity of the XRD peak corresponding to (hkl) planes, n is the number of the diffraction peaks considered, $I_0(hkl)$ denotes the intensity of the XRD peak in the EVA Search–Match software database.

$T(hkl) = 1$ corresponds to films with randomly oriented crystallites, while higher values indicate a large number of grains oriented in a given (hkl) direction. The surface topography images and cross-section images were obtained using the scanning electron microscope (SEM) "Hitachi S-3400N, Hitachi High-Technologies Corporation, Tokyo, Japan". Elemental composition was controlled using an energy-dispersive X-ray spectroscope (EDS) "BrukerXFlash QUAD 5040, Bruker AXS GmbH, Karlsruhe, Germany". Sm concentration was lower in thin films (~13 mol%) than in evaporating powders (~20 mol%) due to the different sublimation temperatures of CeO_2 (2100 °C) and Sm_2O_3 (1950 °C).

Optical transmittance spectra were measured by a "UV-VIS 650, JASCO Deutschland GmbH, Pfungstadt, Germany" spectrometer. The band gap calculations were carried out according to Tauc's relation:

$$\left(h\nu_f\alpha \right)^{\frac{1}{n}} = A\left(h\nu - E_g \right) \qquad (3)$$

where h is the Plank constant, ν_f is the photon frequency, α is the absorption coefficient, E_g is the band gap, A is the proportionality constant, n is $1/2$, $3/2$, 2 or 3.

The total conductivity was measured by the impedance spectrometer "NorECs AS, Oslo, Norway" (EIS). The impedance measurements were performed for thin films deposited on Al_2O_3 because these substrates have a high melting temperature (2345 K) and a low electrical conductivity. The electrodes were made of a platinum paste applied on the top of thin films using a mask of particular geometry. The distance between Pt electrodes was 10 mm and their dimensions were 3 mm × 10 mm. The measurements were carried out in $1 \div 10^6$ Hz frequency range (13 points per decade) and in 200 °C $\div$ 1000 °C temperature interval using the two-probe method. A linear Kramers–Kronig Validity test (KK) was carried out using the "Lin-KK Tool, Karlsruhe Institute of Technology, Karlsruhe, Germany" software [16–19] and analysis of distribution function of relaxation times (DFRT) was performed using the "DRT tools, The Hong Kong University of Science and Technology, Hong Kong, China" software [20]. For the DFRT's analysis, the parameters were chosen as follows:

- discretization Method—Gaussian,
- regularization parameter—$\lambda = 0.01$,
- regularization derivative—first order,
- Radial Basis Function (RBF) Shape control (Coefficient to Full-Width Half-Maximum (FWHM))—0.15.

Finally, the total conductivity was calculated according to:

$$\sigma = \frac{L_e}{R_s A} = \frac{L_e}{R_s h l_e}, \qquad (4)$$

where L_e is the distance between the Pt electrodes, R_s is the resistance obtained from impedance spectra, A is the cross-sectional area, h is the thickness of the thin films, and l_e is the length of the electrodes.

3. Results

3.1. X-Ray Analysis

Patterns of SDC thin films (JCPSD No. 01-075-0157) have peaks belonging to (111), (200), (220), (311), (222), and (400) crystallographic orientations (Figure 1a–c). The positions of the peaks correspond to the fluorite structure, Fm3m space group. However, patterns of SDC thin films do not necessarily have all of these peaks. Peaks might disappear, or their intensities might change by changing the deposition parameters.

All SDC thin films deposited at 50 °C, 150 °C, and 300 °C temperature substrates have a similar structure. No influence of the specific surface area of powders was observed for the thin films deposited at the aforementioned temperatures. A different situation was observed when using 450 °C and higher temperature substrates during deposition. For example, SDC thin films deposited by evaporating powders of different surface areas on SiO_2 substrates at 600 °C had a different crystalline structure (Figure 1a–c). X-ray diffraction patterns of SDC thin films deposited 6.2 m^2/g by evaporating powder had peaks corresponding to (111), (200), (220), (311), (222), and (400) crystallographic orientations. The X-ray diffraction patterns of SDC thin films deposited by evaporating powder (11.3 m^2/g) had peaks which corresponded to (111), (220), (311), and (222) crystallographic orientations. (200) and (400) crystallographic orientations disappeared. An arrangement of diffraction peaks in patterns of SDC thin films deposited by evaporating powder of 201.3 m^2/g specific surface area, was very similar to the arrangement in patterns of SDC thin films deposited by evaporating powder of 11.3 m^2/g specific surface area.

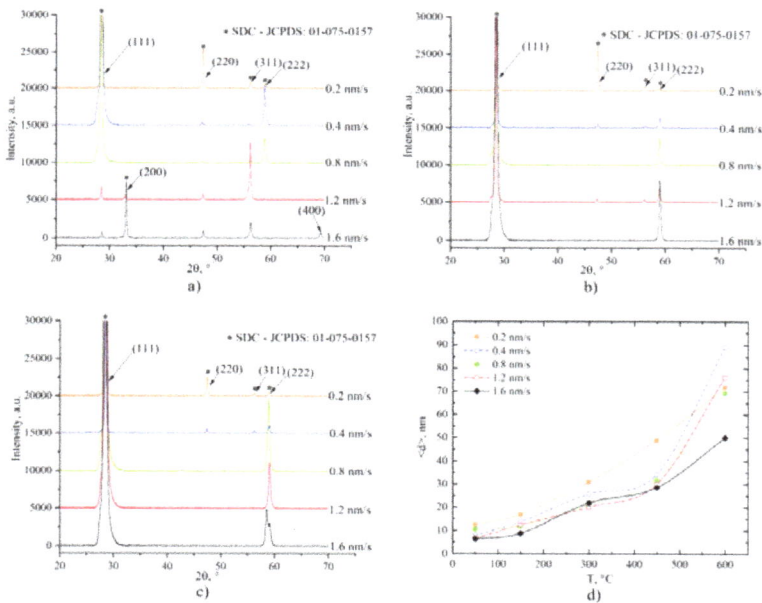

Figure 1. X-Ray Diffraction (XRD) measurements results: XRD patterns of thin Sm-doped cerium (SDC) films deposited on SiO_2 substrate (600 °C) evaporating powders of (**a**) 6.2 m^2/g, (**b**) 11.3 m^2/g, (**c**) 201.3 m^2/g specific surface area, and (**d**) dependences of crystallite size on SiO_2 substrate temperature (11.3 m^2/g powder).

In addition, SDC thin films deposited by evaporating powders of 6.2 m^2/g and 11.3 m^2/g specific surface area changed their preferred orientation from (111) to (311), (200), and (220) (Figure 1a–c, Table 1). The preferred orientation changes to (311) and (200) orientation evaporating 6.2 m^2/g of powder and using 1.2–1.6 nm/s deposition rates, respectively, and to (220) evaporating 11.3 m^2/g of powder and using a 0.2 nm/s deposition rate. SDC thin films deposited by evaporating powder of 201.3 m^2/g specific surface area did not change the preferred orientation.

Table 1. Texture coefficients of SDC thin films deposited by evaporating powders of different specific surface area on SiO$_2$ substrates at 600 °C (v_g—deposition rate).

S_{BET}, m^2/g	v_g, nm/s	$T_{(hkl)\_SiO2}$					
		<111>	<200>	<220>	<311>	<222>	<400>
6.2	1.2	0.2	0.3	0.2	3.2	-	-
6.2	1.6	0.1	3.0	0.3	0.8	1.5	0.1
11.3	0.2	1.3	-	1.7	0.2	0.7	-

The crystalline structure of thin SDC films depends on the composition of the vapor phase during the evaporation process. V. Piacente et. al. investigated the evaporation processes of the Ce-CeO$_2$ mixture and CeO$_2$ [21]. They found that the vapor phase consists of CeO$_2$, CeO, and Ce where concentrations of CeO$_2$ and Ce are 5 and 100 times lower, respectively than the concentration of CeO, if the mixture of Ce-CeO$_2$ evaporates and the concentrations of CeO and Ce are four and six times lower, respectively than the concentrations of CeO$_2$, if CeO$_2$ evaporates. In addition, the reduction of CeO$_2$ occurs on the surface (~7 nm) of CeO$_2$ [22]. This means that the larger surface area of SDC powder is reduced if the powder of a larger specific surface area is used, i.e., there will be a deficiency of oxygen. The experiments of A.F. Orliukas et. al. proved this statement. They found that Ce^{3+} concentration is 14.16 % for powders of 8 m^2/g specific surface area and 19.47 % for powders of 203 m^2/g specific surface area [23]. Therefore, the concentration of CeO and Ce should increase and concentration of CeO$_2$ should decrease in the vapor phase. This means that a higher number of CeO molecules and Ce atoms and lower number CeO$_2$ molecules reach the surface of the substrate using powders with a large specific surface area. Moreover, the diffusion distance and time of CeO$_2$ molecules is shorter than for CeO molecules and Ce adatoms [24]. As a result, the probability to form a crystallographic plane of higher surface energy ($\gamma_{(111)} < \gamma_{(200)} < \gamma_{(220)} < \gamma_{(311)}$) is higher when the concentration of CeO$_2$ is higher in vapor flux. Adatoms do not have enough time to occupy the lowest energy state. Diffusion distance of adatoms and molecules increases in substrates of higher temperature. Hence, more surfaces of low surface energy ((111) and (222)) are formed. However, the probability to form surfaces of (111) and (222) may strongly decrease due to a strong decrease in surface energy at high temperatures. In this case, more surfaces of higher surface energy are formed, i.e., (200), (220), (311) or (400), meaning that the probability of changing preferred orientation increases.

Calculations of crystallite sizes revealed that the crystallite size of SDC thin films is higher for substrates of higher temperature (Figure 1d). The size of the crystallites changes from 6.40 nm to 89.1 nm. The reason is that the diffusion rate of adatoms is higher and the distance is longer for substrates at a higher temperature. Moreover, crystallites grow larger evaporating powders of a larger specific surface area (Table 2). An influence of the specific surface area of powders on the crystallite size was observed only for SDC thin films deposited at high temperatures (600 °C). Such changes in the crystallite size prove the statement that CeO and Ce concentrations are higher for powders with a larger specific surface area. The diffusion rates of CeO dissociation products and Ce adatoms are higher than the diffusion rates of CeO$_2$ molecules or their dissociation products on the surface of the substrate.

Table 2. Crystallite size (<d>$_{SiO2}$) dependence on specific surface area of powders (S_{BET}), 0.4 nm/s deposition rate, 600 °C temperature of SiO$_2$ substrate.

S_{BET}, m^2/g	6.2	11.3	201.3
<d>$_{SiO2}$, nm	66.5 ± 0.3	89.1 ± 0.6	91.8 ± 0.8

3.2. Optical Band Gap Calculation

Ceria becomes conductive for oxygen ions when oxygen vacancies ($2[V_O^{··}] \approx [RE'_{Ce}]$) are created by doping it with lower valence dopants (Sm, Gd) [1]. It is also known that ceria loses oxygen with decreasing partial pressure of oxygen and increasing temperature. This means that a reduction of Ce^{4+} to Ce^{3+} occurs, i.e., oxygen vacancies and electrons $2[V_O^{··}] \approx [e']$ are created [2,22]. X-Ray photoelectron spectroscopy (XPS) measurements of SDC thin films deposited by evaporating powder of 6.2 m^2/g specific surface area proves the hypothesis about the reduction of Ce^{4+} to Ce^{3+}. Concentration of Ce^{3+} varies from 24.5% to 29.1% [25].

In addition, it was found that the optical band gap increases from 3.36 eV to 3.43 eV with increasing crystallite size from 10.2 nm to 91.8 nm (Figure 2). This means that the number of oxygen vacancies and the Ce^{3+} concentration decreases because the surface area where the reduction occurs also decreases with increasing crystallite size [26]. For this reason, fewer donor levels of Ce^{3+} are formed beneath the conduction band. Hence, SDC thin films are mixed ionic-electronic conductors.

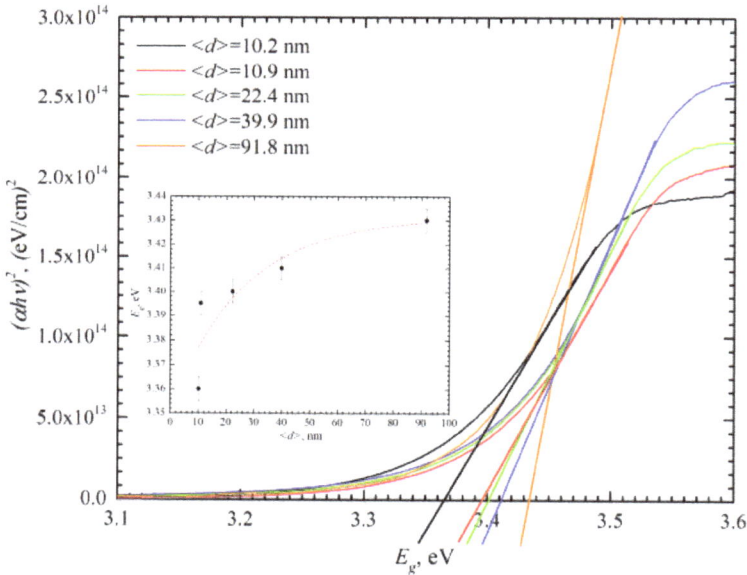

Figure 2. $(h\nu\alpha)^2$ dependence on photon energy and band gap (E_g) dependence on crystallite size ($\langle d \rangle$).

3.3. SEM Results

Thin films consist of grains with the shape of a triangular prism and growing on top of each other in the columnar way (Figure 3). The grain size increases with increasing temperature because the diffusion length of adatoms is longer for substrates of higher temperature. Hence, SDC thin films grow according to the structure zone models of Movchan–Demchishin and Grovenor in Zone 1 and Zone T [27,28]. Moreover, the surface morphology does not depend on substrate type (Figure 3b,c).

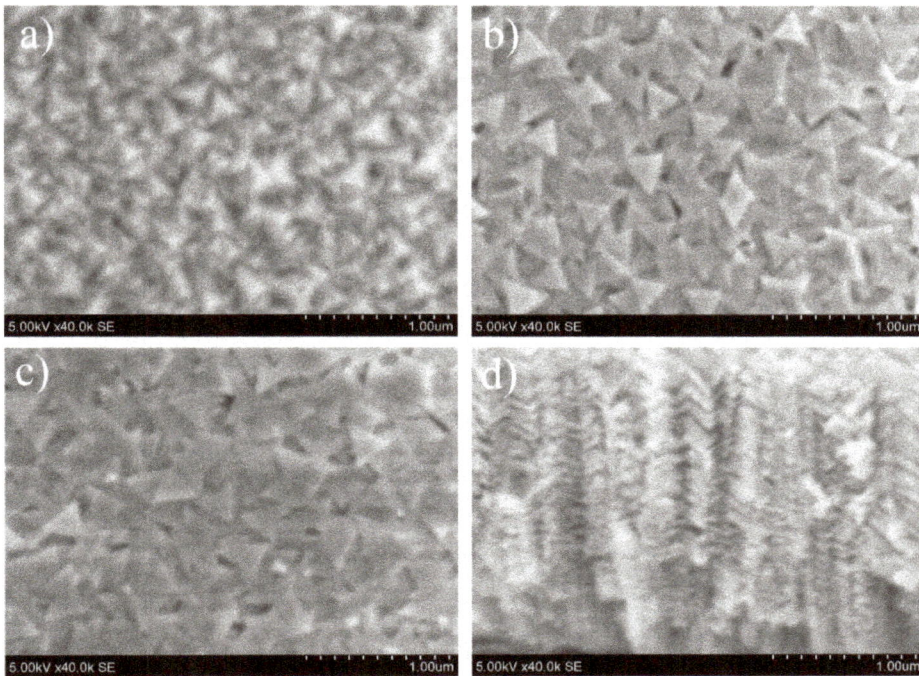

Figure 3. Topography (**a**, **b**, **c**) and cross section (**d**) images of thin SDC films deposited evaporating powder of 201.3 m^2/g specific surface area with a 1.6 nm/s deposition rate. (**a**) 50 °C SiO$_2$ substrate, (**b**) 600 °C SiO$_2$ substrate, (**c**) 600 °C Al$_2$O$_3$ substrate, and (**d**) 600 °C SiO$_2$ substrate.

3.4. EIS Analysis

The activation energy of oxygen vacancies changed from 0.770 eV to 1.042 eV. No relationship between total conductivity and activation energy of oxygen vacancies was observed. In addition, no relationship between total conductivity and changes in preferred crystallographic orientation was determined. Quantitative calculations of total conductivity revealed that, for example, total conductivities of SDC thin films deposited by evaporating powder of 201.3 m^2/g specific surface area on Al$_2$O$_3$ substrates using a 0.4 nm/s deposition rate are 4.4×10^{-7} S/cm, 9.0×10^{-4} S/cm, 1.5×10^{-3} S/cm, 3.3×10^{-3} S/cm, and 7.3×10^{-3} S/cm if 50 °C, 150 °C, 300 °C, 450 °C, and 600 °C substrates are used, respectively (Figure 4a). The values of the total conductivity were calculated for the experimental points measured at 600 °C. It follows that the total conductivity is higher if higher temperature substrates are used during deposition (Figure 4a). In addition, SDC thin films deposited on 50 °C substrates exhibit much lower total conductivity because some SDC thin films deposited on 50 °C–150 °C substrates are nonhomogeneous and have microcracks. These thin films have an extremely low total conductivity (3.4×10^{-8} S/cm–3.9×10^{-6} S/cm). In comparison, thin films deposited on higher temperature substrates are homogeneous (Figure 4c).

A semicircular shape (Figure 4b) indicates that there is more than one component of impedance in the Nyquist plot. However, the plots are very complex, and it is not possible to determine the number of components (RQ elements) and their contribution based on the Nyquist plot. Therefore, it was decided to use the distribution function of relaxation times (DFRT) in the impedance analysis.

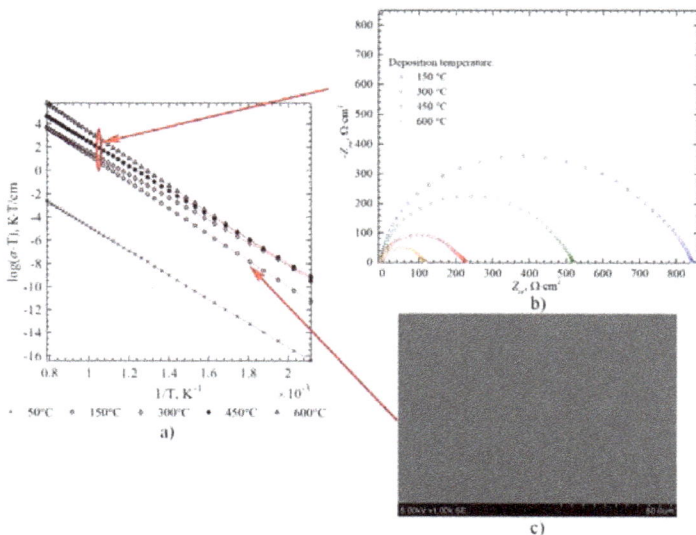

Figure 4. (**a**) Arrhenius plot and (**b**) Nyquist plots for SDC thin film deposited evaporating powder of 201.3 m^2/g specific surface area on Al$_2$O$_3$ substrates with a 0.4 nm/s deposition rate and 150–600 °C substrate temperature, and (**c**) a topography image of a thin SDC film deposited on an Al$_2$O$_3$ substrate at 150 °C with a 0.4 nm/s deposition rate.

Firstly, the Kramers–Kronig (KK) transformations of real and imaginary parts were carried out to check the data validity. Calculations revealed that the relative residuals between the KK transformations and the experimental data are small (± 0.25 %) (Figure 5a). Moreover, residuals do not show systematic deviations from abscissa axis. Therefore, the obtained results confirm the data validity and allow an analysis of the DFRT (Figure 5b). Two peaks are visible in the DFRT plots representing processes in "grains" and grain boundaries, respectively. The relaxation times from 2.4 µs to 12.2 µs represent processes in "gains" and the relaxation times from 80.9 µs to 275.4 µs represent processes in grain boundaries. A similar contribution was observed in all experiments.

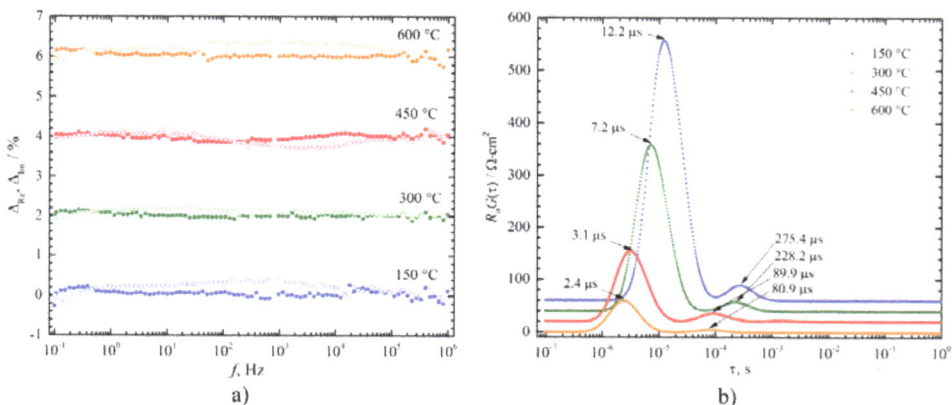

Figure 5. (**a**) Relative residuals of the Kramers–Kronig (KK) test (closed symbols: Δ_{re}, open symbols: Δ_{im}. The data have been offset by steps of 2%) and (**b**) the DRFT's plots for SDC thin films deposited by evaporating powder of 201.3 m^2/g specific surface area and with a 0.4 nm/s deposition rate.

According to Gobel M.C. et al., the increased ratio of grain boundary width (2*b*) to grain width (*L*) is the reason for the decrease in ionic conductivity because the grain boundaries have a blocking effect on the diffusion of oxygen ions [5]. The grain boundaries, which are parallel to the migration path of oxygen ions (perpendicular to the electrodes), affect the ionic conductivity [3]. However, it is not possible to decompose the measured impedance of "grain" into the impedance of the grain boundary and the grain because it is not possible to separate the impedance of circuit elements connected in parallel using EIS [29]. For these reasons, it is assumed that the short relaxation time component in the spectra of DFRT corresponds to the processes in the "grain". The influence of grain boundaries on ionic conductivity (decrease) and electronic conductivity (increase) increase with a decreasing grain size. Moreover, it is reasonable to believe that the influence of substrate temperature and crystallite size on the conductivity of the SDC thin films is similar because the size of the crystallites is smaller when substrates of lower temperature are used during deposition (Figure 1d).

Indeed, calculations revealed that the normalized resistance of "grains" increases from 86.5 Ωcm^2 to 802.1 Ωcm^2 and the normalized resistance of grain boundaries increases from 5.8 Ωcm^2 to 36.2 Ωcm^2 when the crystallite size decreases from 91.8 nm to 10.9 nm (Figure 6a). Further, the resistance of "grains" is much higher than the resistance of grain boundaries (~18 times). The critical frequency of oxygen ions in "grains" increases from 13.1 kHz to 65.5 kHz and the critical frequency of oxygen ions in grain boundaries increases from 0.6 kHz to 2.0 kHz with increasing crystallite size from 10.9 nm to 91.8 nm (Figure 6b).

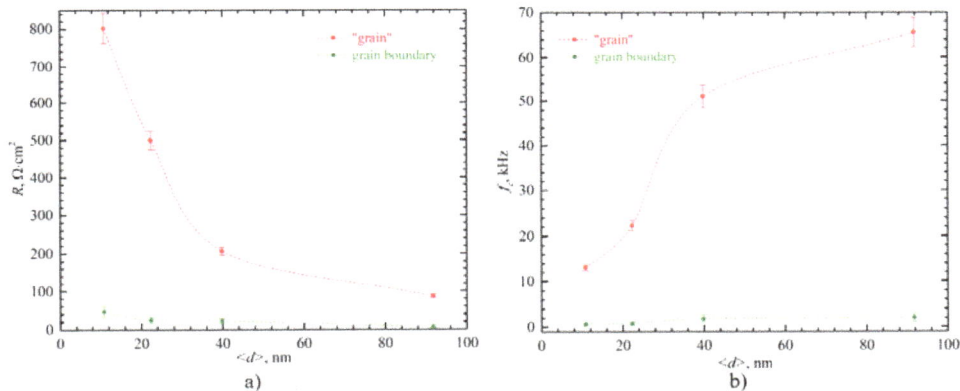

Figure 6. (**a**) Dependencies of normalized resistance (*R*) on crystallite size and (**b**) Dependencies of critical frequency (*f*$_c$) on crystallite size for SDC thin films deposited by evaporating powder of 201.3 m^2/g specific surface area and with a 0.4 nm/s deposition rate.

It follows that the total conductivity is higher if the SDC thin films have larger crystallites (Figure 7). The total conductivity is lower than 0.002 S/cm if the crystallite size is approximately 10–25 nm and more than 0.006 S/cm if the crystallite size is approximately 90 nm. The given values of ionic conductivity are calculated for impedance values measured at 600 °C. The highest value of total conductivity (1.1 $\times$ 10^{-2} S/cm) is similar to the value obtained by other authors: 15SDC–0.013 S/cm [30]. Moreover, the total conductivity of SDC thin films is lower than $Bi_2V_{1.9}Cu_{0.1}O_{5.535}$ (3.3 $\times$ 10^{-1} S/cm) and $La_{0.85}Sr_{0.15}Ga_{0.2}Mg_{0.2}O_{2.825}$ (3.0 $\times$ 10^{-2} S/cm), but higher than Yttrium-Stabilised Zirconia 8YSZ (6.4 $\times$ 10^{-3} S/cm), $La_{1.8}Gd_{0.2}Mo_2O_9$ (7.1 $\times$ 10^{-3} S/cm), $La_{9.75}Sr_{0.25}Si_6O_{26.875}$ (1.0 $\times$ 10^{-2} S/cm) [30–35]. Such ionic conductivity is influenced by crystallite size, dopant concentration, and Ce^{3+} concentration. It is noteworthy that the ionic conductivity may vary greatly even if the thin films are formed with the same crystallite size (e.g., Figure 7, S_{BET} = 11.3 m^2/g and S_{BET} = 201.3 m^2/g). In most cases, this difference occurs due to morphological changes, e.g., increased porosity and different thin film growing kinetics. In the particular case presented in Figure 7, the difference in ionic conductivity appeared due to the lower density of the SDC thin films deposited using 201.3 m^2/g powder.

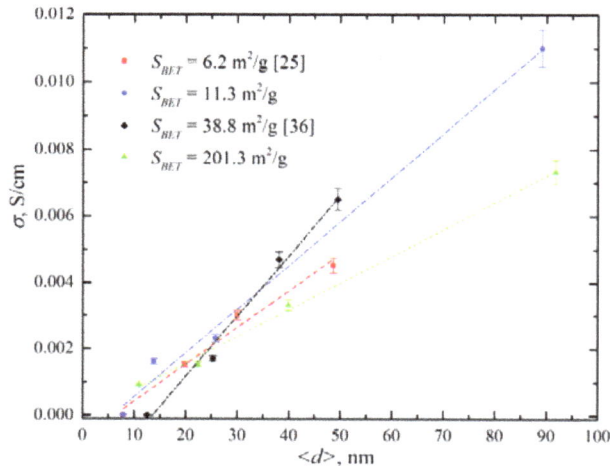

Figure 7. Dependence of total conductivity (σ) on crystallite size ($\langle d \rangle$) (0.4 nm/s deposition rate). Measurements were carried out on an Al$_2$O$_3$ substrates at 600 °C [25,36].

4. Conclusions

An investigation of SDC thin films revealed that preferential crystallographic orientation (111) changes to (200), (220), (311) or (222) if 450–600 °C substrate temperatures and special deposition rates are used during deposition. It was also noticed that there is a relationship between changes in crystallographic orientation and specific surface area of initial powders. Preferential crystallographic orientation (111) tends to change its orientation to (200), (220) or (311) evaporating the powder of 6.2 m^2/g specific surface area to (220) or (311) evaporating the powder of 11.3 m^2/g specific surface area, and to (311) evaporating the powder of 201.3 m^2/g specific surface area. Moreover, a smaller number of SDC thin films changed preferential crystallographic orientation for powders of higher specific surface areas. Crystallite size depended on the temperature of the substrate and specific surface area of the initial powder. Larger crystallites are obtained for substrates of higher temperature and larger specific surface area of powders. The size increases by 1.4 times if the powder of 6.2 m^2/g specific surface area is changed to the powder of 201.3 m^2/g specific surface area.

Changes in the microstructure can be explained by the changes in the gas phase composition. There was more CeO, Ce, and Sm and less CeO$_2$ in the vapor phase when powders of the larger specific surface area were used. Hence, the diffusion distance and time of CeO and adatoms are longer.

It was noticed that the crystallite size has an impact on electrical properties of formed SDC thin films. The dependence of the optical band gap on the crystallite size proved that SDC thin films are mixed conductors and that the Ce^{3+} concentration decreases with increasing crystallite size. The total conductivity also increased with increasing crystallite size. The values of total conductivity are approximately 2.0 × 10^{-3} S/cm when the crystallite size is 10.0–25.0 nm and 6.0 × 10^{-3} S/cm when the crystallite size is 80.0 nm. The highest value of total conductivity was 1.1 × 10^{-2} S/cm (600 °C) when crystallite sizes were 80.6 nm and 89.1 nm.

Author Contributions: Authors contributed equally to the manuscript. More specifically, conceptualization, M.S. and G.L.; methodology, K.B. and N.K.; formal analysis, M.S. and K.B.; investigation, M.S. and N.K.; writing—original draft preparation, M.S., K.B., G.L., and N.K.; writing—review and editing, M.S., K.B., G.L., and N.K.; validation, K.B. and N.K.; supervision, G.L.; funding acquisition, G.L.

Conflicts of Interest: The authors declare no conflict of interest.

References

1. Tuller, H.L.; Bishop, S.R. Point Defects in Oxides: Tailoring Materials Through Defect Engineering. *Annu. Rev. Mater. Res.* **2011**, *41*, 369–398. [CrossRef]

2. Migani, A.; Vayssilov, G.N.; Bromley, S.T.; Illas, F.; Neyman, K.M. Greatly facilitated oxygen vacancy formation in ceria nanocrystallites. *Chem. Commun.* **2010**, *46*, 5936–5938. [CrossRef] [PubMed]

3. Tschöpe, A.; Sommer, E.; Birringer, R. Grain size-dependent electrical conductivity of polycrystalline cerium oxide: I. Experiments. *Solid State Ionics* **2001**, *139*, 255–265. [CrossRef]

4. Daniele, P.; Vladimir, R.; Emiliana, F.; Christof, W.S.; Thomas, L.; Enrico, T.; John, A.K. Probing the bulk ionic conductivity by thin film hetero-epitaxial engineering. *Sci. Technol. Adv. Mater.* **2015**, *16*, 015001.

5. Gobel, M.C.; Gregori, G.; Maier, J. Numerical calculations of space charge layer effects in nanocrystalline ceria. Part I: comparison with the analytical models and derivation of improved analytical solutions. *Phys. Chem. Chem. Phys.* **2014**, *16*, 10214–10231. [CrossRef] [PubMed]

6. Gobel, M.C.; Gregori, G.; Maier, J. Numerical calculations of space charge layer effects in nanocrystalline ceria. Part II: detailed analysis of the space charge layer properties. *Phys. Chem. Chem. Phys.* **2014**, *16*, 10175–10186. [CrossRef] [PubMed]

7. Gobel, M.C.; Gregori, G.; Maier, J. Mixed conductivity in nanocrystalline highly acceptor doped cerium oxide thin films under oxidizing conditions. *Phys. Chem. Chem. Phys.* **2011**, *13*, 10940–10945. [CrossRef] [PubMed]

8. Tschöpe, A. Grain size-dependent electrical conductivity of polycrystalline cerium oxide II: Space charge model. *Solid State Ionics* **2001**, *139*, 267–280. [CrossRef]

9. Bârcă, E.S.; Filipescu, M.; Luculescu, C.; Birjega, R.; Ion, V.; Dumitru, M.; Nistor, L.C.; Stanciu, G.; Abrudeanu, M.; Munteanu, C.; et al. Pyramidal growth of ceria nanostructures by pulsed laser deposition. *Appl. Surf. Sci.* **2016**, *363*, 245–251. [CrossRef]

10. Balakrishnan, G.; Sundari, S.T.; Kuppusami, P.; Mohan, P.C.; Srinivasan, M.P.; Mohandas, E.; Ganesan, V.; Sastikumar, D. A study of microstructural and optical properties of nanocrystalline ceria thin films prepared by pulsed laser deposition. *Thin Solid Films* **2011**, *519*, 2520–2526. [CrossRef]

11. Olding, T.; Sayer, M.; Barrow, D. Ceramic sol–gel composite coatings for electrical insulation. *Thin Solid Films* **2001**, *398*, 581–586. [CrossRef]

12. Virbukas, D.; Laukaitis, G. The structural and electrical properties of samarium doped ceria films formed by e-beam deposition technique. *Solid State Ionics* **2017**, *302*, 107–112. [CrossRef]

13. Laukaitis, G.; Virbukas, D. The structural and electrical properties of GDC10 thin films formed by e-beam technique. *Solid State Ionics* **2013**, *247*, 41–47. [CrossRef]

14. Bail, A.L. Chapter 5 The Profile of a Bragg Reflection for Extracting Intensities. In *Powder Diffraction: Theory and Practice*; Dinnebier, R.E., Billinge, S.J.L., Eds.; The Royal Society of Chemistry: Cambridge, UK, 2008; pp. 134–165.

15. Nagaraju, P.; Vijayakumar, Y.; Ramana Reddy, M.V. Optical and microstructural studies on laser ablated nanocrystalline CeO_2 thin films. *Glass Phys. Chem.* **2015**, *41*, 484–488. [CrossRef]

16. Boukamp, B.A. Linear Kronig-Kramers Transform Test for Immittance Data Validation. *J. Electrochem. Soc.* **1995**, *142*, 1885–1894. [CrossRef]

17. Schönleber, M.; Klotz, D.; Ivers-Tiffée, E. A Method for Improving the Robustness of linear Kramers-Kronig Validity Tests. *Electrochim. Acta* **2014**, *131*, 20–27. [CrossRef]

18. Schönleber, M.; Ivers-Tiffée, E. Approximability of impedance spectra by RC elements and implications for impedance analysis. *Electrochem. Commun.* **2015**, *58*, 15–19. [CrossRef]

19. Tomoyasu, I.; Yasuhiro, Y.; Masataka, S. Electron-beam-assisted evaporation of epitaxial CeO_2 thin films on Si substrates. *J. Vac. Sci. Technol. A* **2001**, *19*.

20. Wan, T.H.; Saccoccio, M.; Chen, C.; Ciucci, F. Influence of the Discretization Methods on the Distribution of Relaxation Times Deconvolution: Implementing Radial Basis Functions with DRTtools. *Electrochim. Acta* **2015**, *184*, 483–499. [CrossRef]

21. Piacente, V.; Bardi, G.; Malaspina, L.; Desideri, A. Dissociation energy of CeO_2 and Ce_2O_2 molecules. *J. Chem. Phys.* **1973**, *59*, 31–36. [CrossRef]

22. Kato, S.; Ammann, M.; Huthwelker, T.; Paun, C.; Lampimaki, M.; Lee, M.-T.; Rothensteiner, M.; van Bokhoven, J.A. Quantitative depth profiling of Ce^{3+} in Pt/CeO_2 by in situ high-energy XPS in a hydrogen atmosphere. *Phys. Chem. Chem. Phys.* **2015**, *17*, 5078–5083. [CrossRef] [PubMed]

23. Orliukas, A.F.; Šalkus, T.; Kežionis, A.; Venckutė, V.; Kazlauskienė, V.; Miškinis, J.; Laukaitis, G.; Dudonis, J. XPS and impedance spectroscopy of some oxygen vacancy conducting solid electrolyte ceramics. *Solid State Ionics* **2011**, *188*, 36–40. [CrossRef]

24. Galdikas, A.; Čerapaitė-Truš, R.; Laukaitis, G.; Dudonis, J. Real-time kinetic modeling of YSZ thin film roughness deposited by e-beam evaporation technique. *Appl. Surf. Sci.* **2008**, *255*, 1929–1933. [CrossRef]

25. Sriubas, M.; Pamakštys, K.; Laukaitis, G. Investigation of microstructure and electrical properties of Sm doped ceria thin films. *Solid State Ionics* **2017**, *302*, 165–172. [CrossRef]

26. Dutta, P.; Pal, S.; Seehra, M.S.; Shi, Y.; Eyring, E.M.; Ernst, R.D. Concentration of Ce^{3+} and Oxygen Vacancies in Cerium Oxide Nanoparticles. *Chem. Mater.* **2006**, *18*, 5144–5146. [CrossRef]

27. Movchan, B.A. Study of the Structure and Properties of Thick Vacuum Condensates of Nickel, Titanium, Tungsten, Aluminium Oxide and Zirconium Oxide. *Fiz Met Metalloved.* **1969**, *28*.

28. Grovenor, C.R.M.; Hentzell, H.T.G.; Smith, D.A. The development of grain structure during growth of metallic films. *Acta Metall.* **1984**, *32*, 773–781. [CrossRef]

29. Maier, J. On the Conductivity of Polycrystalline Materials. *Berichte der Bunsengesellschaft für physikalische Chemie* **1986**, *90*, 26–33. [CrossRef]

30. Acharya, S.A.; Gaikwad, V.M.; D'Souza, S.W.; Barman, S.R. Gd/Sm dopant-modified oxidation state and defect generation in nano-ceria. *Solid State Ionics* **2014**, *260*, 21–29. [CrossRef]

31. Kharton, V.V.; Marques, F.M.B.; Atkinson, A. Transport properties of solid oxide electrolyte ceramics: A brief review. *Solid State Ionics* **2004**, *174*, 135–149. [CrossRef]

32. Guo, X.; Vasco, E.; Mi, S.; Szot, K.; Wachsman, E.; Waser, R. Ionic conduction in zirconia films of nanometer thickness. *Acta Mater.* **2005**, *53*, 5161–5166. [CrossRef]

33. Wei, T.; Singh, P.; Gong, Y.; Goodenough, J.B.; Huang, Y.; Huang, K. $Sr_{3-3x}Na_{3x}Si_3O_{9-1.5x}$ (x = 0.45) as a superior solid oxide-ion electrolyte for intermediate temperature-solid oxide fuel cells. *Energy Environ. Sci.* **2014**, *7*, 1680–1684. [CrossRef]

34. Tsai, D.-S.; Hsieh, M.-J.; Tseng, J.-C.; Lee, H.-Y. Ionic conductivities and phase transitions of lanthanide rare-earth substituted $La_2Mo_2O_9$. *J. Eur. Ceram. Soc.* **2005**, *25*, 481–487. [CrossRef]

35. Belousov, V.V. Oxygen-permeable membrane materials based on solid or liquid Bi_2O_3. *MRS Commun.* **2013**, *3*, 225–233. [CrossRef]

36. Sriubas, M.; Laukaitis, G. The influence of the technological parameters on the ionic conductivity of samarium doped ceria thin films. *Mater. Sci. (Medžiagotyra)* **2015**, *21*. [CrossRef]

Article

Structural and Magnetic Studies of Cr^{3+} Substituted Nickel Ferrite Nanomaterials Prepared by Sol-Gel Auto-Combustion

Jinpei Lin [1,2,†], Yun He [1,3,†], Xianglin Du [1], Qing Lin [1,2,*], Hu Yang [1] and Hongtao Shen [1,4,*]

[1] Guangxi Key Laboratory of Nuclear Physics and Nuclear Technology, Guangxi Normal University, Guilin 541004, China; linjinpei.2007@163.com (J.L.); hy@mailbox.gxnu.edu.cn (Y.H.); linqinglab@126.com(X.D.); yanghu000@126.com (H.Y.)
[2] College of Medical Informatics, Hainan Medical University, Haikou 571199, China
[3] Sate Key Laboratory for Chemistry and Molecular Engineering of Medicinal Resources, Guangxi Normal University, Guilin 541004, China
[4] College of Physics and Technology, Guangxi Normal University, Guilin 541004, China
*   Correspondence: hy@gxnu.edu.cn (Q.L.); heyunlab@163.com (H.S.)
†   These authors contributed equally to this work.

Received: 25 August 2018; Accepted: 17 September 2018; Published: 9 October 2018

Abstract: The present study envisages the preparation of chromium substituted Nickel ferrite NiCr$_x$Fe$_{2-x}$O$_4$ ($x = 0 \sim 1.0$) powders by a sol-gel auto-combustion method. X-ray diffraction analysis (XRD) showed that the specimens with $x > 0.2$ exhibited a single-phase spinel structure, and that more content of Cr within a specimen is favorable for the synthesis of pure Ni-Cr ferrites. The lattice parameter decreased with an increase in the Cr concentration. The sample without calcining exhibited a good crystallinity. Scanning Electron Microscopy (SEM) showed the formation of ferrite powders nano-particles, and that the substitution of Cr weakened the agglomeration between the particles. Mössbauer spectra of NiCr$_x$Fe$_{2-x}$O$_4$ showed two normal Zeeman-split sextets that displayed a ferrimagnetic behavior. Furthermore, the spectra indicated that iron was in the Fe^{3+} state, and the magnetic hyperfine field at the tetrahedral tended to decrease with an increase in the Cr substitution. The saturation magnetization decreased by the Cr^{3+} ions, and reached a minimum value (Ms = 4.46 emu/g). With an increase in the annealing temperature, the coercivity increased initially, which later decreased.

Keywords: Ni-Cr-ferrite; sol-gel; structure; Cr substitution; Mössbauer; magnetic properties

1. Introduction

Nickel ferrite (NiFe$_2$O$_4$) is a typical soft magnetic ferrite, making it one of the most important spinel ferrites [1]. It is widely used in electronic devices, due to its ability to remain permeable at high frequencies, high electrical resistivities, low eddy currents, and low dielectric losses, as well as its continuous chemical stability [2,3]. NiFe$_2$O$_4$ is an inverse spinel ferrite, in which the Ni^{2+} ions occupy octahedral (B) sites, and Fe^{3+} ions are distributed in tetrahedral (A) and octahedral (B) sites [2]. The magnetic and electric properties of the spinel ferrites depend on the distribution of the cations among tetrahedral and octahedral sites. Chromium ions, with an antiferromagnetic nature, are known for achieving control over magnetic parameters, and in the development of technologically important materials. Patange et al. investigated the cation distribution and magnetic properties of chromium-substituted nickel ferrite, and found that Cr^{3+} and Ni^{2+} ions both have strong preference to occupy octahedral (B) site, and the magnetic saturation and Curie temperature all decrease by Cr^{3+} ions substitution [1]. Lee et al. studied the electrical and magnetic properties of ferrite NiCr$_x$Fe$_{2-x}$O$_4$, which established that the magnetic moment and Curie temperature decreased with the chromium substitution, and that this system exhibited n-type semi-conductivity [4]. Prasad et al. prepared nanoparticles $NiCr_xFe_{2-x}O_4$ by sol-gel, and concluded that the magnetization and coercivity decreased with an increase in the chromium ion concentration, due to the depletion of Fe^{3+} ions at the octahedral (B) sites [5]. Patange et al. obtained the cation distribution of nickel chromium ferrites by the IR spectra analysis [6]. The current paper deals with structural and magnetic properties of ferrite $NiCr_xFe_{2-x}O_4$ (x = 0~1.0) powders that were prepared by a sol-gel auto-combustion method. The ferrite powders, prepared by this method exhibited a good sinterability with a homogeneous composition. The other advantages of the synthetic method include the requirement of relatively simple equipment, and low cost of the materials used. From previous studies [4–6], it is evident that increasing the substitution of Cr is favorable for the synthesis of pure Ni-Cr ferrites, and weakens the agglomeration between the particles.

2. Experimental

2.1. Sample Preparation

Chromium-substituted Nickel ferrite $NiCr_xFe_{2-x}O_4$ (x = 0~1.0) powders were synthesized using the sol-gel auto-combustion process. The raw materials were analytical grade $Ni(NO_3)_2 \cdot 6H_2O$, $Cr(NO_3)_3 \cdot 9H_2O$, $Fe(NO_3)_3 \cdot 9H_2O$, $C_6H_8O_7 \cdot H_2O$ (citric acid) and $NH_3 \cdot H_2O$ (ammonia). The molar ratio of metal nitrates to citric acid was taken as 1:1. The metal nitrates and citric acid were weighed and dissolved in deionized water to prepare the test solutions. The pH value of metal nitrate solution was adjusted from 7 to 9 by ammonia addition. The mixed solutions were heated in a thermostat water bath at 80 °C and stirred continuously to form the dried gel. Continuous dropwise addition of citric acid occurred in this process. The gels were dried in an oven at 120 °C for 2 h and were burnt in self-propagating combustion to form loose powder by being ignited in air at room temperature. The powders then were ground and sintered at temperatures of 400 °C and 800 °C.

2.2. Characterization

The crystalline structure was analyzed using X-ray diffraction (D/max-2500V/PC, Rigaku, Tokyo, Japan), with Cu Kα radiation (λ = 0.15405 nm). The micrographs were obtained by scanning electron microscopy (NoVaTM Nano SEM 430, FEI Corporation, Hillsboro, OR, USA). The Mössbauer spectrum was performed at room temperature, using a conventional Mössbauer spectrometer (Fast Com Tec PC-mossII, Oberhaching, Germany), in constant acceleration mode. The γ-rays were provided by a ^{57}Co source in a rhodium matrix. Magnetization measurements were carried out with super conducting quantum interference device (MPMS-XL-7, Quantum Design, San Diego, CA, USA) at room temperature.

3. Results and Discussion

3.1. X-ray Diffraction (XRD) Analysis

Figure 1 shows the XRD patterns of $NiCr_xFe_{2-x}O_4$ (x = 0~1.0) ferrites calcined at 800 °C for 3 h. XRD results show that the specimens with x > 0.2 exhibited a single-phase spinel structure, while an impurity peak of Fe_2O_3 was detected in the samples with x = 0 to 0.2. The results show that increasing the content of Cr was favorable for the synthesis of pure Ni-Cr ferrites. Similar results are evident in earlier studies [7]. Table 1 indicates that the lattice parameter showed a decreased trend, with an increasing substitution of Cr^{3+} ions. The decrease in the lattice parameter was likely due to the replacement of larger Fe^{3+} ions (0.645 Å), by smaller Cr^{3+} ions (0.63 Å) [4,8]. The average crystallite size of the investigated samples that was estimated by the Scherrer's formula [5], were found to be between 25.5 and 57.5 nm (Table 1). It was observed that the average crystallite size decreased by increasing the Cr content, as evident with earlier reports [9].

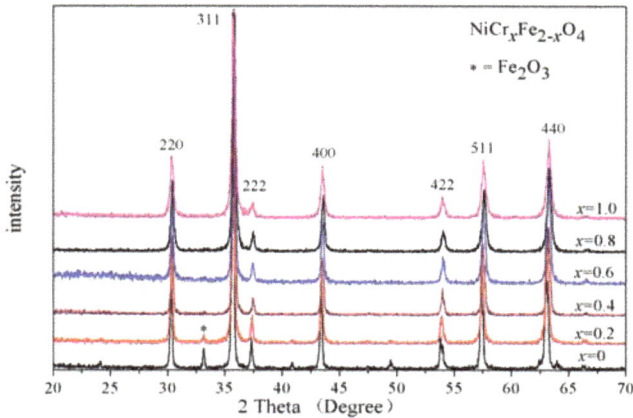

Figure 1. XRD patterns of $NiCr_xFe_{2-x}O_4$ calcined at 800 °C.

The X-ray density was calculated using the relation [10]:

$$\rho_x = \frac{8M}{Na^3} \tag{1}$$

where M is relative molecular mass, N is the Avogadro's number and 'a' is the lattice parameter. Table 1 shows that the X-ray density highlighted a decreasing trend for Cr^{3+} concentration for all samples. The atomic weight of Fe is greater than Cr, so the relative molecular mass decreased with an increase in the Cr concentration. The decrease in the X-ray density is attributed to the fact that the relative molecular mass decreases significantly with a negligible decline in the lattice parameter.

Table 1. XRD data of $NiCrxFe_{2-x}O_4$ calcined at 800 °C.

Content (x)	Lattice Parameter (Å)	Average Crystallite Size (Å)	Density (g/cm³)
0	8.34062	575	5.3664
0.2	8.33600	446	5.3562
0.4	8.31862	453	5.3735
0.6	8.31741	375	5.3581
0.8	8.30992	308	5.3548
1.0	8.31979	255	5.3179

The X-ray patterns of $NiCr_{0.2}Fe_{1.8}O_4$ calcined at different temperature are shown in Figure 2. XRD patterns confirmed the formation of cubic spinel phase as the main phase along with traces of the secondary phase of Fe_2O_3. The intensity of Fe_2O_3 decreased with increasing heating temperatures. Furthermore, it was evident that the heat treatment was favorable for the synthesis of pure Ni-Cr ferrites. The lattice parameter and the X-ray density differed with a change in temperature for the samples. Average crystallite size of $NiCr_{0.2}Fe_{1.8}O_4$ increased with increasing the calcining temperature, as evident from Table 2. In earlier work [11], the diffraction peaks of $Ni_{0.50}Cu_{0.25}Zn_{0.25}Cr_xFe_{2-x}O_4$ calcined at low temperature were not very sharp, but our results indicate that the diffraction peaks of $CoCr_{0.2}Fe_{1.8}O_4$ without burning were very sharp. We prepared chromium substituted cobalt ferrite powders by the sol-gel auto-combustion method, whereas the samples without calcining showed a good crystallinity.

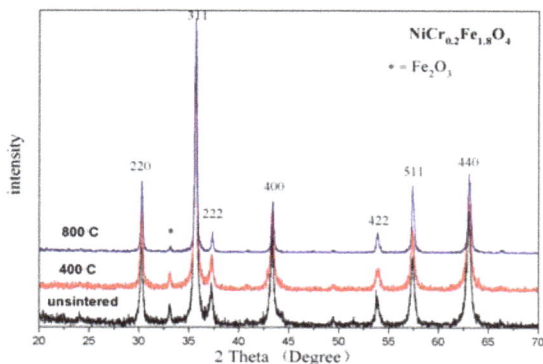

Figure 2. XRD patterns of $NiCr_{0.2}Fe_{1.8}O_4$ sintered at different temperatures.

Table 2. XRD data of $NiCr_{0.2}Fe_{1.8}O_4$ sintered at different temperatures.

Temperature (°C)	Lattice Parameter (Å)	Average Crystallite Size (Å)	Density (g/cm³)
unsintered	8.34459	293	5.3411
400 °C	8.34400	300	5.3424
800 °C	8.36000	446	5.3562

3.2. Structures and Grain Size

The Scanning Electron Microscopy (SEM) results of $NiCr_xFe_{2-x}O_4$ ($x = 0, 0.2$) annealed at 800 °C for 3 h is shown in Figure 3. It can be observed that the distribution of grains was almost uniform in size, and were well crystallized for all samples. The substitution of Cr can weaken the agglomeration between the particles. The histogram of grain size distribution of $NiCr_xFe_{2-x}O_4$ ferrites is shown in Figure 4 shows. The average grain size of $NiFe_2O_4$ and $NiCr_{0.2}Fe_{1.8}O_4$ estimated using a statistical method was approximately 69.51 and 52.63 nm, respectively. This shows that the ferrite powders were nano-particles, and the average grain size decreased significantly with an increase in the Cr content. The average grain size was slightly larger than the average crystallite size, as determined by XRD analysis. The data reveals that every particle was formed by a number of crystallites.

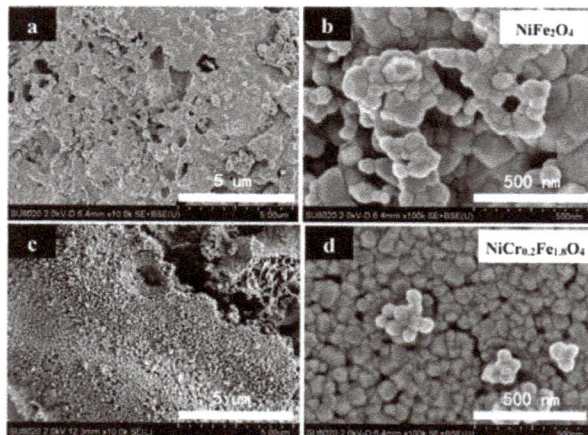

Figure 3. SEM micrographs of $NiFe_2O_4$ and $NiCr_{0.2}Fe_{1.8}O_4$ sintered at 800 °C.

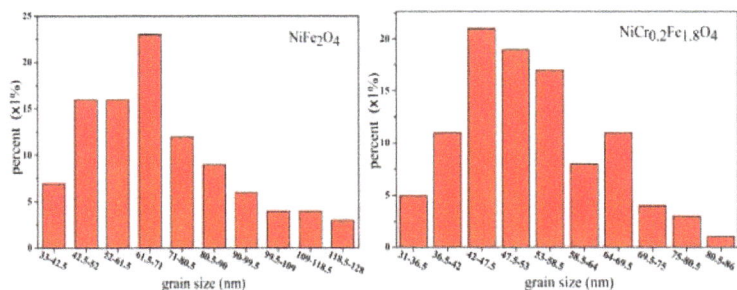

Figure 4. Histogram of grain size distribution of $NiFe_2O_4$ and $NiCr_{0.2}Fe_{1.8}O_4$.

3.3. Mössbauer Spectroscopy

The Mössbauer spectra for $NiCr_xFe_{2-x}O_4$, recorded at room temperature, is shown in Figure 5. All samples were analyzed using the Mösswinn 3.0 program. The spectra exhibited two normal Zeeman-split sextets due to the presence of Fe^{3+} at tetrahedral and octahedral sites, which indicated the ferromagnetic behavior of the samples [12]. The sextet with the larger isomer shift was assigned to the Fe^{3+} ions at the B site, and the one with the smaller isomer shift was assumed to arise from the Fe^{3+} ions, occupying the A site. This may be attributed to the difference in Fe^{3+}–O^{2-} internuclear separations. Since, the bond separation for the B site Fe^{3+} ions, was larger in comparison with the A site ions, small overlapping of the orbits for Fe^{3+} and O^{2+} ions at B site occurred, resulting in the smaller covalency and large isomer shift for B site Fe^{3+} ions [9]. It is reported that the values of Isomer Shift (I.S.) for Fe^{2+} ions lie in the range 0.6–1.7 mm/s, while for Fe^{3+} they lie in the range 0.1–0.5 mm/s [13]. From Table 3, the values for I.S. in our study indicate that iron is in Fe^{3+} state.

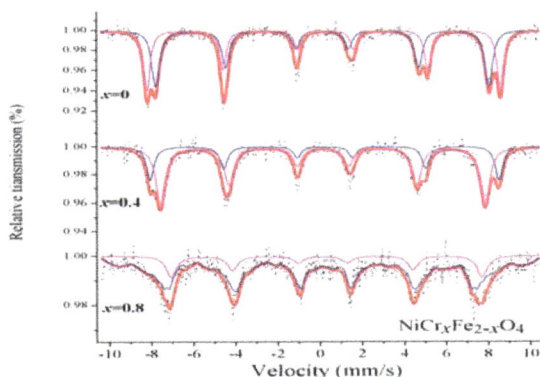

Figure 5. Mössbauer spectra of $NiCr_xFe_{2-x}O_4$ calcined at 800 °C.

Table 3 shows that the magnetic hyperfine field at the tetrahedral A site, and octahedral B site, exhibited a decreasing trend with increasing Cr substitution. The decrease of magnetic hyperfine field with increasing Cr contents is attributed to the decrease of the A-B super-exchange interaction with the magnetic Fe^{3+} ions replaced by magnetic Cr^{3+} ions, which resulted in the decrease of the magnetic hyperfine field [9]. Furthermore, the reduction of magnetic hyperfine field was related to the disappearance of the impurity Fe_2O_3. The decrease of Fe_2O_3 (with valence state − Fe^{3+}) led to the decrease of the superexchange interaction [14]. The quadrupole shift of Mössbauer spectra was nearly zero in our study, which indicated that the ferrites exhibited a cubic symmetry. The decrease in the absorption area ration of B site can be attributed to the Fe^{3+} ions decrease at B site, with the Cr^{3+} ions doping.

Table 3. Mössbauer parameters of NiCr$_x$Fe$_{2-x}$O$_4$ calcined at 800 °C.

Content (x)	Component	Isomer Shift (I.S.) (mm/s)	Quadrupole Shift (Q.S.) (mm/s)	H(T)	Line Width (Γ) (mm/s)	Relative Area (A$_0$) (%)
0	Sextet (A)	0.124	0.003	48.089	0.439	51.7
	Sextet (B)	0.235	−0.087	52.051	0.384	48.3
0.4	Sextet (A)	0.146	−0.003	47.786	0.544	68.3
	Sextet (B)	0.230	−0.014	51.359	0.429	31.7
0.8	Sextet (A)	0.147	−0.219	43.373	0.352	80.2
	Sextet (B)	0.224	0.191	45.820	0.587	19.8

3.4. Magnetic Property of Particles

Figure 6 shows the hysteresis loops of NiCr$_x$Fe$_{2-x}$O$_4$ at room temperature. The magnetization of all samples nearly reached a saturation point at the external field of 5000 Oe. As shown in Table 4, the saturation magnetization decreased with an increase in the Cr content.

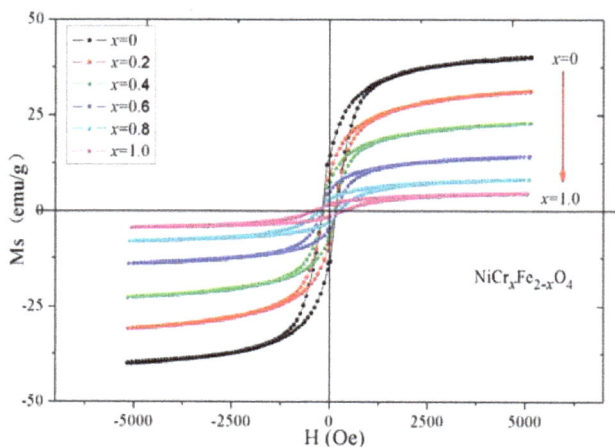

Figure 6. Hysteresis loops of NiCr$_x$Fe$_{2-x}$O$_4$ calcined at 800 °C.

The saturation magnetization can be expressed by means of the following relation [4,14]:

$$\sigma_s = \frac{5585 \times n_B}{M} \tag{2}$$

where n_B is the magnetic moment with Bohr magneton as the unit, and M is the relative molecular mass. The relative molecular mass of NiCr$_x$Fe$_{2-x}$O$_4$ decreased as the Cr content in x increased. The change of magnetic moment nB can be explained with Néel's theory. The magnetic moment μ per ion for Cr^{3+}, Ni^{2+} and Fe^{3+} ions was 3 μ$_B$, 2 μ$_B$ and 5 μ$_B$ [2,3,14], respectively. According to Néel's two sublattice model of ferrimagnetism, the cation distribution of (Fe)A[NiCr$_x$Fe$_{1-x}$]BO$_4$ was used, since Ni^{2+} prefers to occupy the octahedral (B) site in NiFe$_2$O$_4$ ferrite of inverse spinel structure [1,2], and Cr^{3+} ions have strong B-site preference [3,13,14]. The magnetic moment nB is expressed as [3,4,10,14]:

$$n_B = M_B − M_A = 2 + 3x + 5(1 − x) − 5 = 2 − 2x \tag{3}$$

where M_B and M_A are the B and A sublattice magnetic moments. Formula (3) shows that the theoretical magnetic moment decreases when there is an increase in the Cr content. According to the relation of Formula (2), the theoretical saturation magnetization decreased with Cr content x. The variation of the experimental and theoretical saturation magnetization agreed with each other for all samples.

Table 4. Magnetic data for $NiCr_xFe_{2-x}O_4$ calcined at 800 °C.

Content (x)	Ms (emu/g)	Hc (Oe)	Mr (emu/g)
0	40.12	177.25	13.64
0.2	31.02	152.09	9.06
0.4	22.87	154.61	6.94
0.6	14.06	185.70	4.80
0.8	8.11	253.79	2.87
1.0	4.46	361.83	1.55

It is observed from Table 4 that the coercivity of $NiCr_xFe_{2-x}O_4$ initially decreased, then subsequently increased—with an increase in the Cr content x. The coercivity was found to decrease which can be attributed to the decrease of magnetocrystalline anisotropy, as Cr has a negative magnetocrystalline anisotropy [15]. However, the coercivity increased when $x \geq 0.4$, due to influence by many factors, such as impurity phase, crystallinity, microstrain, magnetic particle morphology and size distribution, anisotropy, and magnetic domain size [16–18].

The magnetic hysteresis loops at room temperature for unsintered $NiCr_{0.2}Fe_{1.8}O_4$, and after annealing at 400 °C, 600 °C, and 800 °C, are shown in Figure 7. Table 5 indicates that $NiCr_{0.2}Fe_{1.8}O_4$ after annealing at 800 °C offered a maximum saturation magnetization value, since the particle size increased with an increase in the annealing temperature [17]. The saturation magnetization of $NiCr_{0.2}Fe_{1.8}O_4$ decreased after annealing at 400 °C and 600 °C, which may be due to the presence of an impurity phase [19,20]. Similar studies have been reported in the literature [21,22].

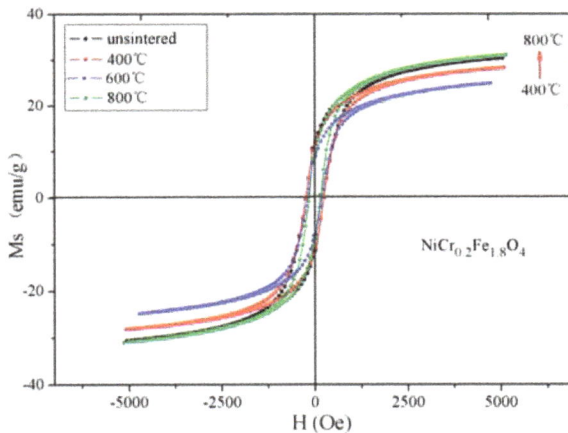

Figure 7. Hysteresis loops of $NiCr_{0.2}Fe_{1.8}O_4$ calcined at different temperatures.

The coercivity of $NiCr_{0.2}Fe_{1.8}O_4$ increased initially and subsequently decreased, with an increase in the annealing temperature. This can be explained by the variation of the grain size. The coercivity in the single domain region is expressed as $H_C = g - h/D^2$. In the multidomain region the variation of the coercivity with grain size can be expressed as in $H_C = a + b/D$, where g, h, a, and b are constants and D is the diameter of the particle [17,18]. Hence, in the single domain region, the coercivity increased with an increase in the grain size, while in the multidomain region the coercivity decreased as the particle diameter increased. In our earlier research, the grain size increased by increasing the sintered temperatures obtained by SEM. The grain size of $NiCr_{0.2}Fe_{1.8}O_4$ calcined at different temperature should be ideally from the single domain region to the multidomain region, so the coercivity increased initially and then decreased with an increase in the annealing temperature [19,20]. Moreover, the impurity phases of Fe_2O_3 in the ferrite nanopowders is also the important factor that results in the decreases of coercivity [23–25].

Table 5. Magnetic data for $NiCr_{0.2}Fe_{1.8}O_4$ calcined at different temperatures.

Temperature (°C)	Ms (emu/g)	Hc (Oe)	Mr (emu/g)
unsintered	30.47	247.03	11.62
400 °C	28.19	247.45	11.10
600 °C	24.87	207.85	6.25
800 °C	31.02	152.09	9.06

4. Conclusions

XRD analysis of the $NiCr_xFe_{2-x}O_4$ ferrite reveals that the specimens with $x > 0.2$ exhibit a single-phase spinel structure, while the impurity peak of Fe_2O_3 was detected in the samples with x = 0 or 0.2. The decrease in the lattice parameter was probably due to the replacement of the larger Fe^{3+} ions by smaller Cr^{3+} ions. The XRD patterns of $NiCr_{0.2}Fe_{1.8}O_4$ calcined at different temperature indicate the intensity of Fe_2O_3 decreased with increasing the heat treatment temperatures, and the diffraction peaks of the sample without burning were very sharp along with a good crystallinity. SEM results indicate that the grains were distributed homogeneously, and the ferrite powers were nano-particles; the average grain size was slightly larger than the average crystallite size determined by XRD, which shows that every particle is formed by a number of crystallites. Room temperature Mössbauer spectra of $NiCr_xFe_{2-x}O_4$ calcined at 800 °C showed the presence of two normal Zeeman-split sextets, and that it exhibited a ferrimagnetic behavior for the all samples. The decrease in the magnetic hyperfine field with increasing Cr contents was attributed to the decrease of the A-B super-exchange interaction. The saturation magnetization decreased with an increase in the Cr content x, which could be explained by Néel's theory. Furthermore, with an increase in the annealing temperature, the coercivity increased initially and subsequently decreased for $NiCr_{0.2}Fe_{1.8}O_4$, since the particle size increased with increasing the annealed temperature.

Author Contributions: J.L. and Y.H. contributed equally to this work. Q.L. and Y.H. conceived and designed the experiments; Q.L. and J.L. performed the experiments; J.L. and H.S. analyzed the data; X.D. and H.Y. contributed reagents/materials/analysis tools. Q.L. and H.S. are co-corresponding authors contributed equally to this study. All authors commented and edited on the manuscript. No potential conflict of interest was reported by the authors.

Acknowledgments: This research was funded by the National Natural Science Foundation of China (NO. 11364004, 11775057, 11747307, 11765004, 11547307, 11647309). The project was funded by Guangxi Key Laboratory of Nuclear Physics and Nuclear Technology.

Conflicts of Interest: The authors declare no conflict of interest.

References

1. Patange, S.M.; Shirsath, S.E.; Jadhav, S.S.; Lohar, K.S.; Mane, D.R.; Jadhav, K.M. Rietveld refinement and switching properties of Cr^{3+} substituted $NiFe_2O_4$ ferrites. *Mater. Lett.* **2010**, *64*, 722–724. [CrossRef]

2. He, Y.; Lei, C.; Lin, Q.; Dong, J.; Yu, Y.; Wang, L. Mössbauer and Structural properties of La-substituted $Ni_{0.4}Cu_{0.2}Zn_{0.4}Fe_2O_4$ nanocrystalline ferrite. *Sci. Adv. Mater.* **2015**, *7*, 1809–1815. [CrossRef]

3. Anh, L.N.; Duong, N.P.; Loan, T.T.; Nguyet, D.T.T.; Hien, T.D. Synchrotron and Magnetic Study of Chromium-Substituted Nickel Ferrites Prepared by Using Sol-Gel Route. *IEEE Trans. Magn.* **2014**, *50*, 1–5.

4. Lee, S.H.; Yoon, S.J.; Lee, G.J.; Kim, H.S.; Yo, C.H.; Ahn, K.; Lee, D.H.; Kim, K.H. Electrical and magnetic properties of $NiCr_xFe_{2-x}O_4$ spinel ($0 \leq x \leq 0.6$). *Mater. Chem. Phys.* **1999**, *61*, 147–152. [CrossRef]

5. Prasad, A.S.; Dolia, S.N.; Pareek, S.P.; Samariya, A.; Sharma, P.K.; Dhawan, M.S. Sol-gel synthesized high anisotropy magnetic nanoparticles of $NiCr_xFe_{2-x}O_4$. *J. Sol-Gel Sci. Technol.* **2013**, *66*, 372–377. [CrossRef]

6. Patange, S.M.; Shirsath, S.E.; Toksha, B.G.; Jadhav, S.S.; Shukla, S.J.; Jadhav, K.M. Cation distribution by Rietveld, spectral and magnetic studies of chromium-substituted nickel ferrites. *Appl. Phys. A* **2009**, *95*, 429–434. [CrossRef]

7. Xia, A.; Liu, S.; Jin, C.; Chen, L.; Lv, Y. Hydrothermal $Mg_{1-x}Zn_xFe_2O_4$ spinel ferrites: Phase formation and mechanism of saturation magnetization. *Mater. Lett.* **2013**, *105*, 199–201. [CrossRef]

8. Iqbal, M.J.; Siddiquah, M.R. Electrical and magnetic properties of chromium-substituted cobalt ferrite nanomaterials. *J. Alloys Compd.* **2008**, *453*, 513–518. [CrossRef]

9. Chae, K.P.; Lee, Y.B.; Lee, J.G.; Lee, S.H. Crystallographic and magnetic properties of $CoCr_xFe_{2-x}O_4$ ferrite powders. *J. Magn. Magn. Mater.* **2000**, *220*, 59–64. [CrossRef]

10. Singhal, S.; Jauhar, S.; Singh, J.; Chandra, K.; Bansal, S. Investigation of structural, magnetic, electrical and optical properties of chromium substituted cobalt ferrites ($CoCr_xFe_{2-x}O_4$, $0 \leq x \leq 1$) synthesized using sol gel auto combustion method. *J. Mol. Struct.* **2012**, *1012*, 182–188. [CrossRef]

11. Bayoumy, W.A.; Gabal, M.A. Synthesis characterization and magnetic properties of Cr-substituted NiCuZn nanocrystalline ferrite. *J. Alloys Compd.* **2010**, *506*, 205–209. [CrossRef]

12. Lin, Q.; Lei, C.; He, Y.; Xu, J.; Wang, R. Mössbauer and XRD studies of $Ni_{0.6}Cu_{0.2}Zn_{0.2}Ce_xFe_{2-x}O_4$ ferrites By Sol-Gel auto-combustion. *J. Nanosci. Nanotechnol.* **2015**, *15*, 2997–3003. [CrossRef] [PubMed]

13. Kumar, S.; Farea, A.M.M.; Batoo, K.M.; Lee, C.G.; Koo, B.H.; Yousef, A.; Alimuddin. Mössbauer studies of $Co_{0.5}Cd_xFe_{2.5-x}O_4$ ($0.0 \leq x \leq 0.5$) ferrite. *Phys. B Condens. Matter.* **2008**, *403*, 3604–3607. [CrossRef]

14. Toksha, B.G.; Shirsath, S.E.; Mane, M.L.; Patange, S.M.; Jadhav, S.S.; Jadhav, K.M. Autocombustion High-Temperature Synthesis, Structural, and Magnetic Properties of $CoCr_xFe_{2-x}O_4$ ($0 \leq x \leq 1.0$). *J. Phys. Chem. C* **2011**, *115*, 20905–20912. [CrossRef]

15. Gabal, M.A.; Al Angari, Y.M.; Al-Agel, F.A. Synthesis, characterization and magnetic properties of Cr-substituted Co-Zn ferrites Nanopowders. *J. Mol. Struct.* **2013**, *1035*, 341–347. [CrossRef]

16. Jiang, T.; Yang, Y.M. Effect of Gd substitution on structural and magnetic properties of Zn-Cu-Cr ferrites prepared by novel rheological technique. *Mater. Sci. Technol.* **2009**, *25*, 415–418. [CrossRef]

17. Singhal, S.; Barthwal, S.K.; Chandra, K. XRD, magnetic and Mössbauer spectral studies of nano size aluminum substituted cobalt ferrites ($CoAl_xFe_{2-x}O_4$). *J. Magn. Magn. Mater.* **2006**, *306*, 233–240. [CrossRef]

18. Harzali, H.; Marzouki, A.; Saida, F.; Megriche, A.; Mgaidi, A. Structural, magnetic and optical properties of nanosized $Ni_{0.4}Cu_{0.2}Zn_{0.4}R_{0.05}Fe_{1.95}O_4$ ($R = Eu^{3+}$, Sm^{3+}, Gd^{3+} and Pr^{3+}) ferrites synthesized by co-precipitation method with ultrasound irradiation. *J. Magn. Magn. Mater.* **2018**, *460*, 89–94. [CrossRef]

19. He, Y.; Yang, X.; Lin, J.; Lin, Q.; Dong, J. Mössbauer spectroscopy, Structural and magnetic studies of Zn^{2+} substituted magnesium ferrite nanomaterials prepared by Sol-Gel method. *J. Nanomater.* **2015**, *2015*, 854840. [CrossRef]

20. Kumar, A.; Shen, J.; Yang, W.; Zhao, H.; Sharma, P.; Varshney, D.; Li, Q. Impact of Rare Earth Gd^{3+} Ions on Structural and Magnetic Properties of $Ni_{0.5}Zn_{0.5}Fe_{2-x}Gd_xO_4$ Spinel Ferrite: Useful for Advanced Spintronic Technologies. *J. Supercond. Novel Magn.* **2018**, *31*, 1173–1182. [CrossRef]

21. Lin, L.Z.; Tu, X.Q.; Wang, R.; Peng, L. Structural and magnetic properties of Cr-substituted NiZnCo ferrite Nanopowders. *J. Magn. Magn. Mater.* **2015**, *381*, 328–331.

22. Cao, C.; Xia, A.; Liu, S.; Tong, L. Synthesis and magnetic properties of hydrothermal magnesium–zinc spinel ferrite powders. *J. Mater. Sci. Mater. Electron.* **2013**, *24*, 4901–4905. [CrossRef]

23. Li, L.Z.; Zhong, X.X.; Wang, R.; Tu, X.Q.; Peng, L. Structural and magnetic properties of Co-substituted NiCu ferrite nanopowders. *J. Magn. Magn. Mater.* **2017**, *433*, 98–103. [CrossRef]

24. Lin, J.; He, Y.; Lin, Q.; Wang, R.; Chen, H. Microstructural and Mössbauer spectroscopy Studies of $Mg_{1-x}Zn_xFe_2O_4$ ($x = 0.5, 0.7$) nanoparticles. *J. Spectrosc.* **2014**, *2014*, 540319. [CrossRef]

25. Lv, H.; Zhang, H.; Zhang, B.; Ji, G.; He, Y.; Lin, Q. A proposed electron transmission mechanism between Fe^{3+}/Co^{2+} and Fe^{3+}/Fe^{3+} in the spinel structure and its practical evidence in quaternary $Fe_{0.5}Ni_{0.5}Co_2S_4$. *J. Mater. Chem. C* **2016**, *4*, 5476–5482. [CrossRef]

crystals

MDPI

Article

Quasi-Equilibrium, Multifoil Platelets of Copper- and Titanium-Substituted Bismuth Vanadate, $Bi_2V_{0.9}(Cu_{0.1-x}Ti_x)O_{5.5-\delta}$, by Molten Salt Synthesis

Kevin Ring and Paul Fuierer *

Materials and Metallurgical Engineering Department, New Mexico Institute of Mining and Technology (New Mexico Tech), 801 LeRoy Place, Socorro, NM 87801, USA; kring@alumni.nmt.edu
* Correspondence: paul.fuierer@nmt.edu; Tel.: +1-575-835-5497

Received: 11 March 2018; Accepted: 12 April 2018; Published: 17 April 2018

Abstract: 10% copper-substituted (BiCUVOX/$Bi_2V_{0.9}Cu_{0.1}O_{5.5-\delta}$) and 5% copper/titanium double-substituted bismuth vanadate (BiCUTIVOX/$Bi_2V_{0.9}(Cu_{0.05}Ti_{0.05})O_{5.5-\delta}$) platelets were formed by molten salt synthesis (MSS) using a eutectic KCl/NaCl salt mixture. The product was phase-pure within the limits of X-ray diffraction. The size and form of the platelets could be controlled by changing the heating temperature and time. The crystallite growth rate at a synthesis temperature of 650 °C and the activation energy for grain growth were determined for BICUTIVOX, which experienced inhibited growth compared to BICUVOX. Quasi-equilibrium, multifoil shapes consisting of lobes around the perimeter of the platelets were observed and explained in the context of relative two-dimensional nucleation and edge growth rates.

Keywords: bismuth vanadate; molten salt synthesis; platelet morphology; multifoil shape; Wulff shape; Ostwald ripening

1. Introduction

In 1988, Abraham et al. published the discovery of a new group of oxide ion-conducting ceramics based on bismuth vanadate ($Bi_4V_2O_{11-\delta}$) [1]. This compound was determined to be a member of the Aurivillius family of crystal structures, and it was described as having alternating layers of $(Bi_2O_2)^{2+}$ and $(VO_{3.5}\square_{0.5})^{2-}$ [1]. Bismuth vanadate has three distinct phases, and from single crystal studies, the highest temperature γ-phase (normally occurring at T > 550 °C) was determined to be tetragonal (space group *I4/mmm*, a = 4.004 Å, c = 15.488 Å). Conductivity was found to be highly anisotropic; two orders of magnitude higher when perpendicular to the c-axis (in the *a*–*b* plane) than parallel to the *c* axis [1–4]. It was later found that a partial substitution of an aliovalent metal cation for vanadium (first was copper at 10 mol %) resulted in the stabilization of the γ-phase at room temperature, and the term BIMEVOX was coined to refer to this and all subsequent formulations, where ME represents one of the many possible substitutive metal cations, typically a transition metal [2]. Having an oxide ion conductivity that is higher than any other ceramic (e.g., yttria-stabilized cubic zirconia ($Zr_{1-x}Y_xO_{2-x/2}$ "YSZ") at moderate temperatures, BIMEVOX compounds have generated interest over the years for their potential applications as solid oxide fuel cells, as oxygen separation pumps, as gas sensors [5–9], and very recently, as catalyst supports [10].

Just as with its conductivity, the anisotropy of the tetragonal phase is expected to result in anisotropy in the other chemical and physical tensor properties of BIMEVOX. While such differences are distinct in single crystals, a random polycrystalline sample exhibits an approximate average of the lattice-specific property coefficients. Enhancing the long-range texture of polycrystalline samples therefore holds promise for enhancing the desired properties of BIMEVOX ceramics; of chief interest is maximizing the bidimensional ionic conductivity [11]. Experimental techniques for enhancing the texture of anisotropic ceramics have included pulsed laser deposition [12], magnetic or electric fields [13,14], and load-assisted sintering or "hot-forging" [14–16]. Shantha et al. [17], used a KCl flux in a two-stage, liquid-phase sintering process which resulted in densities of 97% and textures of up to 79% in BIMEVOX.

Another route for fabricating textured ceramics begins with particles already exhibiting the desired geometry and involves the use of these particles as seeds in "templated grain growth" (TGG). TGG has been used with reasonable success in conjunction with tape-casting for bismuth titanate ($Bi_4Ti_3O_{12}$) and other layered perovskite-type compounds, achieving *(00l)* textures exceeding 90% [18–23]. Due to the crystallographic anisotropy of bismuth titanate and other Aurivillius phases, the surface energy of the *(00l)* planes is less than that of the *(hk0)* planes, and therefore, the growth of anisometric particles (i.e., platelets) is thermodynamically favorable; however, the kinetics in conventional, solid-state synthesis often limit such growth.

Molten salt synthesis (MSS), also called salt melt synthesis, is a favored method for producing such seed particles with a pronounced anisotropic growth; it involves reacting the material in a molten salt flux [24–29]. The molten salt acts as a diffusion-enhancing medium. Phase formation occurs by one of two mechanisms—solution–precipitation or solution–diffusion—depending on the relative solubility of the reactant species [27]. Grain growth by the Ostwald ripening of crystallites is controlled by two-dimensional (2D) nucleation and growth along a crystallographic vector through the generation of ledges on the propagating edge [27,30]. MSS has been used since the 1970s to synthesize binary and complex oxides using a variety of salt systems [24–27], but it is receiving renewed attention in the production of borides, carbides, silicides, and nanomaterials with high crystallinity [28,29]. MSS is high-temperature solution chemistry, but, compared to solid-state synthesis, allows for lower reaction temperatures and much faster mass transport by means of convection and diffusion through the liquid phase. The process is also attractive because it can be scaled up, and, depending on the salt system used, it can be economical and environmentally friendly through the recycling of water-soluble salts.

Roy and Fuierer [31] first published work on the successful synthesis of BICOVOX (cobalt-substituted bismuth vanadate) by MSS. Herein, we report for the first time the molten salt synthesis of BICUVOX and BICUTIVOX using similar techniques. The hypothesis is that the product of the solubility of oxide constituents, Bi_2O_3, V_2O_5, CuO, and TiO_2, is larger than the solubility of the compound $Bi_4(V_{1-x-y}Cu_xTi_y)_2O_{11-\delta}$, leading to its formation under a high degree of supersaturation. Subsequent grain growth is expected to lead to highly anisometric particles. Furthermore, we investigate the influence of temperature, time, and double substitution (Cu plus Ti co-doping) on crystallite growth.

2. Materials and Methods

Stoichiometric amounts of bismuth (III) oxide (Alfa Aesar, 99%), vanadium (V) oxide (Acros Organics 99.6+%), titanium (IV) oxide (JT Baker, ACS grade), and copper (II) oxide (Aldrich, ACS grade) were combined with an equal mass of a eutectic (T_E = 640–657 °C [32]) salt mixture comprised of a 0.506:0.494 molar ratio of sodium chloride (Fisher, 99+%) and potassium chloride (Aldrich 99+%). The batches were milled in a 250 mL, high-density polyethylene jar with 400 grams of 13 mm cylindrical, ceria-stabilized zirconia media, at approximately 130 rpm typically for 5 h. Milling was performed dry, and the media and powder were separated after milling using a #10 sieve (2 mm opening). Reactions were carried out in air using a Thermolyne 47900, a resistive wire heated muffle furnace. The material was placed in a covered alumina crucible and heated at a ramp rate of 10 °C/min to temperatures between 610 °C and 700 °C, and it was then held for 2–10 h. The net loss of mass after heat treatments never exceeded 0.5%. After heating, the charge of the product and the fused salt were broken up by soaking them in deionized water at ~75 °C and then splitting them to expose the interior. The aggregated material was broken up and washed of the salt by stirring in 1.4 L of DI H_2O at 60 °C for periods of 30 min to 18 h. After the material settled, the water was decanted and replaced. This process was repeated a minimum of four times. The final product was oven-dried.

Powder samples for analysis were taken by stirring the entire yielded quantity and taking material from the center. XRD analysis was carried out with a PANalytical X'Pert Pro (Cu$_{K\alpha}$ source, nickel filter, and a 2.122° X'Celerator line detector) over a 2θ of 5.996°–70°, with 0.01671° steps, and the data was analyzed with the accompanying HighScore Plus software. Patterns presented were modified from the raw scan data by dividing the intensity at each 2θ value in a set by the maximum intensity in that set (i.e., normalized to 1), and then those patterns were shifted by integer values to accommodate multiple patterns on the same chart. All patterns were indexed as *I4/mmm* tetragonal, according to ICCD PDF #01-070-9191 [33]. Microscopic analysis was performed using a Hitachi S-3200N scanning electron (SE) microscope. Crystallite platelet dimensions were measured using image digitization of SE micrographs. Particle size distribution analysis was performed by Fraunhoffer laser (780 nm) diffraction using a Beckman Coulter LS 13 320, with the microliquid module filled with deionized water and loaded to 8–12% obscuration.

3. Results and Discussion

The series of XRD patterns for BICUVOX in Figure 1 shows that the γ-phase was successfully synthesized in all heat treatments and that no other phases were present within detectable limits. The enhanced ratio of *(00l)/(013)* peak intensities compared to the random powder pattern peak heights is an indication of an increase in population of *c*-axis-oriented platelets and their alignment under the mild shearing forces involved in the preparation of a powder sample for XRD analysis [34]. Comparing different temperatures, the sample synthesized at 650 °C shows *(00l)* reflections of higher intensity than the material synthesized at 610 °C and at 675 °C. For the purposes of strong crystallite texturing, the result suggests that 650 °C is the best temperature for synthesis.

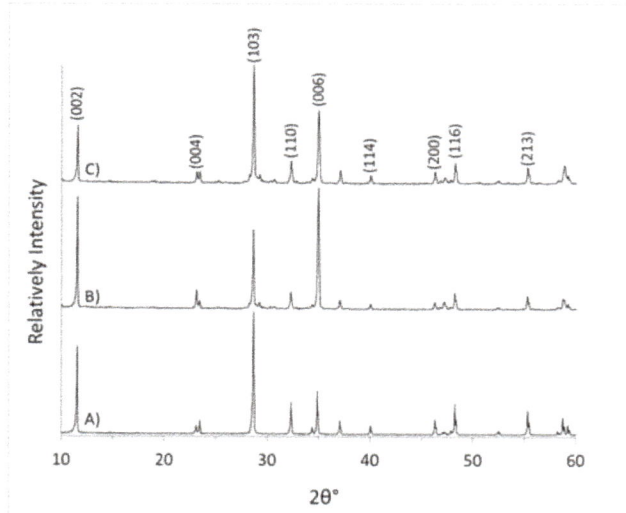

Figure 1. XRD pattern of BICUVOX platelets synthesized by molten salt synthesis (MSS): (A) 610 °C/8 h, (B) 650 °C/8 h, (C) 675 °C/8 h.

Figure 2 shows SEM images of MSS BICUVOX. The 610/8 material appears to consist largely of agglomerates of small platelets. Although 610 °C is substantially lower than the reported eutectic temperature of 640–657 °C [32], the solid salt still facilitates habit growth by enhancing the diffusion of ions [35]. With higher process temperatures and the presence of a fully molten salt, larger platelets result, with a majority of the faces having an irregular shape. An inspection of additional micrographs suggests that the highest temperature (675 °C) also gives a greater population of smaller crystallites, resulting either from fracture or from larger platelets dissolving and re-nucleating as described [26,30,31]. An increase in population of broken and/or re-dissolved and nucleated particles is believed to be the cause of the decrease in *(00l)* reflections for the 675 °C processed material. Figure 3 shows the distribution of particle sizes as measured by laser diffraction. As expected, the major mode shifts to a larger particle size with a process temperature and time, which is in general agreement with the SEM images. For the solid-salt synthesis at 610 °C, the small mode at a large size of about 80 microns is attributed to aggregation. For the 675 °C material, the shoulder located at about 30 µm off of the major peak can be attributed to broken platelets.

Figure 2. SEM images of BICUVOX platelets formed by MSS at: (**A**) 610 °C/8 h, (**B**) 650 °C/8 h, (**C**) 675 °C/8 h.

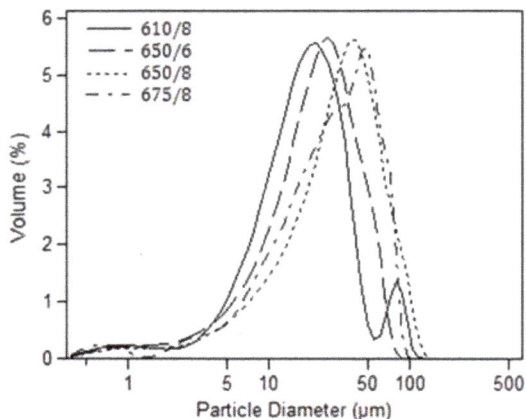

Figure 3. Relative volume distribution of particle sizes for BICUVOX powders synthesized using salt matrix at various treatments T(C°)/t(h).

The diffraction patterns in Figure 4 show that the γ-phase for BICUTIVOX was also synthesized in all heat treatments. Again, all three patterns show higher *(00l)/(103)* intensity ratios than a random powder pattern [33]. In this case, there is a steady decrease in powder texture with an increasing synthesis temperature.

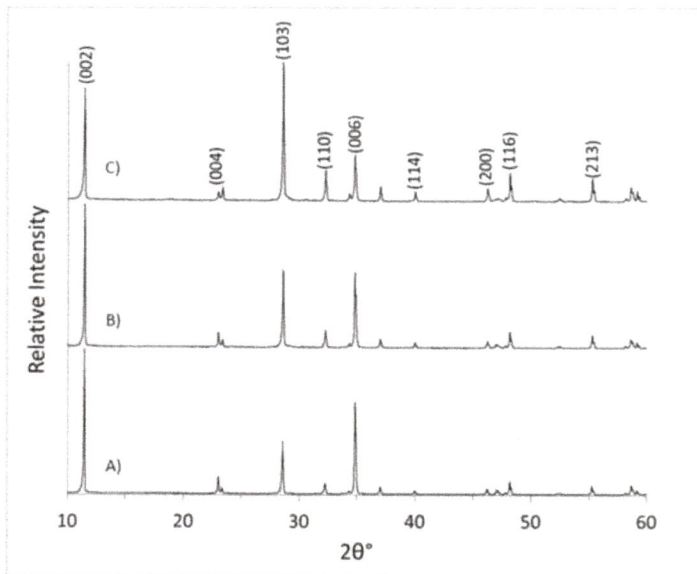

Figure 4. XRD pattern of BICUTIVOX platelets produced via MSS: (A) 650 °C/8 h, (B) 675 °C/8 h, (C) 700 °C/8 h.

The SEM images in Figure 5 show populations of platelets, with some being quite rounded and others being irregular in geometry. The occurrence of platelets with edge lobes and multifoil shapes [36] is evident. A mechanistic explanation for the platelet shape and edge lobes is proposed later in the study. The size of the platelets shows a steady increase with the process temperature. Figure 6 shows the size distribution of particles by laser diffraction, and it also shows the major peak shifting to a smaller size with an increasing temperature. This is contrary to what is expected when compared to the result for BICUVOX, and it is not in agreement with the impression obtained from the SEM micrographs. See Figure 7 for a larger perspective view and also a direct comparison of BICUVOX and BICUTIVOX at equivalent process conditions.

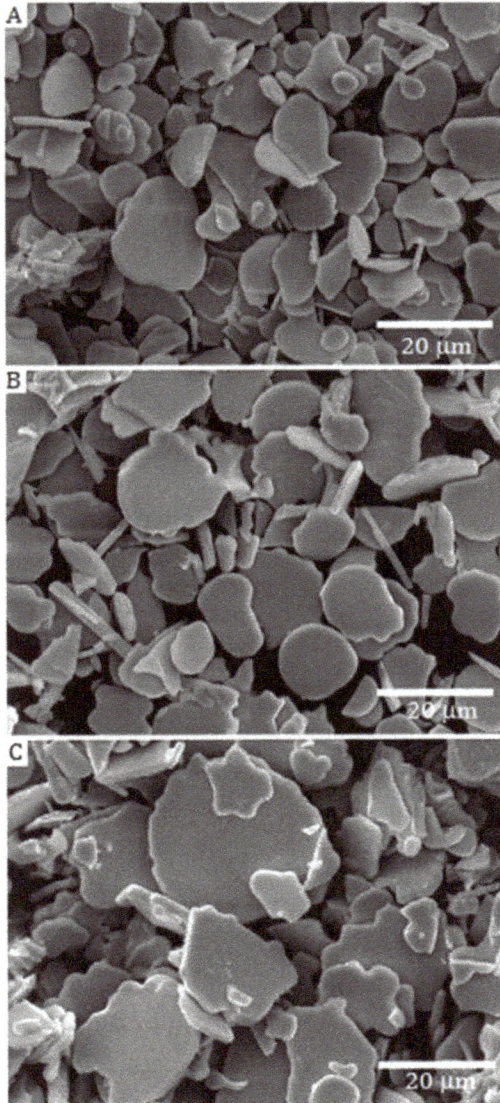

Figure 5. SEM images of BICUTIVOX platelets formed by MSS at: (**A**) 650 °C/8 h, (**B**) 675 °C/8 h, (**C**) 700 °C/8 h.

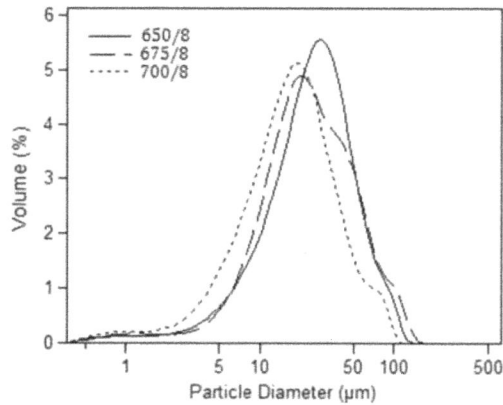

Figure 6. Relative volume distribution of particle sizes for BICUTIVOX platelets synthesized using molten salt at various treatments T(C°)/t(h).

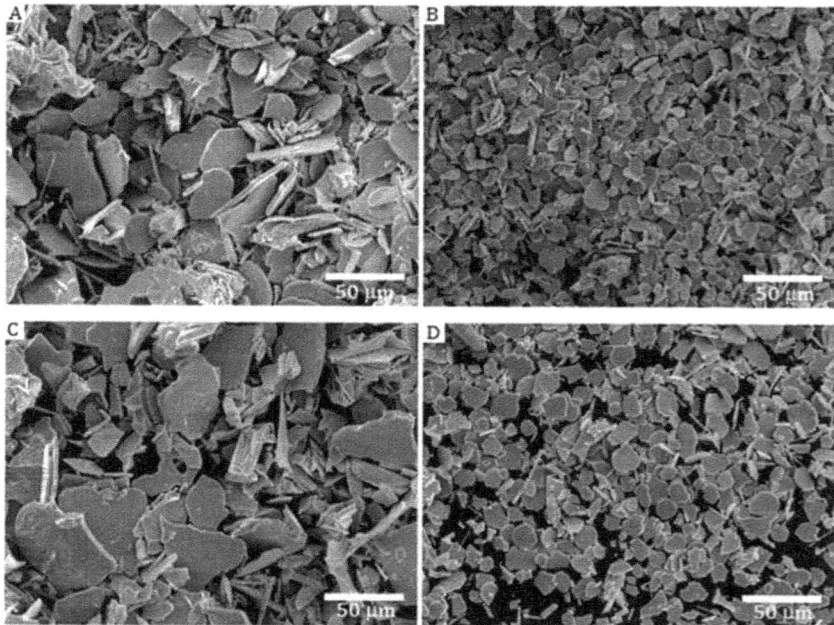

Figure 7. SEM images of BICUVOX (left) and BICUTIVOX (right) synthesized under the same conditions: (**A**) BICUVOX, 650 °C/8 h, (**B**) BICUTIVOX, 650 °C/8 h, (**C**) BICUVOX, 675 °C/8 h, (**D**) BICUTIVOX, 675 °C/8 h.

Particle sizing by laser diffraction is challenging for platelet-type particles and can be misleading because the technique assumes an equivalent spherical diameter, and is subject to the influence of fractured particles and agglomeration. Hence, a direct approach was adopted using SEM images for a more accurate particle size measurement and analytics. Four representative micrographs from each sample, with particles well-dispersed on carbon tape, were carefully analyzed. Dimensions were taken of any particle whose large face was oriented normal to the viewing axis. Platelets which were obviously fractured were not used. Likewise, all edge-on particles were used to measure platelet thickness. Statistical values of these populations (N >100) of particles are summarized in Table 1. Despite the rather large standard deviations in some samples, the table shows a general trend of increased average and maximum platelet sizes with increasing temperatures. At a fixed process temperature of 650 °C, a steady increase in average size with time is apparent. The aspect ratio (AR) undergoes little change except from solid to molten salt (i.e., 610 °C vs. 650 °C). Also given in Table 1 are Lotgering Orientation factors, *F*, obtained from the XRD patterns in Figures 1 and 4, calculated by [37]:

$$F = \frac{p - p_0}{1 - p_0} \text{ and } p = \frac{\Sigma I(00l)}{\Sigma I(hkl)} \tag{1}$$

where p is equal to the sum of the intensities of the selected family of planes divided by the sum of the intensities for all planes within the pattern; p_0 is the factor for a non-textured sample (randomly oriented powder pattern [33]). This is a relative measure of the number and tendency for the orientation of *c*-axis platelets in the packed powder XRD samples. The *F* factor is highest for a process temperature of 650 °C for both materials. Platelet breakage, dissolution/nucleation, and increased edge roughness are all believed to be responsible for the decrease in powder XRD texture with a higher process temperature.

Table 1. Dimensional statistics and orientation factor, *F*, of salt-synthesized BIMEVOX platelets.

Phase	Treatment	$F_{(001)}$	$D_{ave} \pm D_{StDev}$	D_{max}	$AR \pm AR_{StDev}$
	T(°C)/t(h)	(μm)	(μm)		
BICUVOX	610/8	0.14	7.2 ± 1.5	11.7	8 ± 3
	650/2		24.5 ± 2.0	36.3	13 ± 3
	650/4		27.0 ± 4.5	47.5	13 ± 3
	650/6	0.33	26.9 ± 7.2	35.1	12 ± 3
	650/8	0.43	33.5 ±10.3	49.4	15 ± 8
	650/10		29.1 ± 5.5	44.9	9 ± 2
	675/8	0.23	37.3 ± 21.1	65.5	15 ± 9
BICUTIVOX	650/2		9.6 ± 0.6	12.9	8 ± 3
	650/4		11.0 ± 3.2	21.7	11 ± 3
	650/6		13.6 ± 0.5	18.3	9 ± 2
	650/8	0.6	16.1 ± 2.7	29.2	10 ± 2
	650/10		16.3 ± 3.0	24.3	7 ± 1
	675/8	0.47	17.9 ± 2.8	25.0	11 ± 3
	700/8	0.22	21.3 ± 4.2	32.1	10 ± 3

Particle growth can be represented by the following equation [38–40]:

$$D^2 - D_0^2 = k(t - t_0) \exp\left(-\frac{E_a}{RT}\right) \tag{2}$$

where D is the average particle size at time t, D_0 is the initial particle size at initial time t_0, k is the rate constant, E_a is the activation energy for particle growth, R is the gas constant, and T is the synthesis temperature. Figure 8 shows a plot of $2ln(D)$ vs inverse temperature for the two compositions. From the slope of the best fit line, an activation energy of 83 kJ/mol for particle growth for BICUTIVOX is estimated. This value is between 32 kJ/mol for $BaBi_4Ti_4O_{15}$ (another Aurivilius-type phase) synthesized in K_2SO_4–Na_2SO_4 flux [40], and 96 kJ/mol for $Pb(Mg_{.33}Nb_{.67})O_3$ (perovskite) in Li_2SO_4–Na_2SO_4 flux [38], and this value is considered reasonably accurate. With only two molten salt data points and larger deviation for BICUVOX, the activation energy cannot be calculated, but it appears to be lower than for BICUTIVOX. Also plotted on the graph is the single data point for the solid-salt-synthesized BICUVOX, which falls far below the other data points due to a different limiting mechanism for growth in the solid state below the NaCl–KCl eutectic temperature.

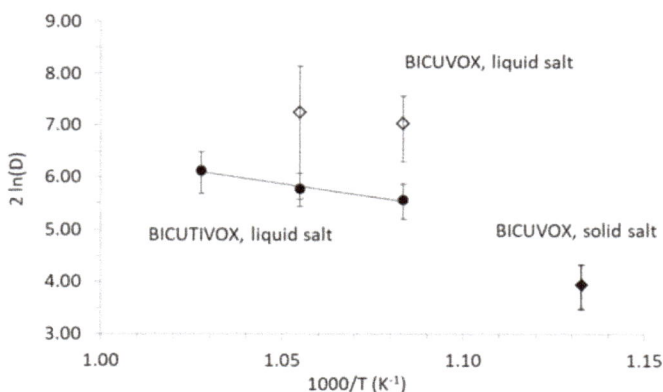

Figure 8. Plot of *2ln(D)* (*D* = platelet diameter) versus inverse temperature for salt-synthesized BICUVOX and BICUTIVOX according to the particle growth model. Error bars show ± one D_{stdev}.

According to Equation (2) and the well-known parabolic rate law, the platelet diameter, *D*, is plotted against the square root of time for a process temperature of 650 °C (Figure 9). BICUTIVOX data appear to be well-behaved. Using the activation energy for growth (83 kJ/mol) in Equation (2) and the slope of the best fit line, one obtains a grain growth rate constant of 9.6×10^{-16} cm^2/s. In this case, the *y*-intercept is 3.2 μm (interpreted as the starting grain size, D_0). If the best fit line is forced through the origin (starting size $D_0 = 0$ μm), then the rate constant increases to 16×10^{-16} cm^2/s. This rate constant is far below the estimated mobility of ionic species in a molten salt (1×10^{-8} cm^2/s), but is not far from the estimated mobility of species in the solid state (1×10^{-18}) cm^2/s [18]. This suggests that the limiting mechanism for grain growth is the diffusion at the surface or near the surface of the platelets. For BICUVOX, the data is less well-behaved, falling further off the best fit line. This is due to the significantly larger standard deviation in the particle size data, due in part to excessive breakage as the maximum platelet size approaches 50 μm (see Table 1). The 650 °C/6 h statistics suggest that this sample in particular was affected. Note that the experimental grain size from the solid salt synthesis is used as the non-zero starting grain size, D_0, which can be justified because BICUVOX formation took place prior to the existence of the liquid salt. Although we do not calculate a value for the growth rate constant due to the absence of a calculated activation energy, the rate constant appears to be larger for BICUVOX than for BICUTIVOX.

An inspection of the micrographs (Figures 2, 5 and 7) reveals the differences between BICUVOX and BICUTIVOX. The most obvious difference is the size difference, but there is also a difference in particle form and edge angularity. BICUTIVOX particles can be described as having a discoidal form with a high circularity, and being sub-angular to well-rounded. In some cases, the particle takes on a multifoil shape reminiscent of a Wulff construction [41–43]. Note the small pentafoil star on top of the larger, nearly round, multifoil platelet in Figure 5B. BICUVOX, on the other hand, can be described as discoidal with a low circularity, and having angular to sub-rounded edges.

Figure 10 is a schematic which aids the description of the platelet growth and provides an explanation for their quasi-equilibrium, multifoil shapes. The top row represents BICUTIVOX. The bottom row represents BICUVOX. In general, particles of Aurivillius-type, bismuth, layered-structure compounds are expected to have atomically smooth *(00l)* faces and atomically rough *(hk0)* faces [44], so high surface area sites nucleate along the propagating edge of a platelet (Figure 10B). During Ostwald ripening, a kinetic restriction (surface diffusion) regulates the growth of the platelets but does not affect the 2D-nucleation rates for the ledge generation. In BICUTIVOX, nucleation of ledges (with fast kinetics) is distributed evenly about the perimeter of the crystallite, and so, the growth of the propagating edge happens evenly in all directions (Figure 10C). The nucleation–growth process repeats itself as the particle grows (Figure 10D–G). For some particles, the process is interrupted at the nucleation stage while others are interrupted at the edge growth stage or somewhere in-between (see especially Figure 5B). The lobes are more pronounced and more numerous in the higher temperature (700 °C) material because the rate of nucleation is very high. In the case of BICUVOX, the mechanism of platelet growth is the same, except that there is a difference in the speed of edge propagation because the rate constant is higher (as seen in Figure 9). In BICUVOX, soon after the nucleation of a ledge site, the edge begins to grow around it as the cusp fills in due to capillary pressure. Because this growth step happens more rapidly than in BICUTIVOX, the growth does not occur evenly in all directions; thus, this leads to the lack of circularity in BICUVOX platelets.

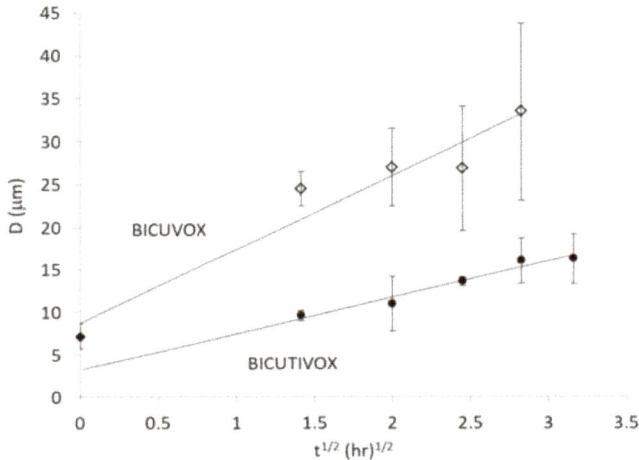

Figure 9. Platelet diameter versus the square root of time plots for BICUVOX and BICUTIVOX at an MSS process temperature of 650 °C. Exception: The solid point at $t = 0$ is the experimental grain size of BICUVOX measured from solid salt synthesis at 610 °C, and it is taken as an approximation of the starting grain size (error bars show $\pm$ one D_{stdev}).

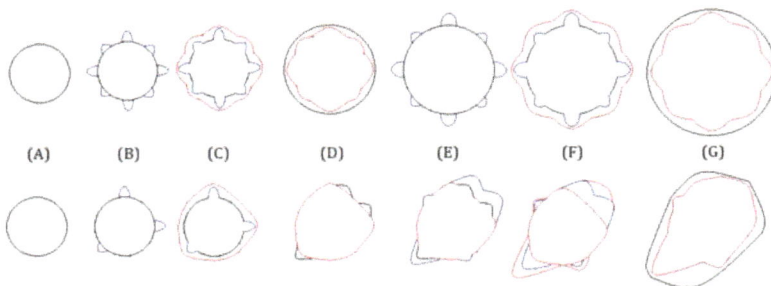

Figure 10. Schematic description of the kinetically limited growth mechanism undergone by BICUTIVOX (**upper row**) as compared to BICUVOX (**bottom row**). Different colors are used to indicate sequential layers of nucleation–growth.

The presence of the Ti^{4+} likely does not have a large effect on molten-salt–liquid-diffusion rates, but it noticeably increases the activation energy and decreases the rate constant for growth. This is consistent with the known effect of solid solution, aliovalent impurities, or substitutions acting as grain growth inhibitors in ceramics [39,45]. In the presence of such solute ions, the activation energy for boundary migration is the sum of the vacancy migration, vacancy creation, impurity solution, and strain energy terms, but it is difficult to separate the relative contribution of each factor. Since doped BICUVOX contains significant oxygen vacancies and is a fast oxide ion conductor particularly in the *a–b* plane, it is conceivable that cations are the rate-limiting diffusing species.

While evidence of edge roughness and lobes on MSS platelets can be found in previous published works [25,46,47], they generally do not appear as pronounced or as regular multifoil shapes like the BICUTIVOX material presented here. The closest example is found in Nd/V co-doped $Bi_4Ti_3O_{12}$ by Tang et al. [47]. To our knowledge, the shapes have not been explained or even addressed. The BICUTIVOX multifoil and lobed platelets synthesized in this work are similar to Wulff-construction polar plots of the anisotropic surface energies of crystal planes. They can be described as quasi-equilibrium shapes which result from the combination of both surface energy anisotropy and growth rate anisotropy, not just between *(00l)* and *(hk0)* planes, but also within the family of *(hk0)* planes. The growth rate anisotropy occurs when a particular crystal face affords easy atom/ion attachment and grows rapidly, and it can result in shapes which exaggerate actual surface energy anisotropies [48]. The identification of the crystal planes within a multifoil platelet, their correlation with edge lobes, and the role of aliovalent dopants are the subjects of future work.

4. Conclusions

The γ-phase of both BICUVOX and BICUTIVOX can be synthesized from simple metal oxides using a NaCl–KCl eutectic mix in an MSS process. The salt flux enhances grain growth with platelet habit, even with a process temperature that is 40 °C below the eutectic temperature, indicating that whether solid or liquid, the salt increases the diffusion rates of ions and facilitates anisotropic growth. The average size of platelets increases with the process temperature as expected, but when the platelets grow to a diameter in excess of about 50 μm, significant fracturing can be expected due to the rigors of post-processing (washing/filtering), and therefore, there is a practical limit to how large the platelets can be synthesized and recovered. Optimal results with respect to achieving high grain (*c*-axis or *(00l)*) orientation factors are achieved at 650 °C—very near the salt eutectic temperature—for both compounds.

Titanium ions, Ti^{4+}, as a second, aliovalent substitution for V^{5+}, acts as a particle growth inhibitor during MSS. The activation energy for platelet growth of BICUTIVOX in the molten salt is estimated to be 83 kJ/mol, which is consistent with MSS values reported in the literature for other simple and complex perovskites. The (growth) reaction rate constant at 650 °C is estimated to be 1.0×10^{-15} cm^2/s, indicating that the limiting mechanism for platelet growth is the (near) surface reaction and diffusion. Excessive breakage of large platelets of BICUVOX causes greater uncertainty in particle size measurement, and precludes the calculation of grain growth activation energy and rate constant, but evidence suggests that these are lower and higher, respectively, than for BICUTIVOX. Due to the significant difference in 2D nucleation and edge propagation rates during Ostwald ripening, BICUVOX shows a tendency for multifoil platelet shapes with pronounced lobes extending from platelet edges. Faster volume and/or surface diffusion in BICUVOX allows for a higher overall growth rate and causes the platelet shape to be noncircular and more angular. Future interest may lie in producing lobed platelets with a higher surface area to take advantage of the high oxide ion mobility in the *(00l)* (*a–b* crystal) plane and the functionality of the *(hk0)* surfaces in BICUTIVOX.

Acknowledgments: This research was funded by the National Science Foundation (Materials World Network, Division of Materials Research grant # 1108466). The authors also acknowledge the assistance of Riley Reprogle with image analysis and particle size measurement.

Author Contributions: Kevin Ring performed the synthesis and characterization of the samples, and wrote the first draft of the manuscript. Paul Fuierer conceived the original idea, managed the overall project, and wrote and edited the final manuscript.

Conflicts of Interest: The authors declare no conflict of interest.

References

1. Abraham, K.; Debreuille, M.; Mairesse, G.; Nowogrocki, G. Phase transitions and ionic conductivity in $Bi_4V_2O_{11}$ an oxide with a layered structure. *Solid State Ion.* **1988**, *28–30*, 529–532. [CrossRef]
2. Abraham, F.; Boivin, J.; Mairesse, G.; Nowogrocki, G. The BIMEVOX Series: A New Family of High Performance Oxide Ion Conductors. *Solid State Ion.* **1990**, *40–41*, 934–937. [CrossRef]
3. Yaremchenko, A.; Avdeev, M.; Kharton, K.; Kovalevsky, A.; Naumovich, E.; Marques, F. Structure and electronic conductivity of $Bi_{2-x}La_xV_{0.9}Cu_{0.1}O_{5.5-\delta}$. *Mater. Chem. Phys.* **2002**, *77*, 552–558. [CrossRef]
4. Kurek, P.; Dygas, J.; Breiter, M. Impedance measurements on single crystals of the oxygen ion conductor BICUVOX. *J. Electroanal. Chem.* **1994**, *378*, 77–83. [CrossRef]
5. Goodenough, J. Oxide-Ion Electrolytes, a review. *Ann. Rev. Mater. Res.* **2003**, *33*, 91–128. [CrossRef]
6. Kendall, K.; Navas, C.; Thomas, J.; Loye, H. Recent Developments in Oxide Ion Conductors: Aurivillius Phases. *Chem. Mater.* **1996**, *8*, 642–649. [CrossRef]
7. Cho, H.; Sakai, G.; Shimanoe, K.; Yamazoe, N. Preparation of BiMeVOx (Me=Cu, Ti, Zr, Nb, Ta) compounds as solid electrolyte and behavior of their oxygen concentration cells. *Sens. Actuators B Chem.* **2005**, *109*, 307–314. [CrossRef]
8. Kida, T.; Minami, T.; Kishi, S.; Yuasa, M.; Shimanoe, K.; Yamazoe, N. Planar-type BiCuVOx solid electrolyte sensor for the detection of volatile organic compounds. *Sens. Actuators B Chem.* **2009**, *137*, 147–153. [CrossRef]
9. Kida, T.; Harano, H.; Minami, T.; Kishi, S.; Morinaga, N.; Yamazoe, N.; Shimanoe, K. Control of electrode reactions in a mixed-potential-type gas sensor based on a BiCuVOx solid electrolyte. *J. Phys. Chem. C* **2010**, *114*, 15141–15148. [CrossRef]
10. Patil, B.; Sharma, S.; Mohanta, H.; Roy, B. BINIVOX catalyst for hydrogen production from ethanol by low temperature steam reforming (LTSR). *J. Chem. Sci.* **2017**, *129*, 1741–1746. [CrossRef]
11. Fuierer, P.; Maier, R.; Roder-Roith, U.; Moos, R. Processing Issues Related to the Bi-dimensional Ionic Conductivity of BIMEVOX Ceramics. *J. Mater. Sci.* **2011**, *46*, 5447–5453. [CrossRef]
12. Sant, C.; Contour, J. Pulsed laser deposition of $Bi_4Cu_{2x}V_{2(1-x)}O_{11}$ thin films. *J. Cryst. Growth* **1995**, *153*, 63–67. [CrossRef]
13. Muller, C.; Chateigner, D.; Anne, M.; Bacmann, M.; Fouletier, J.; Rango, P. Pressure and magnetic field effects on the crystallographic texture and electrical conductivity of the compound. *J. Phys. D Appl. Phys.* **1996**, *29*, 3106–3111. [CrossRef]
14. Fuierer, P.; Nichtawitz, A. Electric Field Assisted Hot Forging of Bismuth Titanate. In Proceedings of the IEEE International Symposium on Applications of Ferroelectrics, University Park, PA, USA, 7–10 August 1994; pp. 126–129.
15. Fuierer, P.; Newnham, R. Newnham, $La_2Ti_2O_7$ Ceramics. *J. Am. Ceram. Soc.* **1991**, *74*, 2876–2881. [CrossRef]
16. Fuierer, P.; Maier, M.; Exner, J.; Moos, R. Anisotropy and thermal stability of hot-forged BICUTIVOX oxygen ion conducting ceramics. *J. Eur. Ceram. Soc.* **2014**, *34*, 943–951. [CrossRef]
17. Shantha, K.; Varma, K. Fabrication and characterization of grain-oriented bismuth vanadate ceramics. *Mater. Res. Bull.* **1997**, *32*, 1581–1591. [CrossRef]
18. Seth, V.; Schulze, W. Grain-Oriented Fabrication of Bismuth Titanate Ceramics and Its Electrical Properties. *IEEE Trans. Ultrason. Ferroelectr. Freq. Control* **1989**, *36*, 41–42. [CrossRef] [PubMed]
19. Watanabe, H.; Kimura, T.; Yamaguchi, T. Particle Orientation during Tape Casting in the Fabrication of Grain-Oriented Bismuth Titanate. *J. Am. Ceram. Soc.* **1989**, *72*, 289–293. [CrossRef]
20. Horn, J.; Zhang, S.; Selvaraj, U.; Messing, G.; Trolier, S. Templated Grain Growth of Textured Bismuth Titanate. *J. Am. Ceram. Soc.* **1999**, *82*, 921–926. [CrossRef]
21. Yilmaz, H.; Messing, G.; Trolier, S. (Reactive) Templated Grain Growth of Textured Sodium Bismuth Titanate ($Na_{0.5}Bi_{0.5}TiO_3$-$BaTiO_3$) Ceramics-I Processing. *J. Electroceram.* **2003**, *11*, 207–215. [CrossRef]
22. West, D.; Payne, D. Reactive-Templated Grain Growth of $Bi_{0.5}(Na,K)_{0.5}TiO_3$: Effects of Formulation on Texture Development. *J. Am. Ceram. Soc.* **2003**, *86*, 1132–1137. [CrossRef]
23. Kimura, T.; Yoshida, Y. Origin of Texture Development in Barium Bismuth Titanate Prepared by the Templated Grain Growth Method. *J. Am. Ceram. Soc.* **2006**, *89*, 869–874. [CrossRef]
24. Arendt, R.; Rosolowski, J.; Szymaszek, J. Lead zirconate titanate ceramics from molten salt solvent synthesized powders. *Mater. Res. Bull.* **1979**, *15*, 703–709. [CrossRef]
25. Kimura, T.; Yamaguchi, T. Fused Salt Synthesis of $Bi_4Ti_3O_{12}$. *Ceram. Int.* **1983**, *9*, 13–17. [CrossRef]

26. Li, Z.; Zhang, X.; Hou, J.; Zhou, K. Molten salt synthesis of anisometric $Sr_3Ti_2O_7$ particles. *J. Cryst. Growth* **2007**, *305*, 265–270. [CrossRef]

27. Kimura, T. Chapter 4. Molten Salt Synthesis of Ceramic Powders. In *Advances in Ceramics—Synthesis and Characterization, Processing and Specific Applications*; Sikalidis, C., Ed.; INTECH: Rijeka, Croatia, 2011; pp. 75–100, ISBN 978-953-307-505-1.

28. Liu, X.; Fechler, N.; Antonietti, M. Salt melt synthesis of ceramics, semiconductors and carbon nanostructures. *Chem. Soc. Rev.* **2013**, *42*, 8237–8265. [CrossRef] [PubMed]

29. Preethi, G.; Ninan, A.; Kumar, K.; Balan, R.; Nagaswarupa, H. Molten Salt Synthesis of Nanocrystalline $ZnFe_2O_4$ and Its Photocatalytic Dye Degradation Studies. *Mater. Today Proc.* **2017**, *4*, 11816–11819. [CrossRef]

30. Kang, M.; Kim, D.; Hwang, N. Ostwald ripening kinetics of angular grains dispersed in a liquid phase by two dimensional nucleation and abnormal grain growth. *J. Eur. Ceram. Soc.* **2002**, *22*, 603–612. [CrossRef]

31. Roy, B.; Fuierer, P. Molten Salt Synthesis of $Bi_4(V_{0.85}Co_{0.15})_2O_{11-\delta}$ (BICOVOX) Ceramic Powders. *J. Am. Ceram. Soc.* **2009**, *92*, 520–523. [CrossRef]

32. Sangster, J.; Pelton, A. Phase Diagrams and Thermodynamic Properties of the 70 Binary Alkali Halide Systems Having Common Ions. *J. Phys. Chem. Ref. Data* **1987**, *16*, 509. [CrossRef]

33. International Center for Diffraction Data. *Powder Diffraction File #01-070-9191, Bismuth Vanadium Copper Oxide*; International Center for Diffraction Data: Newtown Square, PA, USA.

34. Inoue, M.; Hirasawa, I. The relationship between crystal morphology and XRD peak intensity on $CaSO_4 \cdot 2H_2O$. *J. Cryst. Growth* **2013**, *380*, 169–175. [CrossRef]

35. Roy, B.; Scott, P.; Ahrenkiel, Y.; Fuierer, P. Controlling the Size and Morphology of TiO_2 Powder by Molten and Solid Salt Synthesis. *J. Am. Ceram. Soc.* **2008**, *91*, 2455–2463. [CrossRef]

36. Betz, W.; Webb, H. *Plane Geometry, Book, V. Regular Polygons and Circles*; Ginn & Co.: Cambridge, UK, 1921; p. 321.

37. Lotgering, F. Topotactical Reaction with Ferrimagnetic Oxides Having Hexagonal Structures. *J. Inorg. Nucl. Chem.* **1959**, *9*, 113–123. [CrossRef]

38. Yoon, K.; Yong, S.; Kang, D. Review: Molten Salt synthesis of lead-based relaxors. *J. Mater. Sci.* **1998**, *33*, 2977–2984. [CrossRef]

39. Kapadia, L. *The Mechanism of Grain Growth in Ceramics*; NASA Report; NASA: Washington, DC, USA, 1973; pp. 6–34.

40. Gu, Y.; Huang, J.; Li, L.; Zhang, K.; Wang, X.; Li, Q.; Tan, X.; Xu, H. Molten salt synthesis of anisotropy $BaBi_4Ti_4O_{15}$ powders in K_2SO_4-Na_2SO_4 flux. *Mater. Sci. Forum* **2011**, *687*, 333–338. [CrossRef]

41. Wulff, G. Zur Frage der Geschwindigkeit des Wachsthums und der Auflösung der Krystallflächen. *Z. Kristallogr.* **1901**, *34*, 449–530. [CrossRef]

42. Herring, C. Some Theorems on the Free Energies of Crystal Surfaces. *Phys Rev.* **1951**, *82*, 87–93. [CrossRef]

43. Kang, S.C. 15 Grain Shape and Grain Growth in a Liquid Matrix. In *Sintering*; Elsevier: London, UK, 2005; p. 218.

44. Kimura, T.; Tani, T. Chapt 15, Processing and Properties of Textured Bismuth Layer-Structured Ferroelectrics. In *Lead Free Piezoelectrics*; Priya, S., Nahm, S., Eds.; Springer: New York, NY, USA, 2012; p. 465, ISBN 978-1-4419-9598-8.

45. Hollenberg, G.; Gordon, R. Origin of Anomalously High Activation Energies in Sintering and Creep of Impure Refractory. *J. Am. Ceram. Soc.* **1973**, *56*, 109–110. [CrossRef]

46. Afanasiev, P. Preparation of Mixed Phosphates in Molten Alkali Metal Nitrates. *Chem. Mater.* **1999**, *11*, 1999–2007. [CrossRef]

47. Tang, Q.Y.; Kan, Y.M.; Wang, P.L.; Li, Y.G.; Zhang, G.J. Nd/V Co-Doped $Bi_4Ti_3O_{12}$ Powder Prepared by Molten Salt Synthesis. *J. Am. Ceram. Soc.* **2007**, *90*, 3353–3356. [CrossRef]

48. Chiang, Y.; Birnie, D.; Kingery, W. Chapter 5, Microstructure. In *Physical Ceramics*; J. Wiley & Sons: New York, NY, USA, 1997; p. 354, ISBN 0-471-59873-9.

crystals

MDPI

Article

Crystal Structure, Hydration, and Two-Fold/Single-Fold Diffusion Kinetics in Proton-Conducting Ba$_{0.9}$La$_{0.1}$Zr$_{0.25}$Sn$_{0.25}$In$_{0.5}$O$_{3-a}$ Oxide

Wojciech Skubida [1], Anna Niemczyk [1], Kun Zheng [1], Xin Liu [2] and Konrad Świeczek [1,3,*]

[1] Faculty of Energy and Fuels, Department of Hydrogen Energy, AGH University of Science and Technology, al. A. Mickiewicza 30, 30-059 Krakow, Poland; wskubida@agh.edu.pl (W.S.); niemczyk@agh.edu.pl (A.N.); zheng@agh.edu.pl (K.Z.)
[2] Contemporary Amperex Technology Co., Limited, Fujian 352100, China; rzxhygr@163.com
[3] AGH Centre of Energy, AGH University of Science and Technology, ul. Czarnowiejska 36, 30-054 Krakow, Poland
[*] Correspondence: xi@agh.edu.pl

Received: 15 February 2018; Accepted: 13 March 2018; Published: 16 March 2018

Abstract: In this work, hydration kinetics related to the incorporation of water into proton-conducting Ba$_{0.9}$La$_{0.1}$Zr$_{0.25}$Sn$_{0.25}$In$_{0.5}$O$_{3-a}$ perovskite-type oxide are presented, with a recorded transition on temperature from a single-fold to a two-fold behavior. This can be correlated with an appearance of the electronic hole component of the conductivity at high temperatures. The collected electrical conductivity relaxation data allowed to calculate chemical diffusion coefficient D and surface exchange reaction coefficient k, as well as respective activation energies of their changes on temperature. Presented results are supplemented with a systematic characterization of the structural properties of materials synthesized at different temperatures, amount of incorporated water after hydration in different conditions, influence of water content on the crystal structure, as well as electrical conductivity in dry, H$_2$O- and D$_2$O-containing air, which enabled to evaluate proton (deuterium) conductivity.

Keywords: perovskite oxides; substituted barium indate; hydration; proton conductivity; relaxation experiments; coupled/decoupled ionic transport

1. Introduction

High-temperature proton conductors, used e.g., as electrolytes in Protonic Ceramic Fuel Cell (PCFC) technology, are characterized by a high ionic conductivity at the intermediate temperatures (ca. 500–700 °C), which may surpass the one of commonly used oxide ion conductors [1,2]. The first reports on proton-conducting ceramics were presented by Iwahara and co-workers in the 1980s [3,4]. Since that time, many acceptor-doped AB$_{1-x}$M$_x$O$_{3-\delta}$ (A: Ba, Sr, Ca; B: Zr, Ce, In, Sn; M: different +3/+2 elements) perovskite-type oxides were discovered and characterized as efficient proton conductors [5–10].

One of the highest values of H$^+$ conductivity in wet atmospheres, on the order of 10^{-2} S cm^{-1} at ca. 600 °C, show materials based on BaCeO$_3$. At higher temperatures, their total ionic conductivity increases even more, but oxygen ion component becomes predominant, which results in a significant decrease of the proton transfer number t$_H$ [11,12]. Unfortunately, cerium-based perovskites react with acidic oxides (e.g., CO$_2$ or SO$_3$) and steam, forming carbonates and hydroxides, which limits usage of these materials in fuel cells or high temperature steam electrolyzers [13,14]. Zr-based oxides exhibit improved stability, but their proton conductivity is lower and much higher temperatures are required for a successful sintering of dense pellets/membranes [15,16]. Recently, other materials having different crystal structures are also considered as high-temperature proton conductors [17–24].

Proton, due to a lack of electrons on its electron shell, strongly polarizes surrounding ions in a host lattice, leading to the formation of a relatively high-energy bonds nearby. Consequently, protons in oxides are located in the electron cloud of the oxygen anions, forming $OH_O^{\bullet}$ defects [17]. Depending on the atmosphere, protons can be absorbed into the oxide structure according to the following reactions [25]:

- in dry hydrogen atmosphere,

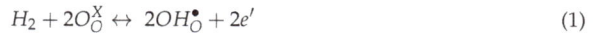

$$H_2 + 2O_O^X \leftrightarrow 2OH_O^{\bullet} + 2e' \tag{1}$$

- in humidified atmosphere,

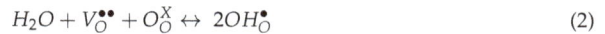

$$H_2O + V_O^{\bullet\bullet} + O_O^X \leftrightarrow 2OH_O^{\bullet} \tag{2}$$

For the second reaction, commonly observed for perovskite-type oxides, the presence of the oxygen vacancies in the crystal structure is required. In the $A^{2+}B^{4+}O_3$ perovskite oxides, partial substitution of the B-site cation by e.g., trivalent one (M^{3+}) is used to form the oxygen vacancies [13].

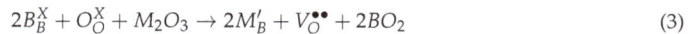

$$2B_B^X + O_O^X + M_2O_3 \rightarrow 2M_B' + V_O^{\bullet\bullet} + 2BO_2 \tag{3}$$

The amount of the oxygen vacancies is therefore half of the introduced M^{3+}. Simulations and isotopic studies have suggested that the Grotthuss mechanism can be generally considered as the proton transport mechanism, occurring at elevated temperatures in perovskite oxides [26]. It is known that the activation energy of proton hopping from one oxygen anion to the neighboring one depends on the oxygen-oxygen distance, but regardless of the crystal structure of the oxides, the value is about 2/3 of the activation energy of the oxygen transport through the vacancy mechanism. This can be explained by the fact that the migration of protons is completely dependent on the oscillation of the oxygen network. For both types of conductivity, the charge carriers need to overcome the same energy barrier, but while the oxygen anion must go through the so-called saddle point between the energy minima, the proton can jump (or tunnel) when the ion on which it is located will overcome 2/3 of the height of the energy barrier [27].

It is also known that there is a correlation between a difference of electronegativity of the A- and the B-site cations and the enthalpy of the hydration reaction. If the electronegativity difference is higher, it limits the amount of incorporated protons, lowering the proton conductivity [28].

Coupled and Decoupled Diffusion of Ionic Charge Carriers

Considering Equation (2), the respective equation for the reaction's constant can be written as

$$K_W = \frac{[OH_O^{\bullet}]^2}{p_{H_2O}[V_O^{\bullet\bullet}][O_O^X]} \tag{4}$$

Also, Equations (3) and (4) can be re-written in the following form [29]:

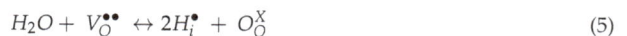

$$H_2O + V_O^{\bullet\bullet} \leftrightarrow 2H_i^{\bullet} + O_O^X \tag{5}$$

$$K'_W = \frac{[H_i^{\bullet}]^2[O_O^X]}{p_{H_2O}[V_O^{\bullet\bullet}]} \tag{6}$$

As noticed in work [29], this should correspond to the situation that during incorporation (or release) of water into (from) the oxide, the diffusion of the oxygen anions and protons is ambipolar. In other words, respective fluxes are correlated to each other. Consequently, in e.g., the electrical relaxation experiments conducted during change of the water vapor partial pressure in the sample's surrounding atmosphere, the respective relaxation curves should be of a single fold nature. This is in contrast to numerous literature reports showing a two-fold character of the relaxation curves recorded in specific conditions [30–36].

In a new approach presented in work [29], the two-fold relaxation kinetics are explained by a decoupled transport of hydrogen and oxygen ions, which initially occurs on the surface and is followed by diffusion in the bulk.

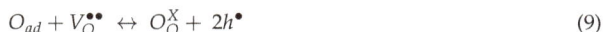

$$H_2O + V_O^{\bullet\bullet} \leftrightarrow 2H_i^{\bullet} + O_O^X \tag{7}$$

$$H_{ad} + h^{\bullet} \leftrightarrow H_i^{\bullet} \tag{8}$$

$$O_{ad} + V_O^{\bullet\bullet} \leftrightarrow O_O^X + 2h^{\bullet} \tag{9}$$

This corresponds to the situation of decoupled chemical diffusion of hydrogen and oxygen. Both processes can be described, respectively, as ambipolar diffusion of protons and electronic holes, as well as ambipolar diffusion of oxygen anions and electronic holes. Furthermore, as shown in work [30], the observed change in the total conductivity ($\Delta\sigma$) as a function of time (i.e., in relaxation experiments) is a sum of the H-related and O-related conductivities, which proceed in opposite directions: $\Delta\sigma$(H) is decreasing with time, while $\Delta\sigma$(O) is increasing.

Crucial for the relaxation behavior is the presence of electronic holes, generated according to the Equation (9) and consumed in Equation (8). If their concentration is negligible, the corresponding transference number of the hole conductivity approaches zero. In this case, the decoupled transport (two-fold relaxation kinetics) reduces to the single-fold diffusion of water (single-fold relaxation) [29].

In this paper, comprehensive results of structural, thermogravimetric, and transport properties of proton-conducting $Ba_{0.9}La_{0.1}Zr_{0.25}Sn_{0.25}In_{0.5}O_{3-a}$ oxide are presented. Notwithstanding that the chosen chemical composition is not characterized by a very high proton conductivity, it enabled us to observe the temperature-dependent transition from the single-fold to the two-fold relaxation kinetics associated to the appearance of the electronic hole-related conductivity. Apart from the characterization of the basic physicochemical properties, the conducted studies with humidified atmosphere allowed us to determine the values of the activation energy of the surface exchange coefficient k and the bulk diffusion coefficient D for the coupled and decoupled ionic transport.

2. Materials and Methods

Analyzed $Ba_{0.9}La_{0.1}Zr_{0.25}Sn_{0.25}In_{0.5}O_{3-a}$ was synthesized by a standard high-temperature solid state route. Stoichiometric amounts of the respective oxides (La_2O_3, ZrO_2, SnO_2, In_2O_3) and barium carbonate ($BaCO_3$), all with ≥99.9% purity, were homogenized in a high-efficiency mill. The used La_2O_3 was earlier preheated up to 800 °C in order to remove water and ensure proper stoichiometry. The obtained mixture was pressed into pellets and then annealed in a temperature range of 1200–1450 °C. For preparation of dense pellets (ca. 98%), the initial calcination performed at 950 °C in synthetic air (decomposition of $BaCO_3$) was required, and also, the sintering temperature was increased up to 1600 °C. In order to minimize evaporation of barium or other cations, the pressed pellets were covered with a layer of a powder having the same composition and were placed in lid-covered crucibles. Chemical composition of the materials was established by inductively coupled plasma-optical emission spectroscopy (ICP-OES, Optima 2100, Perkin Elmer, Wellesley, MA, USA).

Structural data were obtained by X-ray diffraction (XRD) measurements conducted in 10°–110° range using CuK_{α} radiation on Panalytical Empyrean diffractometer (Empyrean, Panalytical, Almelo, Netherlands) equipped with PIXcel3D detector. An Anton Paar HTK 1200N (Panalytical, Almelo, The Netherlands) oven chamber was used for high-temperatures XRD studies, which were carried out in 25–900 °C temperature range in a synthetic air flow (H_2O content < 3 ppm). Structural parameters were refined using GSAS/EXPGUI set of software, based on Rietveld method (a least squares approach, which enable to refine a theoretical XRD profile until it matches the measured data) [37,38]. In the fittings done using cubic *Pm-3m* space group, unit call parameter a, profile parameters, zero shift, and thermal U_{iso} factors were refined. It should be emphasized that (Ba,La)-O distance can be derived as equal to $a\sqrt{2}/2$, while (Zr,Sn,In)-O bond length equals $a/2$ in the cubic perovskite structure. Stored and dried at 400 °C for 24 h (in synthetic air), the materials were studied during XRD measurements. Also, Scherrer equation was utilized to obtain crystallite size.

Microstructure of the as-synthesized materials was analyzed by scanning electron microscopy (SEM) technique. SEM images were taken on FEI Nova NanoSEM 200 apparatus (FEI EUROPE COMPANY, Eindhoven, The Netherlands).

In order to evaluate the amount of the incorporated water, thermogravimetric (TG) measurements were carried out on TA Q5000 IR thermobalance (TA Instruments, New Castle, DE, USA). Tests were performed up to 900 °C (heating/cooling rate 5 °C min^{-1}) in synthetic air on previously hydrated and dried samples. The hydration process was conducted by annealing of $Ba_{0.9}La_{0.1}Zr_{0.25}Sn_{0.25}In_{0.5}O_{3-a}$ powder at 250 °C and 400 °C in humidified air (ca. 3% vol. H_2O), while drying was done by similar annealing at 400 °C and 800 °C in synthetic air. TG measurements were accompanied by mass spectroscopy analysis of the out-going gas in order to establish H_2O and CO_2 presence in the studied samples. For the studies, Pfeiffer Vacuum ThermoStar mass spectrometer (GSD 301 T1 (2008/08), Pfeiffer Vacuum, Aßlar, Germany) was used. The initial and relatively high water signal recorded in the studies likely comes from the apparatus-related affect (presence of residual moisture). This line should be treated as a background one.

Electrical properties of a dense $Ba_{0.9}La_{0.1}Zr_{0.75}Sn_{0.75}In_{0.5}O_{3-a}$ sinter were measured by the AC electrochemical impedance spectroscopy (EIS) technique using Solartron 1260 frequency response analyzer (1260, Solartron, Bognor Regis, Great Britain). Samples were mounted on Probostat holder, with Pt electrodes attached. The applied disturbance amplitude was 25 mV and the frequency range was 0.1–10^6 Hz. Tests were performed in 350–800 °C temperature range in dry synthetic air, and also in humidified synthetic air with H_2O and D_2O (ca. 3% vol.) in order to establish proton (deuterium) conductivity component of the total conductivity. Additional measurements were conducted in dry and H_2O-containing argon to evaluate pO_2 influence on conductivity. Gathered data were analyzed using Scribner Associates Inc. ZView 2.9 software (Scribner Associates Inc., Southern Pines, NC, USA) by selection of a proper equivalent electrical circuit [39]. An approximate value of proton conductivity (σ_H) was obtained as a difference between the recorded conductivity in wet and dry atmospheres: $\sigma_H = \sigma_{wet} - \sigma_{dry}$. Consequently, the transference number $t_H = \sigma_H/\sigma_{wet}$. Similar calculations were carried out in the case of measurements in D_2O-containing atmosphere in order to determine σ_D and t_D [40].

Electrical conductivity relaxation (ECR) measurements were performed during an abrupt gas change from dry to humidified synthetic air at the constant temperature in a range of 350–800 °C. The recorded ECR profiles were normalized and refined using custom-made Matlab code. Considering geometry of the pellets, the one-dimensional diffusion model was found to be the correct approximation [41]. Interestingly, a coupled and decoupled transport phenomena related to the ionic conductivity were observed. While these effects were previously reported in several studies [34,36], for the presented results, a clear transition occurs as a function of temperature.

3. Results and Discussion

3.1. Crystal Structure at Room and Elevated Temperatures

As shown in Figure 1a, $Ba_{0.9}La_{0.1}Zr_{0.25}Sn_{0.25}In_{0.5}O_{3-a}$ oxide can be obtained as a single phase, with no major impurities present in a wide range of temperatures (1200–1600 °C). Figure 1b shows an exemplary diffractogram recorded at room temperature (RT) of a sample after heat treatment at 1400 °C, together with the performed structural refinement. It can be seen that all reflections correspond well with the cubic *Pm-3m* space group. This structure was present in all samples sintered at different temperatures. It is also visible (Figure 1a) that, with the increasing sintering temperature, the recorded peaks shift toward higher angles, suggesting a decrease of the unit cell parameter. This was confirmed by Rietveld refinements, and the calculated dependence is shown in Figure 2. In order to establish chemical composition of the samples, ICP studies were conducted, with results being in agreement with the assumed content of the elements (e.g., material annealed at

1400 °C has $Ba_{0.93}La_{0.10}Zr_{0.22}Sn_{0.29}In_{0.49}O_{2.84}$ composition). In addition, these measurements indicated somewhat increased evaporation of barium, as well as tin and indium for materials annealed at higher temperatures, but the changes were found be rather too small to fully explain the decrease of the *a* parameter with the sintering temperature (both Ba^{2+} and In^{3+} are large cations). Alternatively, it seems that the observed variation of the unit cell parameter mainly originates from a high affinity of the samples for water incorporation. For comparison, XRD measurements were also conducted immediately after drying of the samples in synthetic air at 400 °C for 24 h. The refined unit cell parameters are significantly lower (except for the material sintered at 1600 °C), suggesting that the compounds stored under ambient conditions reacted with moisture from air.

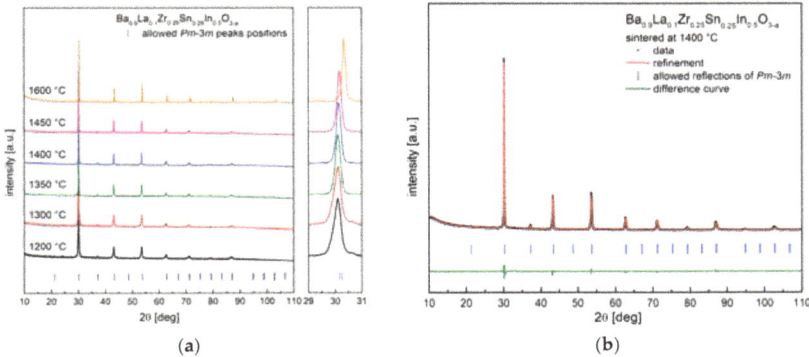

Figure 1. (**a**) X-ray diffraction (XRD) data recorded at room temperature (RT) for samples sintered at different temperatures; (**b**) XRD diffractogram with Rietveld refinement for $Ba_{0.9}La_{0.1}Zr_{0.25}Sn_{0.25}In_{0.5}O_{3-a}$ sintered at 1400 °C.

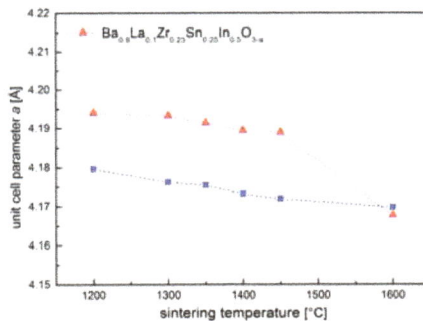

Figure 2. Unit cell parameter *a* dependence on the sintering temperature for $Ba_{0.9}La_{0.1}Zr_{0.25}Sn_{0.25}In_{0.5}O_{3-a}$.

As the temperature of the synthesis increases, the average crystallite size was found to grow in the expected manner (Figure 3). Nevertheless, the pellets obtained at 1450 °C and below are not well-sintered, with a relative density not exceeding 70%. On the contrary, material prepared at 1600 °C was very dense (ca. 98%). Consequently, a large surface area of contact between the material and atmospheric air of the porous specimens (Figure 4) intensifies reactivity with the water vapor, while for the dense material, only the pellet's surface has direct contact with the atmosphere, suppressing the reaction. This is of a crucial consequence because $Ba_{0.9}La_{0.1}Zr_{0.25}Sn_{0.25}In_{0.5}O_{3-a}$ in a powder form is reactive toward H_2O at very low temperatures, even at RT.

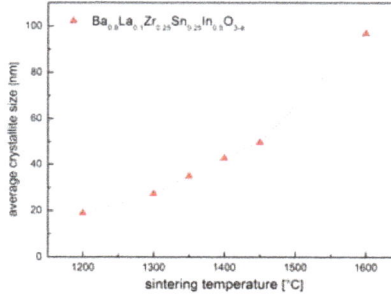

Figure 3. Calculated average crystallite size of $Ba_{0.9}La_{0.1}Zr_{0.25}Sn_{0.25}In_{0.5}O_{3-a}$ as a function of the sintering temperature.

Figure 4. Scanning electron microscopy (SEM) micrograph of $Ba_{0.9}La_{0.1}Zr_{0.25}Sn_{0.25}In_{0.5}O_{3-a}$ material (crushed and grinded pellet) sintered at 1400 °C.

As shown in Figure 5, material sintered at 1400 °C and dried at 400 °C in synthetic air, which was protected from the ambient atmosphere and studied in a short time after the synthesis, exhibits much lower unit cell parameter (4.1742 Å), comparing to the one stored under ambient conditions (4.1894 Å). The effect of the unit cell parameter (and volume) increase on hydration of the material was previously documented in works [42–44].

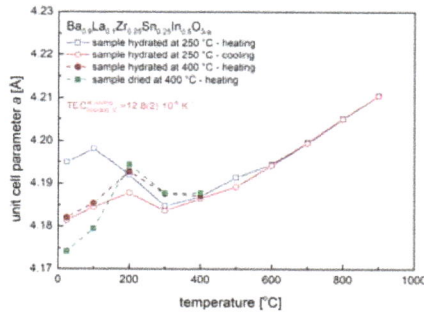

Figure 5. Unit cell parameter dependence on temperature, as calculated from high-temperature XRD data, recorded in synthetic air for samples, which were previously hydrated at 250 °C and 400 °C, as well as dried at 400 °C.

With an initial increase of temperature above RT, the material dried at 400 °C was found to increase its unit cell parameter (and volume) up to 200 °C, above which temperature the recorded unit cell parameter decreased and then increased again above 400 °C (Figure 5). Essentially the same behavior was observed for samples previously hydrated in humidified synthetic air (ca. 3 vol. % H_2O) at 250 °C and 400 °C. The quantitative difference between the materials results from the value of the initial unit cell parameter, which was found to the highest for the oxide hydrated at 250 °C. As documented below in TG studies, this corresponds to the highest amount of the incorporated water into the crystal structure. Above ca. 400 °C all studied samples were found to behave practically the same. During cooling from 900 °C to 500 °C the relative changes of the unit call parameter scaled linearly with temperature, which allowed to evaluate thermal expansion coefficient (TEC). The TEC value was found to be moderate, $12.8(2) \times 10^{-6}$ K^{-1}, a typical one for this class of materials. Interestingly, on cooling below 300 °C, the unit cell parameter of $Ba_{0.9}La_{0.1}Zr_{0.25}Sn_{0.25}In_{0.5}O_{3-a}$ increased and then decreased again. This effect can be explained as due to incorporation of residual water vapor present in synthetic air, and while H_2O content is below 3 ppm, a constant gas flow on the order of 100 cm^3 min^{-1} supplied into the oven-chamber was enough for the sample to absorb relatively large quantity of water [45–47].

Selected data concerning structural refinements performed at RT and at elevated temperatures are gathered in Table 1 below.

Table 1. Refined structural data for $Ba_{0.9}La_{0.1}Zr_{0.25}Sn_{0.25}In_{0.5}O_{3-a}$ material obtained in different conditions.

Conditions	Unit Cell Parameter a [Å]	Unit Cell Volume V [Å^3]	(Ba,La)-O Distance [Å]	(Zr,Sn,In)-O Distance [Å]	x^2/R_{wp} [%]	$U_{iso} \cdot 100$ [Å^2] CationsO
stored sample	4.1893(1)	73.53(1)	2.9623(1)	2.0947(1)	4.125/3.53	2.39(2)/4.6(1)
material hydrated at 250 °C (data at RT)	4.1949(1)	73.82(1)	2.9662(1)	2.0974(1)	1.548/2.79	2.08(2)/3.1(1)
data recorded at 600 °C in synthetic air	4.1944(1)	73.79(1)	2.9659(1)	2.0972(1)	2.291/3.45	2.09(2)/4.3(2)
data recorded at 900 °C in synthetic air	4.2105(1)	74.64(1)	2.9773(1)	2.1053(1)	1.959/3.21	2.83(2)/5.7(2)

All of the presented above results can be explained as originating from a presence of protons in the studied material, which incorporation/release (in a form of water) on heating/cooling is visible in structural changes. What is more, strong effects recorded for the dried sample prove a very strong affinity of the material toward water uptake, even at very low water partial pressures.

3.2. Thermogravimetric Measurements and Presence of Protons

The results of thermogravimetric studies conducted in synthetic air for $Ba_{0.9}La_{0.1}Zr_{0.25}Sn_{0.25}In_{0.5}O_{3-a}$ material hydrated at 250 °C are shown in Figure 6a. TG measurements were accompanied by mass spectroscopy studies, and the recorded (relative) ionic current data of H_2O- and CO_2-related signals are presented in Figure 6b,c. As can be seen, during first heating, a considerable weight loss occurs at above 200 °C, matching well the behavior seen in the high-temperature XRD studies for the hydrated material (Figure 5). The maximum signal of the H_2O occurs at lower temperature than the CO_2-related peak (Figure 6b,c), and consequently, it can be assumed that majority of the weight change up to ca. 300 °C is related to the water release. Using TG data the amount of the incorporated water during hydration at 250 °C can be estimated as ca. 0.13 mole of water per mole of material, corresponding to the $Ba_{0.9}La_{0.1}Zr_{0.25}Sn_{0.25}In_{0.5}O_{2.93}H_{0.26}$ composition. For the material hydrated at 400 °C, the water content was found to be smaller, ca. 0.11. Interestingly, on the second TG cycle some remaining water signal was also recorded. For the CO_2 signal, while the first maximum can be linked to a desorption process from the powder's surface, the peak at ca. 600 °C can be related to a decomposition of residual carbonates. Apart from the first heating, practically no CO_2-related signal was recorded in the following steps of the TG studies. Importantly, in additional studies of materials previously dried in synthetic air at 400 °C and 800 °C, qualitatively similar behavior was observed, i.e., presence of H_2O- and CO_2-related peaks was evident. This strongly indicates that samples are highly moisture- and CO_2-sensitive, even during contact with atmospheric air at ambient temperature for a relatively short time.

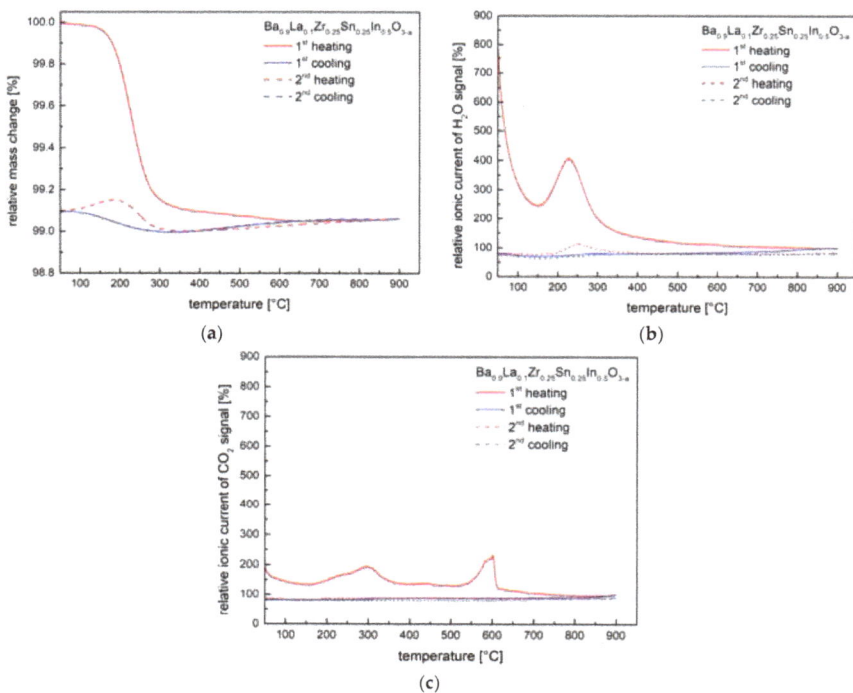

Figure 6. (**a**) Thermogravimetric characteristics for the hydrated at 250 °C $Ba_{0.9}La_{0.1}Zr_{0.25}Sn_{0.25}In_{0.5}O_{3-a}$ material, recorded in synthetic air during two cycles; (**b**) Relative ionic current of H_2O-related signal recorded by mass spectrometer, corresponding to the performed TG studies; (**c**) Relative ionic current of CO_2-related signal recorded by mass spectrometer, corresponding to the performed TG studies.

3.3. Transport Properties in Dry and Humidified Air

Figure 7a,b show exemplary impedance data with fitting, measured for the $Ba_{0.9}La_{0.1}Zr_{0.25}Sn_{0.25}In_{0.5}O_{3-a}$ material in different atmospheres at 450 °C and 750 °C. The spectra at the lower temperature contain three semi-circles. However, the refinements suggest the presence of the R_0 shift from the origin of coordinates (see inset with the equivalent circuit). Such results can be interpreted that the bulk and the grain boundary conductivities correspond to the shift and the high-frequency arc visible on the left, while charge transfer between the sample and the electrodes is visible in a form of the middle- and the low-frequency arc on the right [43,48]. There is a noticeable decrease of the high frequency arc and the shift upon change of the atmosphere from dry to H_2O (D_2O)-containing air, which is an evidence for the appearance of the proton (deuterium) conductivity. It is also expected that the relative increase of the conductivity related to D_2O is smaller [45]. EIS data recorded at higher temperatures, e.g., 750 °C (Figure 7b) are different, and strongly suggest presence of the electronic component of the conductivity, with R_0 shift from the origin of coordinates likely comprising also the ohmic resistance. This issue is discussed in more details below.

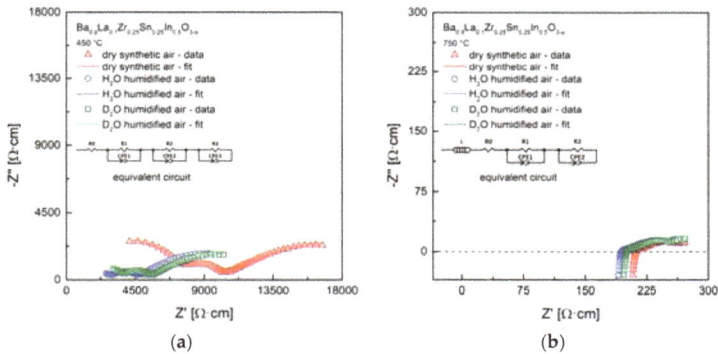

Figure 7. (a) Exemplary impedance data with fitting, measured in dry, H_2O-, and D_2O-containing synthetic air at 450 °C; (b) Exemplary impedance data with fitting, measured in dry, H_2O-, and D_2O-containing synthetic air at 750 °C.

Gathered impedance data allowed us to establish the temperature dependence of the total electrical conductivity, as presented in Arrhenius-type coordinates (Figure 8a). The results are expected, with the highest conductivity values and lowest activation energy appearing in H_2O-containing atmosphere. It can be noticed that the relative increase of the conductivity in wet atmospheres is not very high, and consequently the calculated proton (deuterium) conductivities and the respective transference numbers are not high (Table 2). The exemplary proton conductivities are ca. 1.1×10^{-4} S cm^{-1} at 350 °C and increase to 6.4×10^{-4} S cm^{-1} at 550 °C. A very similar dependence of total conductivity was observed in dry and wet Ar (Figure 8b), which suggest negligible influence of the oxygen partial pressure in this range, making the approximations about determination of proton and deuterium conductivities viable (electronic component is insignificant in such conditions).

At the highest temperatures of 700–800 °C, practically no change of the conductivity of the material during change of the atmosphere was recorded (Figure 8a,b). Considering the shape of the impedance data (Figure 7b), a large part of the resistance can be related to the electronic hole component. This was further supported by polarization-type (Hebb-Wagner) conductivity experiment, which allowed us to establish that, at 800 °C in dry air, the electronic component of the conductivity seems to be dominant. However, due to intrinsic limitations of the method, the exact evaluation of the electronic transference number was hindered.

Table 2. Calculated proton (σ_H) and deuterium (σ_D) conductivities together with respective transference numbers (t_H, t_D) for measurement conducted in synthetic air and argon.

Temperature [°C]	$\sigma_H = \sigma_{wet} - \sigma_{dry}$ [S cm^{-1}]		$\sigma_D = \sigma_{wet} - \sigma_{dry}$ [S cm^{-1}]	$t_H = \sigma_H/\sigma_{wet}$		$t_D = \sigma_D/\sigma_{wet}$
	Air	Ar	Air	Air	Ar	Air
650	5.0×10^{-4}	1.3×10^{-3}	3.3×10^{-4}	0.05	0.16	0.04
600	6.2×10^{-4}	9.5×10^{-4}	2.3×10^{-4}	0.08	0.16	0.03
550	6.4×10^{-4}	9.0×10^{-4}	1.5×10^{-4}	0.14	0.20	0.04
500	4.8×10^{-4}	7.1×10^{-4}	3.9×10^{-5}	0.17	0.25	0.02
450	4.1×10^{-4}	5.1×10^{-4}	1.9×10^{-4}	0.25	0.31	0.13
400	2.4×10^{-4}	2.9×10^{-4}	3.8×10^{-5}	0.30	0.33	0.06
350	1.1×10^{-4}	1.4×10^{-4}	2.6×10^{-5}	0.30	0.32	0.10

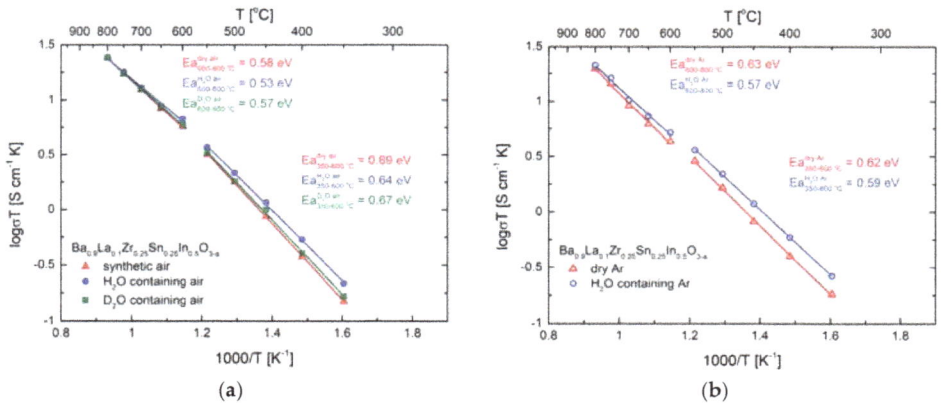

Figure 8. (a) Temperature dependence of the electrical conductivity of $Ba_{0.9}La_{0.1}Zr_{0.25}Sn_{0.25}In_{0.5}O_{3-a}$ in dry, H_2O-, and D_2O-containing air, with calculated values of the activation energy. Data shown in Arrhenius-type coordinates; (b) Temperature dependence of the electrical conductivity of $Ba_{0.9}La_{0.1}Zr_{0.25}Sn_{0.25}In_{0.5}O_{3-a}$ in dry and H_2O- containing Ar, with calculated values of the activation energy. Data shown in Arrhenius-type coordinates.

Overall, such electrical properties of $Ba_{0.9}La_{0.1}Zr_{0.25}Sn_{0.25}In_{0.5}O_{3-a}$ can be linked with a relatively low initial concentration of the oxygen vacancies, as well as with presence of Sn and in cations having relatively high electronegativity, and thus limiting proton (deuterium) conductivity at lower temperatures. Comparing to materials having higher initial oxygen vacancy concentration, the reported in Table 2 values of the proton conductivity are lower [43]. While from a point of view of application as PCFC solid electrolyte the material seems not suitable, for this compound, it was possible to record transition on temperature from a single-fold to a two-fold behavior on the electrical conductivity relaxation kinetics, indicating qualitative change of diffusion of the charge carriers, as presented in the next sub-chapter.

3.4. Conductivity Relaxations Studies upon Hydration

Figure 9a,b shows exemplary ECR data measured for the $Ba_{0.9}La_{0.1}Zr_{0.25}Sn_{0.25}In_{0.5}O_{3-a}$ material during an abrupt change of the surrounding atmospheres from dry to humidified air, as recorded at 450 °C and 650 °C. As can be seen, while the results from the lower temperature show single-fold character, data from the higher temperature are evidently pointing to a decoupled transport of hydrogen and oxygen. Using similar reasoning as presented in work [33], this corresponds to a qualitative change of the diffusion process, which from the coupled, ambipolar transfer of protons and oxygen (i.e., water) becomes decoupled, and both charge carries move independently. Such change implies the presence of the electronic hole conductivity, as can be expected from the discussed characteristics of the EIS data recorded at higher ($\geq$600 °C) temperatures. It should be clarified that, in the case of ECR profiles fitted at 600 °C and above, only R_0 component was considered, which is in accordance with the direct current measurements presented in the literature [33,36].

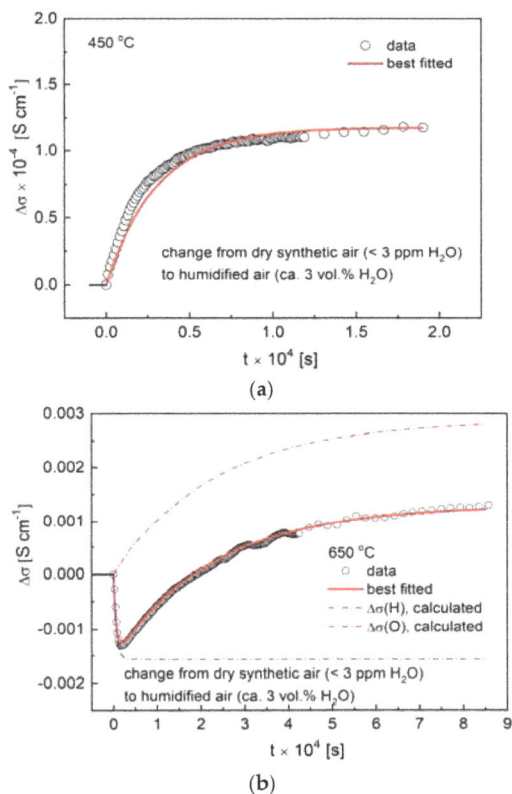

Figure 9. (a) Electrical conductivity relaxation (ECR) data with fit recorded at 450 °C during change of the atmosphere from dry to humidified air; (b) ECR data with fit recorded at 650 °C during change of the atmosphere from dry to humidified air.

All the measured ECR data allowed to prepare the Arrhenius-type graph (Figure 10), in which calculated values of the chemical diffusion coefficient D and the surface exchange coefficient k are presented as a function of temperature. In a range up to 450 °C, in which the single fold relaxation curves were measured, the refined D and k values can be related directly to water, and both coefficients increase with temperature. The activation energy of D_{H2O} in this range was evaluated as 0.77(1) eV. The reported values are comparable in magnitude to the published data for doped BaCeO$_3$ and similar materials ([30,35] and references therein).

In the 550–700 °C range, the ECR data show decoupled behavior, and transport coefficients can be fitted separately for hydrogen and oxygen, with much higher values (faster diffusion) observed for protons. Such decoupled diffusion was reported previously, e.g., for SrCe$_{0.95}$Yb$_{0.05}$O$_{3-a}$ in work [29]. Interested reader can find a detailed theoretical explanation of the nature of decoupling in the cited paper. Interestingly, it seems there is a discontinuity of the D and k values in the intermediate range (a vicinity of 500 °C), which is in accordance with a qualitative change of the nature of the charge carriers diffusion, which takes place when electronic hole-related conductivity becomes significant. Also, as almost no proton-related conductivity component was observed at the highest temperatures (Figure 8a), it is not surprising that ECR at the temperatures above 700 °C do not show an H-related component. Again, the D and k values in this range (corresponding likely to the diffusion of oxygen only) do not follow linearly values from lower temperatures. In the considered 550–700 °C range, values of the hydrogen- and oxygen-related transport coefficients D and k can be calculated separately for both charge carries (Figure 10). In agreement with literature reports, hydrogen is found to diffuse faster than oxygen during decoupled transport, which manifests itself in a fact that the value of hydrogen-related D is about two orders of magnitudes faster that the oxygen-related one. Similar differences, as well as respective orders of magnitude of the transport coefficients were reported before [29,31].

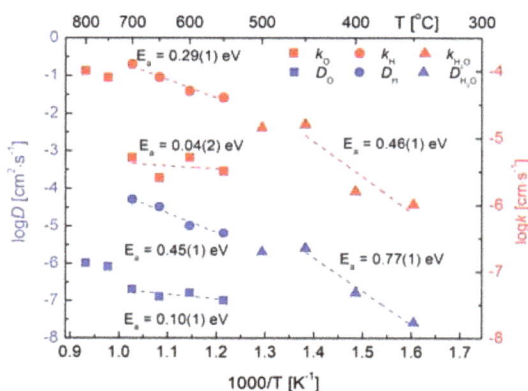

Figure 10. Temperature dependence of the transport coefficients D (blue) and k (red) with calculated values of the activation energy. Please notice description of the used symbols. Data shown in Arrhenius-type coordinates.

As presented above, the nature of the ionic conductivity in $Ba_{0.9}La_{0.1}Zr_{0.25}Sn_{0.25}In_{0.5}O_{3-a}$ is complex and changes with temperature. It depends on the proton content in the material and also on the appearance of the electronic hole-related conductivity at elevated temperatures. As can be expected, diffusion of protons is faster than that of oxygens. However, in the intermediate range at which the decoupling of the ionic charge carriers occurs, the temperature behavior seems non-monotonous.

4. Conclusions

$Ba_{0.9}La_{0.1}Zr_{0.25}Sn_{0.25}In_{0.5}O_{3-a}$ perovskite-type oxide was synthesized and studied in terms of its crystal structure, with RT and high-temperature XRD measurements, its ability for water uptake in different conditions as documented by thermogravimetric measurements with supplementary mass spectroscopy studies, as well as studies of the transport properties in dry, H_2O-, and D_2O-containing air. Single-phase samples were obtained in a wide temperature range of 1200–1600 °C. However, only at the highest temperature was it possible to acquire the dense sinter needed for the electrical measurements. The materials in a powdered form are strongly reacting with moisture from air, which results in a significantly increased unit cell volume. Thermogravimetric studies with analysis of H_2O- and CO_2-related signals, as well as the high-temperature structural evaluation, confirmed incorporation of water during hydration and presence of protons in the bulk of the material. EIS measurements allowed us to evaluate total electrical conductivity and calculate the values of the activation energy. Also, it was possible to determine proton and deuterium conductivities in wet atmospheres and calculate transference numbers. Importantly, ECR measurements were performed during an abrupt gas change from dry to humidified synthetic air at the constant temperature in a range of 350–800 °C, showing a transition from the single- to two-fold relaxation kinetics. At lower temperatures up to 450 °C, the gathered data allowed us to estimate transport coefficients D and k of water (ambipolar diffusion of H^+ and O^{2-}), while in the 550–700 °C range, it was possible to determine values of D and k separately for both charge carriers. The coupled/decoupled transition occurring on temperature seems complex, with non-monotonous behavior in the intermediate range observed.

Acknowledgments: This project was funded by the National Science Centre, Poland, on the basis of the decision number DEC-2012/05/E/ST5/03772.

Author Contributions: A.N. and X.L. prepared samples for all measurements. A.N. conducted XRD and TG experiments. W.S. conducted conductivity and relaxation measurements. K.Z. calculated k and D values. K.Ś. analyzed obtained data and wrote the paper.

Conflicts of Interest: The authors declare no conflict of interest.

References

1. Radenahmad, N.; Afif, A.; Petra, P.I.; Rahman, S.M.H.; Eriksson, S.-G.; Azad, A.K. Proton-conducting electrolytes for direct methanol and direct urea fuel cells—A state-of-the-art review. *Renew. Sustain. Energy Rev.* **2016**, *57*, 1347–1358. [CrossRef]

2. Duan, C.; Tong, J.; Shang, M.; Nikodemski, S.; Sanders, M.; Ricote, S.; Almansoori, A.; O'Hayre, R. Readily processed protonic ceramic fuel cells with high performance at low temperatures. *Science* **2015**, *349*, 1321–1327. [CrossRef] [PubMed]

3. Takahashi, T.; Iwahara, H. Proton conduction in perovskite type oxide solid solution. *Rev. Chim. Miner.* **1980**, *17*, 243–253.

4. Iwahara, H.; Uchida, H.; Maeda, N. High temperature fuel and steam electrolysis cells using proton conductive solid electrolytes. *J. Power Sources* **1982**, *7*, 293–301. [CrossRef]

5. Bonano, N.; Ellis, B.; Mahmood, M.N. Construction and operation of fuel cells based on the solid electrolyte $BaCeO_3$:Gd. *Solid State Ion.* **1991**, *44*, 305–311. [CrossRef]

6. Iwahara, H.; Uchida, H.; Yamasaki, I. High-temperature steam electrolysis using $SrCeO_3$-based proton conductive solid electrolyte. *Int. J. Hydrog. Energy* **1987**, *12*, 73–77. [CrossRef]

7. Iwahara, H.; Mori, T.; Hibino, T. Electrochemical studies on ionic conduction in Ca-doped $BaCeO_3$. *Solid State Ion.* **1995**, *79*, 177–182. [CrossRef]

8. Kreuer, K.D.; Fuchs, A.; Maier, J. HD isotope effect of proton conductivity and proton conduction mechanism in oxides. *Solid State Ion.* **1995**, *77*, 157–162. [CrossRef]

9. Zając, W.; Hanc, E.; Gorzkowska-Sobas, A.; Świerczek, K.; Molenda, J. Nd-doped $Ba(Ce,Zr)O_{3-\delta}$ proton conductors for application in conversion of CO_2 into liquid fuels. *Solid State Ion.* **2012**, *225*, 297–303. [CrossRef]

10. Haile, S.M.; West, D.L.; Campbell, J. The role of microstructure and processing on the proton conducting properties of gadolinium-doped barium cerate. *J. Mater. Res.* **1998**, *13*, 1576–1595. [CrossRef]

11. Iwahara, H.; Yajima, T.; Hibino, T.; Ushida, H. Performance of Solid Oxide Fuel Cell Using Proton and Oxide Ion Mixed Conductors Based on $BaCe_{1-x}Sm_xO_{3-\alpha}$. *J. Electrochem. Soc.* **1993**, *140*, 1687–1691. [CrossRef]

12. Liang, K.C.; Nowick, A.S. High-temperature protonic conduction in mixed perovskite ceramics. *Solid State Ion.* **1993**, *61*, 77–81. [CrossRef]

13. Fabbri, E.; Pergolesi, D.; Traversa, E. Materials challenges toward proton-conducting oxide fuel cells: A critical review. *Chem. Soc. Rev.* **2010**, *39*, 4355–4369. [CrossRef] [PubMed]

14. Haile, S.M.; Staneff, G.; Ryu, K.H. Non-stoichiometry, grain boundary transport and chemical stability of proton conducting perovskites. *J. Mater. Sci.* **2001**, *36*, 1149–1160. [CrossRef]

15. Traversa, E.; Fabbri, E. *Functional Materials for Sustainable Energy Applications*; Kilner, J.A., Skinner, S.J., Irvine, S.J.C., Edwards, P.P., Eds.; Woodhead Publishing Series in Energy, Chapter 16; Woodhead Publishing: Sawston, UK, 2012; pp. 515–537, ISBN 9780857096371.

16. Azad, A.M.; Subramaniam, S.; Dung, T.W. On the development of high density barium metazirconate ($BaZrO_3$) ceramics. *J. Alloys Compd.* **2002**, *334*, 118–130. [CrossRef]

17. Norby, T. *Perovskite Oxide for Solid Oxide Fuel Cells*; Ishihara, T., Ed.; Chapter 11; Springer Science Business Media: Berlin, Germany, 2009; pp. 217–241, ISBN 978-0-387-77708-5.

18. Kreuer, K.D. Proton-Conducting Oxides. *Annu. Rev. Mater. Rev.* **2003**, *33*, 333–359. [CrossRef]

19. Zhu, Z.; Liu, B.; Shen, J.; Lou, Y.; Ji, Y. $La_2Ce_2O_7$: A promising proton ceramic conductor in hydrogen economy. *J. Alloys Compd.* **2015**, *659*, 232–239. [CrossRef]

20. Malavasi, L.; Fisherb, C.; Saiful Islam, A. Oxide-ion and proton conducting electrolyte materials for clean energy applications: Structural and mechanistic features. *Chem. Soc. Rev.* **2010**, *39*, 4370–4387. [CrossRef] [PubMed]

21. Marrony, M. (Ed.) *Proton-Conducting Ceramics: From Fundamentals to Applied Research*; PAN Stanford Publishing: Singapore, 2016; ISBN 9789814613842.

22. Haugsrud, R.; Norby, T. High-temperature proton conductivity in acceptor-doped LaNbO$_4$. *Solid State Ion.* **2006**, *177*, 1129–1135. [CrossRef]

23. Quarez, E.; Noirault, S.; Caldes, M.T.; Joubert, O. Water incorporation and proton conductivity in titanium substituted barium indate. *J. Power Sources* **2010**, *195*, 1136–1141. [CrossRef]

24. Jankovic, J.; Wilkinson, D.P.; Hui, R. Proton conductivity and stability of Ba$_2$In$_2$O$_5$ in hydrogen containing atmospheres. *J. Electrochem. Soc.* **2011**, *158*, B61–B68. [CrossRef]

25. Geffroy, P.M.; Pons, A.; Bechade, E.; Masson, O.; Fouletier, J. Characterization of electrical conduction and nature of charge carriers in mixed and ionic conductors. *J. Power Sources* **2017**, *360*, 70–79. [CrossRef]

26. Agmon, N. The Grotthuss mechanism. *Chem. Phys. Lett.* **1995**, *244*, 456–462. [CrossRef]

27. Kreuer, K.D. *Perovskite Oxide for Solid Oxide Fuel Cells*; Ishihara, T., Ed.; Chapter 12; Springer Science Business Media: Berlin, Germany, 2009; pp. 261–272, ISBN 978-0-387-77708-5.

28. Norby, T.; Widerøe, M.; Glöckner, R.; Larring, Y. Hydrogen in oxides. *Dalton Trans.* **2004**, *19*, 3012–3018. [CrossRef] [PubMed]

29. Kim, E.; Yoo, H.I. Two-fold-to-single-fold transition of the conductivity relaxation patterns of proton-conducting oxides upon hydration/dehydration. *Solid State Ion.* **2013**, *252*, 132–139. [CrossRef]

30. Yoo, H.I.; Yoon, J.Y.; Ha, J.S.; Lee, C.E. Hydration and oxidation kinetics of a proton conductor oxide, SrCe$_{0.95}$Yb$_{0.05}$O$_{2.975}$. *Phys. Chem. Chem. Phys.* **2008**, *10*, 974–982. [CrossRef] [PubMed]

31. Yoo, H.I.; Yeon, J.I.; Kim, J.K. Mass relaxation vs. electrical conductivity relaxation of a proton conducting oxide upon hydration and dehydration. *Solid State Ion.* **2009**, *180*, 1443–1447. [CrossRef]

32. Lee, D.K.; Yoo, H.I. Unusual oxygen re-equilibration kinetics of TiO$_{2-\delta}$. *Solid State Ion.* **2006**, *177*, 1–9. [CrossRef]

33. Yoo, H.I.; Lee, C.E. Two-Fold Diffusion Kinetics of Oxygen Re-Equilibration in Donor-Doped BaTiO$_3$. *J. Am. Ceram. Soc.* **2005**, *88*, 617–623. [CrossRef]

34. Yu, J.H.; Lee, J.S.; Maier, J. Peculiar nonmonotonic water incorporation in oxides detected by local in situ optical absorption spectroscopy. *Angew. Chem. Int. Ed.* **2007**, *46*, 8992–8994. [CrossRef] [PubMed]

35. Kreuer, K.D.; Schönherr, E.; Maier, J. Proton and oxygen diffusion in BaCeO$_3$ based compounds: A combined thermal gravimetric analysis and conductivity study. *Solid State Ion.* **1994**, *70*, 278–284. [CrossRef]

36. Lim, D.K.; Choi, M.B.; Lee, K.T.; Yoon, H.S.; Wachsman, E.D.; Song, S.J. Conductivity Relaxation of Proton-Conducting BaCe$_{0.85}$Y$_{0.15}$O$_{3-\delta}$ Upon Oxidation and Reduction. *J. Electrochem. Soc.* **2011**, *158*, B852–B856. [CrossRef]

37. Singh, K.K.; Ganguly, P.; Rao, C.N. Structural transitions in (La,Ln)$_2$CuO$_4$ and La$_2$(Cu,Ni)O$_4$ systems. *Mater. Res. Bull.* **1982**, *17*, 493–500. [CrossRef]

38. Kanai, H.; Mizusaki, J.; Tagawa, H.; Hoshiyama, S.; Hirano, K.; Fujita, K.; Tezuka, M.; Hashimoto, T. Defect Chemistry of La$_{2-x}$Sr$_x$CuO$_{4-\delta}$: Oxygen Nonstoichiometry and Thermodynamic Stability. *J. Solid State Chem.* **1997**, *131*, 150–159. [CrossRef]

39. Colomban, P.; Tran, C.; Zaafrani, O.; Slodczyk, A. Aqua oxyhydroxycarbonate second phases at the surface of Ba/Sr-based proton conducting perovskites: A source of confusion in the understanding of proton conduction. *J. Raman Spectrosc.* **2013**, *44*, 312–321. [CrossRef]

40. Zhang, G.B.; Smyth, D.M. Protonic conduction in Ba$_2$In$_2$O$_5$. *Solid State Ion.* **1995**, *82*, 153–160. [CrossRef]

41. Crank, J. *The Mathematics of Diffusion*, 2nd ed.; Oxford University Press: New York, NY, USA, 1975; ISBN 0198533446.

42. Han, D.; Majima, M.; Uda, T. Structure Analysis of BaCe0.8Y0.2O3-δ in Dry and Wet Atmospheres by High-Temperature X-ray Diffraction Measurement. *J. Solid State Chem.* **2013**, *205*, 122–128. [CrossRef]

43. Świerczek, K.; Skubida, W.; Niemczyk, A.; Olszewska, A.; Zheng, K. Structure and transport properties of proton-conducting BaSn$_{0.5}$In$_{0.5}$O$_{2.75}$ and A-site substituted Ba$_{0.9}$Ln$_{0.1}$Sn$_{0.5}$In$_{0.5}$O$_{2.8}$ (Ln = La, Gd) oxides. *Solid State Ion.* **2017**, *307*, 44–50. [CrossRef]

44. Skubida, W.; Świerczek, K. Structural properties and presence of protons in Ba$_{0.9}$Gd$_{0.1}$Zr$_{1-x-y}$Sn$_x$In$_y$O$_{3-(y-0.1)/2}$ perovskites. *Funct. Mater. Lett.* **2016**, *9*, 1641005. [CrossRef]

45. Andersson, A.K.E.; Selbach, S.M.; Knee, S.C.; Grande, T. Chemical Expansion Due to Hydration of Proton-Conducting Perovskite Oxide Ceramics. *J. Am. Ceram. Soc.* **2014**, *97*, 2654–2661. [CrossRef]
46. Han, D.; Shinoda, K.; Uda, T. Dopant Site Occupancy and Chemical Expansion in Rare Earth-Doped Barium Zirconate. *J. Am. Ceram. Soc.* **2014**, *97*, 643–650. [CrossRef]
47. Kreuer, K.D. On the complexity of proton conduction phenomena. *Solid State Ion.* **2000**, *136*, 149–160. [CrossRef]
48. Babilo, P.; Uda, T.; Haile, S.M. Processing of yttrium-doped barium zirconate for high proton conductivity. *J. Appl. Phys.* **2007**, *22*, 1322–1330. [CrossRef]

crystals

MDPI

Article

Application of La-Doped SrTiO$_3$ in Advanced Metal-Supported Solid Oxide Fuel Cells

Sabrina Presto [1], **Antonio Barbucci** [1,2], **Maria Paola Carpanese** [1,2] , **Feng Han** [3], **Rémi Costa** [3] and **Massimo Viviani** [1,*]

[1] Institute of Condensed Matter Chemistry and Energy Technologies (ICMATE), National Council of Research (CNR), c/o DICCA-UNIGE, Via all'Opera Pia 15, 16145 Genova, Italy; sabrina.presto@cnr.it (S.P.); barbucci@unige.it (A.B.); carpanese@unige.it (M.P.C.)

[2] Department of Chemical, Civil and Environmental Engineering, DICCA-UNIGE, Via all'Opera Pia 15, 16145 Genova, Italy

[3] Deutsches Zentrum für Luft- und Raumfahrt e.V. (DLR), German Aerospace Center, Institute of Engineering Thermodynamics—Electrochemical Energy Technology, Pfaffenwaldring 38-40, 70569 Stuttgart, Germany; Feng.Han@dlr.de (F.H.); remi.costa@dlr.de (R.C.)

* Correspondence: massimo.viviani@cnr.it; Tel.: +39-010-353-6025

Received: 31 January 2018; Accepted: 10 March 2018; Published: 13 March 2018

Abstract: Composite materials frequently allow the drawbacks of single components to be overcome thanks to a synergistic combination of material- and structure-specific features, leading to enhanced and also new properties. This is the case of a metallic-ceramic composite, a nickel-chromium-aluminum (NiCrAl) foam impregnated with La-doped Strontium Titanate (LST). This particular cermet has very interesting properties that can be used in different fields of application, namely: mechanical robustness provided by the metal foam; and chemical stability in harsh conditions of temperature and atmosphere by promotion of a thin protective layer of alumina (Al$_2$O$_3$); high electronic conductivity given by a percolating ceramic conducting phase, i.e., La-doped Strontium Titanate. In this paper, its application as a current collector in a metal-supported Solid Oxide Fuel Cells (SOFC) was studied. Firstly, the electronic properties of different compositions, stoichiometric and under stoichiometric, of LST were analyzed to choose the best one in terms of conductivity and phase purity. Then, LST chemical stability was studied in the presence of Al$_2$O$_3$ at different temperatures, gas compositions and aging times. Finally, stability and conductivity of LST-impregnated NiCrAl foam composite materials were measured, and LST was found to be fully compatible with the NiCrAl foam, as no reactions were detected in oxidizing and reducing atmosphere after up to 300 h operation at 750 °C and 900 °C between the Al$_2$O$_3$ layer and LST. Results showed that the composite is suitable as a current collector in innovative designs of metal-supported SOFC, like the Evolve cell, in which the metallic part is supposed not only to provide the structural stability to the cell, but also to play the role of current collector due to the impregnation of ceramic material.

Keywords: La-doped SrTiO$_3$; Solid Oxide Fuel Cells; electronic conductivity; impregnation; redox cycle; current collector; metal foam

---

1. Introduction

Solid Oxide Fuel Cells (SOFC) are one of the most efficient and environmentally friendly technologies available for generating power from hydrogen, natural gas, and other renewable fuels [1,2]. The challenges in the commercialization of SOFC are to reduce cost and increase the reliability of the system. Enormous research efforts are being directed towards lowering the temperature of SOFC from 1000 °C to below 800 °C and to reducing the startup time and reliability for portable power applications and transportation [3]. To achieve good performances, attention must be paid to optimization strategies [4], and among these, to materials, geometrical aspects and design. SOFC are categorized into two major types, planar and tubular [5], the former being more promising at the moment. Technical parameters that can influence the processes of production of planar SOFC and, indirectly, their performance, have also been reported in literature [6]. Several designs of planar fuel cells have been reported: anode- or cathode-supported [7]; dual membrane [8,9], which is based on properties of a mixed anionic and protonic conductor used as a central membrane; electrolyte-supported cells in symmetrical configuration [10], which are very useful as reversible solid oxide cells; and metal-supported [11].

Metal-supported SOFC offer an alternative to conventional electrode- and electrolyte-supported cells, with some important advantages like the use of cheaper materials, enhanced mechanical stability to thermal cycling and temperature gradients, and manufacturability benefits.

The main obstacles in SOFC fabrication on porous Fe-Cr steel alloy supports are the constraints on sintering conditions due to high-temperature corrosion of the metal support, decomposition of the cathode materials in inert or reducing atmosphere, and the diffusion of Fe, Cr, and Ni between the ferritic Fe-Cr steel and nickel-containing anode.

To overcome these issues, several processing techniques have been proposed. The infiltration of active materials into pre-sintered porous metallic structures at relatively low temperature has been reported to ensure good performance and protection of the metallic structure. Ni-$Ce_{0.8}Sm_{0.2}O_{2-\delta}$ mixtures [12], Nb-doped $SrTiO_3$ [13], and Mo-doped $SrFeO_{3-\delta}$ [14] have been successfully employed as alternative electrocatalysts to the standard Ni-YSZ anodic composite, which exhibits excellent catalytic activity and current collection, but also suffers from serious limitations, such as Ni sintering, carbon deposition, sulphur poisoning, and low tolerance to redox cycles [15,16]. Additionally, alternative designs have been proposed, including the deposition of a protective La-doped $SrTiO_3$ layer between the metal support-active anode and the electrolyte layers, minimizing the Ni reaction with YSZ at sintering temperature [17].

An innovative approach was developed [18,19] in which the so called "Evolve" cell relies on a supporting current collector made of a ceramic-metal composite in the form of a metallic foam (0.6–1 mm thickness) impregnated with conductive ceramics. The metallic foam is a Ni-Cr-Al, and is supposed to provide structural stability to the cell and also to promote the formation of a thin continuous layer of Al_2O_3 during thermal processing in air. The coating is expected to enhance chemical stability, particularly avoiding Cr migration towards the active anode layer. Robustness against sulphur poisoning and thermal cycling is expected to increase significantly.

The main advantages of this oxidized structure are in its thermal stability and its ability to slow down the oxidation kinetics of the underlying metal in case of contact with an oxidant environment during the stack operation. The backbone of oxidized metal is then impregnated with conducting ceramics, such as La-doped Strontium Titanate, $La_{1-x}Sr_xTiO_3$ (LST), which leads to a percolating electron conducting phase.

Among the numerous other perovskite systems that have been explored, those containing Titanium continue to attract attention, because Ti remains mixed-valent Ti^{4+}/Ti^{3+} in the reducing atmosphere at the anode, and this redox couple can accept electrons from a hydrocarbon or H_2, thus promoting their dissociation. Because the formation of Ti^{3+} is favored by the reducing conditions, the atmosphere used during the sintering stage plays a crucial role in determining the final properties of these materials. Usually, the higher the sintering temperature and the stronger the reducing conditions, the higher the final conductivity will be, because of the higher concentration of Ti^{4+}/Ti^{3+} couples created. LST exhibits n-type semiconducting behavior when it is donor-doped and/or exposed to a reducing atmosphere. Lanthanum is a good donor dopant because the La^{3+} ionic radius (0.132 nm, CN XII) is similar to that of Sr^{2+} (0.140 nm, CN XII). Under oxidizing conditions, the compensation occurs due to the formation of Sr vacancies in the lattice, coupled with the formation of SrO layers within the structure. Under reducing conditions, Sr vacancies and SrO layers are eliminated, and the charge compensation for La^{3+} ions becomes electronic through the formation of electrons in the conduction band, i.e., conversion of Ti^{4+} to Ti^{3+} [20]. Because of its electrical properties and chemical stability, this material is also considered promising for use in SOFC anodes [21], when it is either A- or B-site donor-doped [22–25], when mixed with other materials [26–28], under-stoichiometric compositions on site A [29] or A and B co-doped [30].

Oxygen-deficient compositions can be obtained by doping on the B-site with a cation of lower valence than the valence of the nominal B-site ion or by reducing a stoichiometric composition. The first approach led to the discovery of the compound with the highest known oxide ionic conductivity, $La_{1-x}Sr_xGa_{1-y}Mg_yO_{3-\delta}$ [31,32]. A-site deficiency has recently been used in various systems to prepare materials that are suitable for SOFC anodes. $La_xSr_{1-(3x/2)}TiO_3$ compositions sintered in air have been shown to exhibit conductivity values up to 7 S cm^{-1} upon reduction at 930 °C and pO$_2$ = 10^{-20} atm [33,34]. The optimized composition of $La_{0.08}Sr_{0.86}TiO_{3-\delta}$ presents a conductivity of 82 S cm^{-1} at pO$_2$ = 10^{-19} atm and 800 °C, after being pre-reduced in 7% H$_2$/Ar at 1400 °C, [35,36], which does not greatly increase when further doped on the B-site [37]. The effect of A-site deficiency has also been found to increase the electronic conduction of Sr$_{1-x}$TiO$_3$ under reducing conditions [38]; however, the increase is not significant, and phase segregation was also observed [39].

The aim of this paper is to study the thermal/chemical stability of LST-impregnated NiCrAl foam, a potential candidate as current collector for a metal-supported SOFC, under typical operative temperature and atmosphere conditions after up to 300 h operation and to measure its conductivity. For this purpose, the reduction of LST-based electron conducting ceramics was studied for several compositions, both stoichiometric and under-stoichiometric. LST-impregnated foams were also investigated in terms of their microstructural and chemical stability. LST was found to be fully compatible with the NiCrAl foam, as no reactions were detected between the Al$_2$O$_3$ layer and LST in oxidizing and reducing atmosphere after up to 300 h operation at 750 °C and 900 °C. Results showed that the composite is suitable as support in the innovative design of metal-supported Solid Oxide Fuel Cells (SOFC), in which the metallic part is supposed not only to provide structural stability to the cell, but also to play the role of current collector due to the impregnation of ceramic material.

2. Materials and Methods

In Table 1, the composition, synthesis method, and processing of the prepared and tested LST ceramics and impregnated foams are reported, for clarity. Details of each sample are also given in the text.

Table 1. List of prepared and tested LST samples. SSR = solid state reaction, SP = Spray Pyrolysis.

Name	Nominal Composition	Synthesis Route	Processing	Testing
LST185	$La_{0.1}Sr_{0.85}TiO_3$	SSR, 900 °C Air	Sintering: 1450 °C 12 h Air	DC conductivity and Redox
			Reduction:1450 °C 24 h Ar/H2	
LST27	$La_{0.2}Sr_{0.7}TiO_3$	SSR, 900 °C Air	Sintering: 1450 °C 12 h Air	DC conductivity and Redox
			Reduction:1450 °C 24 h Ar/H2	
LST-c	$La_{0.1}Sr_{0.9}TiO_3$	SP	Sintering: 1450 °C 12 h H$_2$	DC conductivity
LST-p-120	$La_{0.1}Sr_{0.9}TiO_3$ + graphite (40 vol %)	SP	Sintering: 1200 °C 12 h Air	DC conductivity and Redox
			Reduction: 1000 °C 24 h H$_2$	
LST-Al-mix	$La_{0.1}Sr_{0.9}TiO_3$ + Al$_2$O$_3$ (1:1 mol.)	SP	Calcination: 700-800-900-1000 °C 24 h	-
LST-Al-core	$La_{0.1}Sr_{0.9}TiO_3$ + Al$_2$O$_3$ pre-sintered (1600 °C) pellet	SP	Calcination: 700-800-900-1000 °C 24 h	-
Current collector	LST-c impregnated NiCrAl foam	SP	Reported in ref 18	DC conductivity and Redox

2.1. Powders and Pellets

A-site deficient LST powders, namely LST185 and LST27, as reported in Table 1, were prepared by solid-state reaction of the following precursors: $SrCO_3$ (grade HP, Solvay Bario e Derivati, Massa, Italy), TiO_2 (grade Aeroxide TiO_2 P 25, Evonik Degussa GmbH, Hanau, Germany), $La(OH)_3$, prepared by hydrolysis of an aqueous solution of $La(NO_3)_3 \cdot 4H_2O$ (Aldrich, 99.9%) with concentrated ammonia (Fluka, 25 wt %) and a solution of ammonium polyacrylate (Acros Chimica, MW 2000) as dispersant. The method consisted of the following steps: mixing and homogenizing for 24 h in stabilized zirconia media, freeze drying, sieving, and finally calcination of powder for 2 h at 900 °C in air. This temperature was chosen after various attempts to obtain a single LST phase powder.

A commercial LST powder (Ceramic Powder Technology AS, Heimdal, Norway), LST-c, with 10% at of La and prepared by spray pyrolysis, was also studied for comparison. LST powders were pelletized by a uniaxial pressing procedure and sintered at 1450 °C, in Ar-H_2 (5% of H_2) or air and then reduced at different temperatures as reported in Table 1. In some cases, graphite (Timcal Timerex, KS6) was added to the LST powder before pressing to obtain a porous material. Pt (Metalor, Birmingham-UK Pt ink) electrodes were brushed on the plane surfaces of sintered pellets and then cured at 1000 °C under the same atmosphere used for sintering.

To study the inter-material reactivity and compatibility of LST and Al_2O_3 materials, two different types of pellets were prepared, starting from LST-c and Alumina (Baikowski, batch 1134N) powders:

1. LST-Al-mix, obtained by mixing powders by water-based ball milling in stabilized zirconia balls, freeze drying, sieving, uni-axial pressing of pellet and calcination at 700–800–900–1000 °C for 24 h.
2. LST-Al-core, obtained by incorporating (uni-axial pressing) a pre-sintered (1600 °C) fully dense Alumina core in a LST pellet, and then calcination at 700–800–900–1000 °C for 24 h.

2.2. LST-Impregnated NiCrAl Foams

A NiCrAl standard foam (NiCrAl #1) from Alantum Europe GmbH, with an open cell size of about 450 µm was used as an alumina-forming alloy, in the form of sheets with 1.6 mm thickness. Square samples of 50 × 50 mm were cut from these sheets and impregnated with the LST-c slurry and fired in air at 1000 °C to obtain a current collector. The pore size of the substrates was reduced to 200 nm through the impregnation of LST ceramic into the metal foam. The details of impregnation are reported elsewhere [19].

2.3. Characterizations

Powders, pellets and impregnated foams (current collector) before and after reduction, and before and after thermal treatment and electrical characterization were analyzed by X-ray diffraction (CubiX, Panalytical) and Scanning Electron Microscopy (1450VP, LEO) to observe phase purity, microstructure and any degradation effect. XRD profiles were analyzed by the Rietveld method, as implemented in the FullProf software. The chemical composition was also confirmed, and element migration was studied by an energy-dispersive electron microprobe (INCA 300, Oxford Instruments). The electrical properties were studied with a 2-electrodes–4-wires set-up (ProboStat, NorECs AS, interfaced with a Solartron SI 1286 Potentiostat) both in air and in H_2 at different operative temperature and during redox cycles.

3. Results

3.1. LST Ceramics

3.1.1. Structural and Microstructural Properties

Phase purity of samples LST27, LST185 and LST-c was checked by XRD on ceramics sintered at 1450 °C for 12 h. In order to study the structural stability of A-site deficient compositions in reducing conditions, XRD was repeated after treatment at 1450 °C for 24 h under Ar-H$_2$ (5% of H$_2$). Results are reported in Figure 1 for all compositions, with Rietveld Refinement.

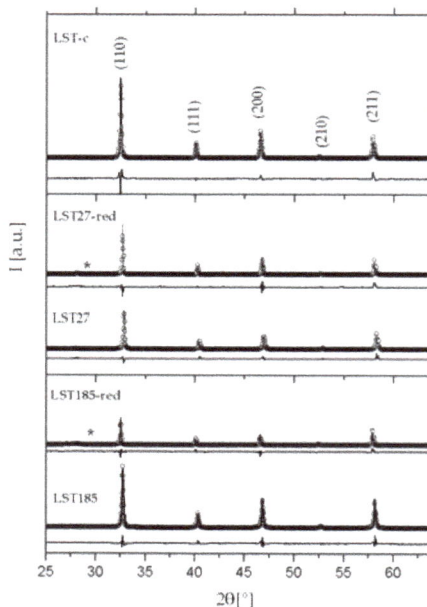

Figure 1. XRD patterns and Rietveld refinement of as-sintered and reduced LST samples, as indicated in labels. Residuals and Miller indexes of principal phase are also reported. (*) indicates major reflection of the La$_2$Ti$_2$O$_7$ secondary phase.

As-sintered materials are single cubic-phase, SrTiO$_3$-type (JCPDS card No. 86-179). Nevertheless, after reduction, some secondary phases, mostly La$_2$Ti$_2$O$_7$, are formed in LST27 and LST185 samples. The lattice parameters, and some reliability factors of Rietveld Refinement, are summarized in Table 2. Refinement, for reduced samples, was performed by also considering the secondary phase. From this table, it can be observed that lattice parameter increases with reduction. This behavior is in good agreement with previous results [24] and is due to partial reduction of Ti^{4+} to Ti^{3+}.

Table 2. Rietveld refined parameters of patterns reported in Figure 1.

Sample	Lattice Parameter (Å)	χ^2	R$_B$	R$_F$
LST-c	3.89767	1.72	6.11	3.01
LST185	3.90038	4.20	6.28	3.30
LST185-red	3.90712	8.12	17.5	11.2
LST27	3.89503	5.5	8.15	4.15
LST27-red	3.90551	7.53	12.7	6.48

Such phases were also detected by SEM, for LST185 and LST27 as separate darker intergranular areas (Figure 2b–d). EDX analysis, reported in Table 3, confirmed that dark grey zones correspond to Ti-rich phases, in which there is a depletion of Sr compared to the main phase. No defects were visible in the sample LST-c.

Table 3. Results of EDX analysis performed on reduced samples, as reported in Figure 2.

LST185-Red	O/Ti	Ti/(Sr+La)	La/Sr
Spectrum 1	3.44	0.97	41
Spectrum 2	2.92	1.01	0.12
LST27-Red	**O/Ti**	**Ti/(Sr+La)**	**La/Sr**
Spectrum 1	3.58	0.94	30
Spectrum 2	2.95	0.97	0.12
LST-c	**O/Ti**	**Ti/(Sr+La)**	**La/Sr**
Spectrum 1	3.10	0.94	0.12
Spectrum 2	2.96	0.98	0.13

Grain growth is coherently more evident in the sample containing 20 at % of La. In fact, average grain size is about 10 μm in LST27 and about 1 μm in LST185. LST-c showed a broader grain size distribution between about 1–4 μm which can be attributed to the different synthesis method used. In general, all samples show a high-density microstructure, even if, in the A-site-deficient samples, some intergranular voids are present, due to the fast grain growth process; such defects are typical in this type of sample and can be reduced by a fine tuning of the sintering process. Density was also measured by Archimedes' method, and a relative value of 96–97% for all ceramic materials was obtained.

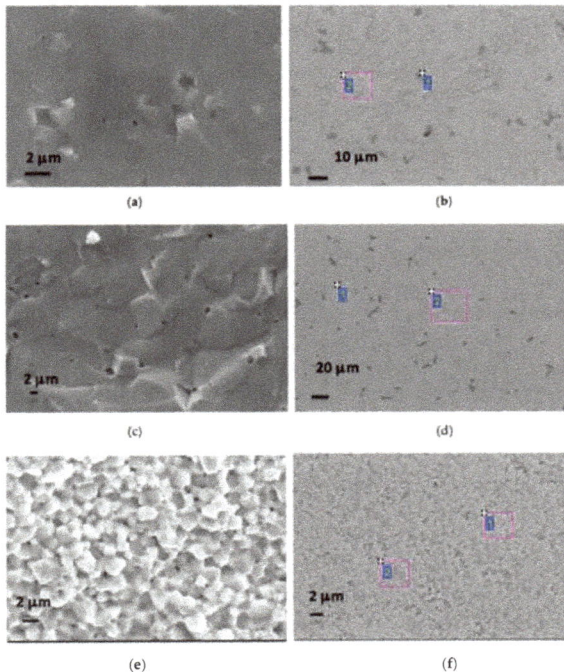

Figure 2. Microstructures obtained by SEM: SE image on fracture of as-sintered LST185 (**a**), LST27 (**c**) and LST-c (**e**); BSE image on cross section of reduced LST185 (**b**), LST27 (**d**) and of as-sintered LST-c (**f**).

3.1.2. Electrical Properties

In Figure 3, conductivity of oxidized ceramics, LST185 and LST27, measured in air is reported. LST-c was not measured in oxygen because it was sintered in Ar-H$_2$ to avoid strontium oxide segregation at the grain boundary, which could limit conductivity processes. A-site-deficient samples show thermally activated conduction as indicated by Arrhenius plot and low conductivity. Activation energy is 1.1 eV in LST185 sample, higher than 0.7 eV of LST27, possibly due to the prevailing grain boundary resistance given by the higher density of grain boundaries in the structure.

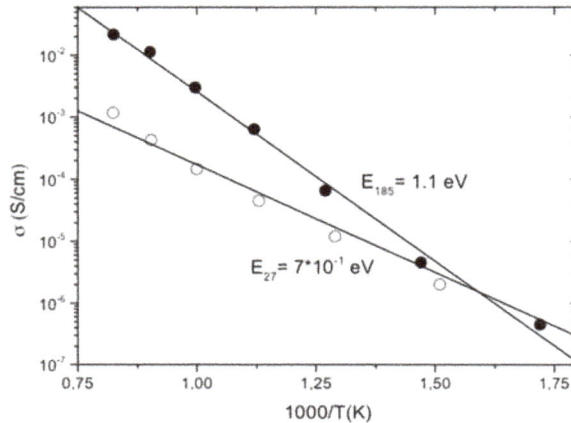

Figure 3. Arrhenius plot of total conductivity measured in air on oxidized samples (o) LST27, and (●) LST185.

After reduction, the conductivity was increased for both A-site-deficient samples (Figure 4), being higher for the LST27 sample. However, when compared to Sr$_{0.9}$La$_{0.1}$TiO$_3$ prepared by spray pyrolysis, conductivity appeared considerably limited. This result might be attributed to the secondary phases, which partially deplete the La concentration in the perovskite and also introduce a resistive phase in the volume. In addition, differences can be introduced by synthesis routes, i.e., by chemical homogeneity and the particle size of powders.

The effect of atmospheric composition on conducting properties was investigated by simulating RedOx cycles at fixed temperature on porous sample of LST-c, which showed higher conductivity and was chosen as the best powder to impregnate the NiCrAl foams and to study the composite material as a possible candidate for being a current collector in a metal-supported fuel cell.

As the main function of LST ceramics is to act as electronic conductor within the current collector and the active anode, its conductivity under reducing conditions is the most important quantity to check for practical use. However, the kinetics of transition from the oxidized non-conducting to the reduced conducting state is also of great importance and must be taken into account for the optimization of the current collector functionality.

At microscopic level, the oxidation (reduction) of LST involves two phenomena, which are the inbound (outbound) solid state diffusion of oxide ions and the surface exchange between solid and gas phases. The former is the limiting step; therefore, the kinetics of RedOx transitions is macroscopically determined by the microstructure, i.e., by the gas permeation through the LST network structure. The apparent conductivity is reported in Figure 5 for the sample sintered at 1200 °C (LST-p-120).

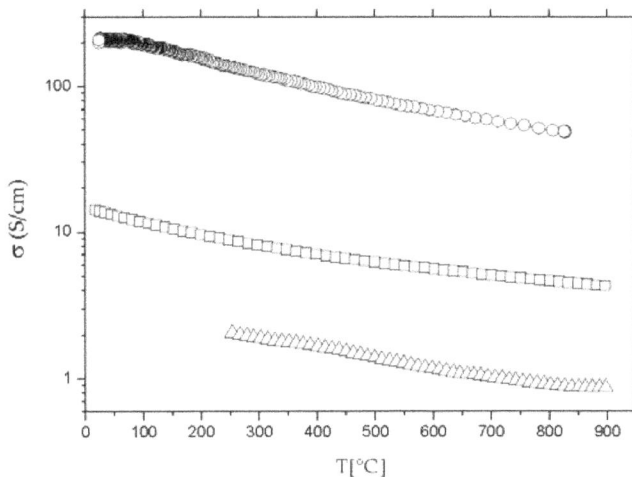

Figure 4. Conductivity as function of temperature, measured in hydrogen on sample (o) LST-c, and after reduction on samples (□) LST27 and (Δ) LST185.

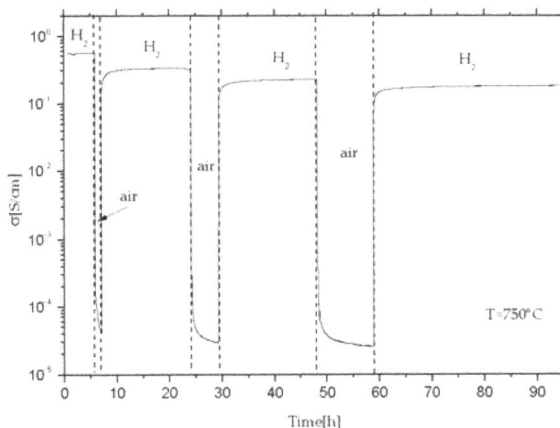

Figure 5. Conductivity under redox cycling (air-H$_2$) at fixed temperature of sample LST-p-120.

In the beginning, when the sample is reduced at 1000 °C for 24 h, the apparent conductivity is 0.5 S/cm at 750 °C, and the sample showed a metallic behavior. Such a low value of conductivity was expected and was due to the high porosity introduced by graphite. The response to the atmosphere switch decreased the conductivity by about 4 orders of magnitude, and equilibrium was hardly reached, as suggested by hysteresis in the conductivity value after several cycles.

3.2. LST-Impregnated NiCrAl Foams

As mentioned in the previous paragraph, LST-c was chosen as the best powder to impregnate the NiCrAl foams and to study the composite material as a possible candidate for being a current collector in a metal-supported fuel cell. The study of LST-impregnated NiCrAl foams regarded two aspects of current collector: its chemical stability and its conductivity.

3.2.1. Chemical Stability

Before testing the chemical stability of LST-impregnated NiCrAl foams in real operative conditions as current collector, two intermediate experiments were performed to systematically study the chemical stability of the system, due to the formation of a layer of alumina on NiCrAl foams. The experiments are described in the following together with the results:

1.   Study of pellets obtained by mixing of powders LST-c and Al_2O_3 after treatment at several different temperatures in air
2.   Study of pellets obtained as a pre-sintered (1600 °C) fully dense Alumina core in a LST-c pellet after treatment at several different temperatures in air

The actual microstructure of the composite current collector is better described by experiment (2); nevertheless, in experiment (1), any possible reactivity phenomenon is emphasized because of higher surface density of powders. From experiment type (1) formation of new phases (due to reactivity of two powders) was not detected by XRD at any temperature. Peaks of measured patterns on mixed powders were compared with those of pure materials and reported in Figure 6, with Rietveld refinement. The lattice parameters, and some reliability factors of Rietveld Refinement are summarized in Table 4. Refinement was performed by considering both the two phases. From this table, it can be observed that the lattice parameter remains quite constant. Stability was confirmed by experiment (2), and no migration of elements was detected in LST or in alumina for all studied temperature at interface by EDAX analysis.

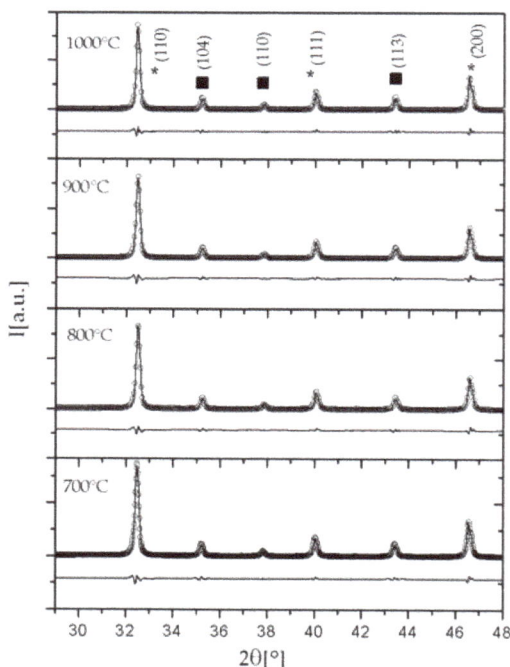

Figure 6. XRD patterns and Rietveld refinement of mixing of powders LST-c and Al_2O_3 after treatment in air at several different temperatures, as reported in the labels. Residuals and Miller indexes of Alumina (■) and LST (*) phases are also reported.

Table 4. Rietveld refined parameters of patterns reported in Figure 6.

Sample	Lattice Parameter (Å)	χ^2	R_B	R_F
Mixing at 700 °C	3.90051	3.8	2.52	1.78
Mixing at 800 °C	3.90026	3.5	2.68	2.13
Mixing at 900 °C	3.90058	2.8	2.22	1.84
Mixing at 1000 °C	3.90086	2.4	2.15	1.72

The same findings were detected by post-measurement analysis on the current collector composite, and a good stability, of both the ceramic impregnating material and the NiCrAl foam, was obtained in all conditions. In fact, as shown in Figure 7, after treatment at 750 °C for 300 h in H_2 the Al_2O_3 layer at the surface of the foam is still present, covering the entirety of the foam. Ni and Cr are confined inside the foam, although some segregation of Cr at the surface can be detected at a few points.

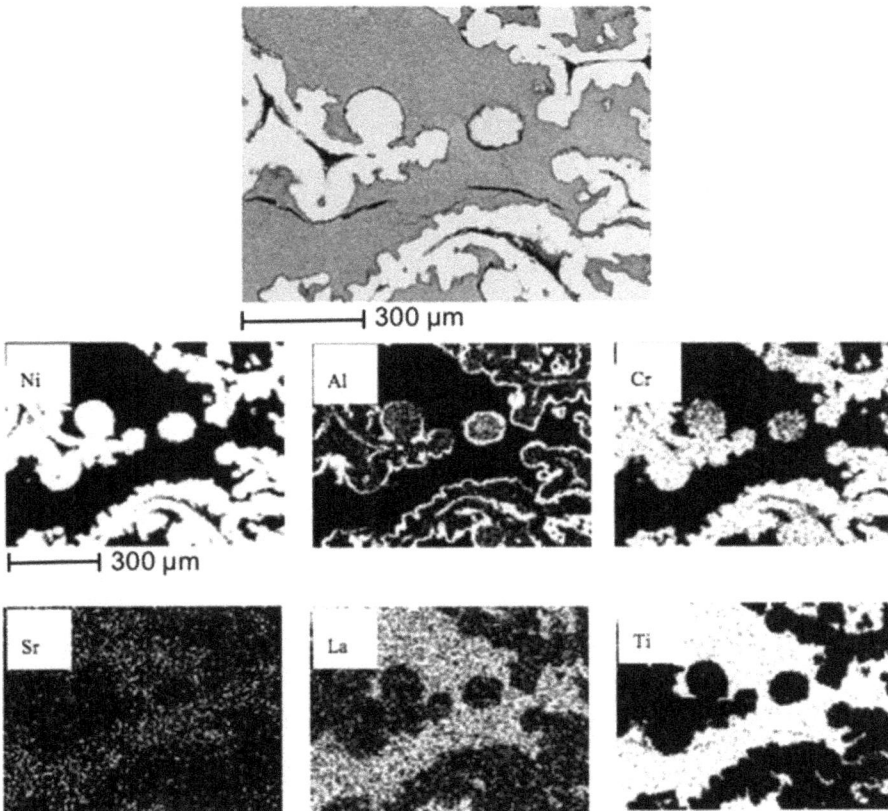

Figure 7. SEM+EDX maps of LST-NiCrAl current collector after reduction at 750 °C for 300 h.

One of these is shown in Figure 8, with spots of Cr oxide visible within the alumina layer. Diffusion of cations from ceramic components towards the foam was not detected, as Sr, La and Ti are only present in the core of the ceramic and do not react with Al.

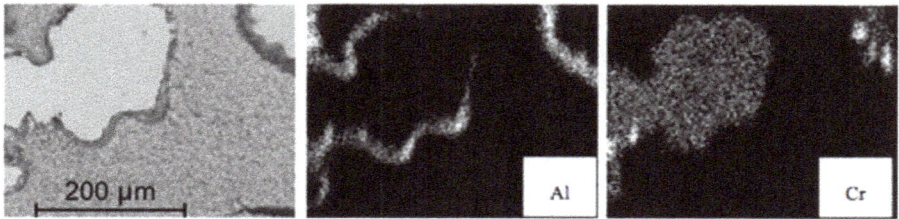

Figure 8. SEM+EDX maps of LST-NiCrAl current collector after reduction at 750 °C for 300 h: chromium segregation.

Very similar features (Figure 9) can be found after treatment at 900 °C and RedOx cycling (detailed treatment is reported in Figure 10b). In this case, no formation of Cr oxides is detected. From the above results on ceramics and composites, it can be concluded that LST is highly compatible with NiCrAl foams, and long-term operation without significant degradation can be expected.

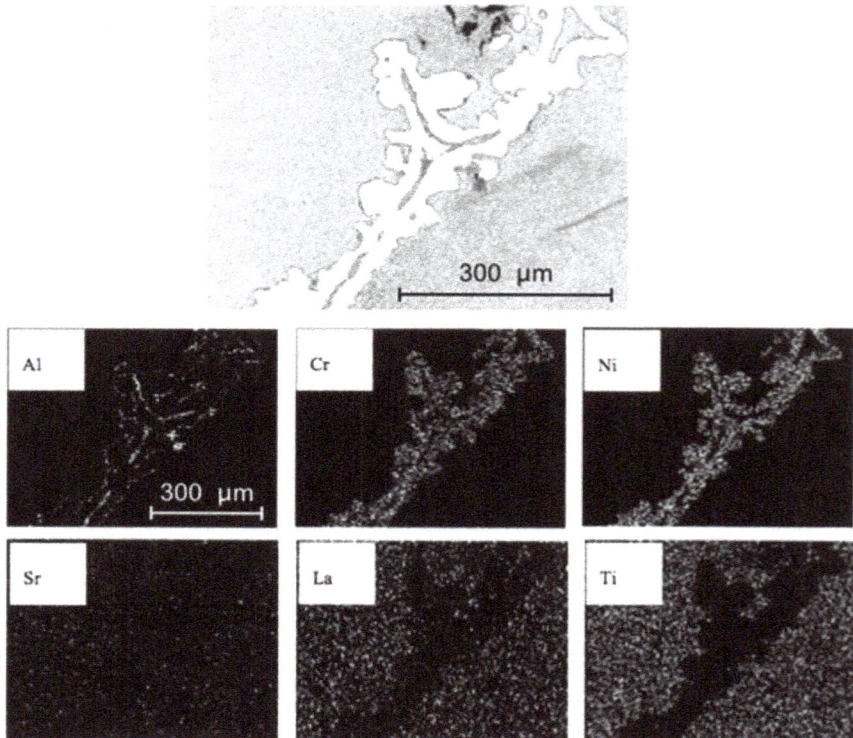

Figure 9. SEM+EDX maps of LST-NiCrAl current collector after treatment at 900 °C and RedOx cycling.

However, the reduction of LST requires high temperature—at least 900 °C—and high porosity—at least 30 vol %—as it was in composites presented here. Such requirements need to be carefully matched with constraints imposed by other parts of the cell and when robustness against RedOX cycling is required.

3.2.2. Conductivity

Conductivity of LST-impregnated NiCrAl foams reduced under H$_2$, was measured at two different temperatures, 750 and 900 °C, i.e., the operating temperature of the EVOLVE cell and the maximum temperature used for the cell fabrication. Results are reported in Figure 10. In general, the apparent conductivity is quite low due to the porosity (the current collector microstructure is reported in literature, ref [18]) and the lack of metallic electrodes. This was intentional in order to avoid short-circuiting with the metallic foam. At 750 °C (Figure 10a), reduction of the ceramic component is not yet completed after 300 h, and the behavior is substantially semiconducting as the conductivity drops by at least one order of magnitude on cooling. At 900 °C (Figure 10b), the sample reduces faster, but not completely after 20 h. Switching from H$_2$ to synthetic air produces huge conductivity change, by more than 3 orders, although equilibration appears very slow. Additionally, in this case an hysteresis in conductivity value after few cycles is observed.

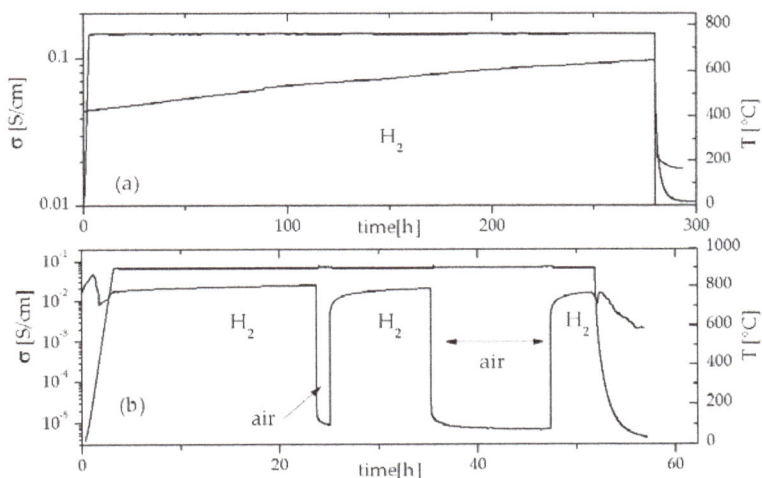

Figure 10. Conductivity of LST-impregnated NiCrAl foam at 750 °C (**a**) and 900 °C (**b**) under different atmospheres.

4. Discussion and Conclusions

Two different mechanisms of defect compensation, i.e., *electronic* at (*i*) low P(O$_2$) and *cation vacancy type* at (*ii*) high P(O$_2$), are assumed in donor-doped SrTiO$_3$ [38]. Incorporation mechanisms in La-doped SrTiO$_3$ can be described by the following reactions, respectively relevant for case (*i*) and (*ii*):

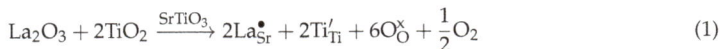

$$La_2O_3 + 2TiO_2 \xrightarrow{\text{SrTiO}_3} 2La^{\bullet}_{Sr} + 2Ti'_{Ti} + 6O^x_O + \frac{1}{2}O_2 \tag{1}$$

$$La_2O_3 + 3TiO_2 \xrightarrow{\text{SrTiO}_3} 2La^{\bullet}_{Sr} + V''_{Sr} + 3Ti^x_{Ti} + 9O^x_O \tag{2}$$

where Kröger-Vink notation was used. As a result, the thermodynamically stable phase at low P(O$_2$) is the conductive stoichiometric Sr$_{1-x}$La$_x$TiO$_3$, while at high P(O$_2$) it is the insulating Sr-deficient Sr$_{1-1.5x}$La$_x$TiO$_3$. When the stoichiometric phase is exposed to oxidizing atmosphere, charge compensation changes from electronic to cation vacancy, causing the formation of secondary phases accommodating the Sr excess. Such phases were identified as SrO either precipitated at grain boundaries or present inside grains as shear planes [38,40] and can be hardly revealed at lower doping levels and by standard-resolution techniques. On the contrary, when the Sr-deficient phase is subjected to reducing atmosphere, establishment of the electronic compensation imply accommodation of Ti excess by formation of TiO_2 Magneli phases either embedded in $SrTiO_3$ or as precipitate [40]. The process can be described as:

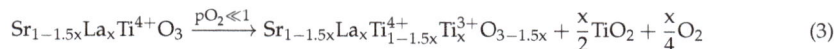

$$Sr_{1-1.5x}La_xTi^{4+}O_3 \xrightarrow{pO_2 \ll 1} Sr_{1-1.5x}La_xTi^{4+}_{1-1.5x}Ti^{3+}_xO_{3-1.5x} + \frac{x}{2}TiO_2 + \frac{x}{4}O_2 \tag{3}$$

The results presented about LST ceramics are in good agreement with the above-presented mechanisms: A-site deficient compositions (LST185 and LST27) did not show any secondary phase after sintering in air, and neither did LST-c after sintering in H_2. The application of reducing atmosphere promoted in LST185 and LST27 the formation of Ti-rich phase as indicated by Equation (3). Oxidation of LST-c did not reveal appearance of SrO-rich phases (see Figure 6) possibly due to kinetic limitations at moderate temperatures (T < 1000 °C).

Conductivity and activation energy values reported in Figure 3 suggest the existence of some ionic conduction, that could be explained by a mixed incorporation mode, i.e., by simultaneous formation of Sr and O vacancies, recently proposed for RE-doped $SrTiO_3$ [41].

Reduced materials showed metallic-type behavior (Figure 4) in any case. However, of A-site-deficient composition resulted in lower conductivity than in stoichiometric material, which is contrary to several reports. This might be explained by the presence of secondary phases in A-site-deficient LST and by the high temperature of reduction applied, as already reported in the literature [24]. The apparent inconsistency could be also ascribed to the different synthesis method used for stoichiometric (spray pyrolysis) and A-site-deficient (solid-state reaction) compositions, resulting in much finer and highly reactive powders in the first case.

Values of conductivity found in composites were limited by contact, which was intentionally not optimized in order to avoid possible side effects from metallic pastes.

In conclusion, the application of a composite material, the LST-impregnated NiCrAl foam, as a current collector in a metal-supported solid-oxide fuel cell was studied. For this purpose, the electronic properties of different compositions, stoichiometric and under stoichiometric, of LST were analyzed to choose the best one in terms of conductivity and phase purity. Then, LST chemical stability was studied in the presence of Al_2O_3 at different temperatures, gas compositions and aging times. Finally, stability and conductivity of the composite materials LST-impregnated NiCrAl foam were measured, and LST was found fully compatible with the NiCrAl foam, as no reactions were detected in oxidizing and reducing atmosphere after up to 300 h operation at 750 °C and 900 °C between the Al_2O_3 layer and LST. Results showed that the composite is suitable as current collector in the innovative design of metal-supported Solid Oxide Fuel Cells (SOFC), EVOLVE cells, in which the metallic part is supposed not only to provide the structural stability to the cell, but also to play the role of current collector due to the impregnation of the ceramic material.

Improvement in the layer performances might be obtained by using a LST-GDC-impregnated foam. The choice is motivated by the catalytic activity of GDC in the reduction of hydrogen, which should improve the total operation of the EVOLVE cell and by a better mechanical compatibility between the anodic part of the cell, the LST-GDC layer, and the GDC electrolyte.

Acknowledgments: The research leading to these results received funding from the European Union's Seventh Framework Programme (FP7/2007-2013) for the Fuel Cells and Hydrogen Joint Technology Initiative under grant agreement NO. 303429—EVOLVE. The authors are grateful to Alantum Europe GmbH, for supplying foams, and to Ceramic Powder Technology AS, for supplying commercial LST powders. Finally, the authors wish to thank F. Perrozzi for helping in doing measurements and Maria Teresa Buscaglia for synthesis of LST powders.

Author Contributions: All authors conceived, designed and performed the experiments; Sabrina Presto and Massimo Viviani analyzed the data and wrote the paper. Rémi Costa and Feng Han carried out foams impregnation.

Conflicts of Interest: The authors declare no conflict of interest.

References

1. Choudhury, A.; Chandra, H.; Arora, A. Application of solid oxide fuel cell technology for power generation—A review. *Renew. Sustain. Energy Rev.* **2013**, *20*, 430–442. [CrossRef]

2. Andersson, M.; Sundén, B. *Technology Review—Solid Oxide Fuel Cell*; Report 2017:359; Energiforsk: Stockholm, Sweden, 2017; ISBN 978-91-7673-359-2. Available online: https://energiforskmedia.blob.core.windows.net/media/22411/technology-review-solid-oxide-fuel-cell-energiforskrapport-2017-359.pdf (accessed on 26 January 2018).

3. Massardo, A.F.; Lubelli, F. Internal Reforming Solid Oxide Fuel Cell-Gas Turbine Combined Cycles (IRSOFC-GT): Part A—Cell Model and Cycle Thermodynamic Analysis. *J. Eng. Gas Turbines Power* **1999**, *122*, 27–35. [CrossRef]

4. Ramadhani, F.; Hussain, M.A.; Mokhlis, H.; Hajimolana, S. Optimization strategies for Solid Oxide Fuel Cell (SOFC) application: A literature survey. *Renew. Sustain. Energy Rev.* **2017**, *76*, 460–484. [CrossRef]

5. Kendall, K.; Minh, N.Q.; Singhal, S.C. Chapter 8 Cell and Stack Designs. In *High-temperature Solid Oxide Fuel Cells: Fundamentals, Design and Applications*, 1st ed.; Singhal, S.C., Kendall, K., Eds.; Elsevier Science: London, UK, 2003; pp. 197–228; ISBN 9780080508085.

6. Mahmud, L.S.; Muchtar, A.; Somalu, M.R. Challenges in fabricating planar solid oxide fuel cells: A review. *Renew. Sustain. Energy Rev.* **2017**, *72*, 105–116. [CrossRef]

7. Su, S.; Gao, X.; Zhang, Q.; Kong, W.; Chen, D. Anode-Versus Cathode-Supported Solid Oxide Fuel Cell: Effect of Cell Design on the Stack Performance. *Int. J. Electrochem. Sci.* **2015**, *10*, 2487–2503.

8. Thorel, A.S.; Abreu, J.; Ansar, S.-A.; Barbucci, A.; Brylewski, T.; Chesnaud, A.; Ilhan, Z.; Piccardo, P.; Prazuch, J.; Presto, S.; et al. Proof of concept for the dual membrane cell I. Fabrication and electrochemical testing of first prototypes. *J. Electrochem. Soc.* **2013**, *160*, F360–F366. [CrossRef]

9. Presto, S.; Barbucci, A.; Viviani, M.; Ilhan, Z.; Ansar, S.-A.; Soysal, D.; Thorel, A.S.; Abreu, J.; Chesnaud, A.; Politova, T.; et al. IDEAL-Cell, innovative dual membrane fuel-cell: Fabrication and electrochemical testing of first prototypes. *ECS Trans.* **2009**, *25*, 773–782.

10. Xu, J.; Zhou, X.; Cheng, J.; Pan, L.; Wu, M.; Dong, X.; Sun, K. Electrochemical performance of highly active ceramic symmetrical electrode $La_{0.3}Sr_{0.7}Ti_{0.3}Fe_{0.7}O_{3-\delta}$-$CeO_2$ for reversible solid oxide cells. *Electrochim. Acta* **2017**, *257*, 64–72. [CrossRef]

11. Krishnan, V.V. Recent developments in metal-supported solid oxide fuel cells. *WIREs Energy Environ.* **2017**, *6*, e246. [CrossRef]

12. Zhou, Y.; Yuan, C.; Chen, T.; Meng, X.; Ye, X.; Li, J.; Wang, S.; Zhan, Z.J. Evaluation of Ni and Ni–$Ce_{0.8}Sm_{0.2}O_{2-\delta}$ (SDC) impregnated 430L anodes for metal-supported solid oxide fuel. *J. Power Sources* **2014**, *267*, 117–122. [CrossRef]

13. Blennow, P.; Sudireddy, B.R.; Persson, Å.H.; Klemensø, T.; Nielsen, J.; Thydén, K. Infiltrated $SrTiO_3$:FeCr-based Anodes for Metal-Supported SOFC. *Fuel Cells* **2013**, *13*, 494–505. [CrossRef]

14. Zhou, Y.; Meng, X.; Liu, X.; Pan, X.; Li, J.; Ye, X.; Nie, H.; Xia, C.; Wang, S.; Zhan, Z. Novel architectured metal-supported solid oxide fuel cells with Mo-doped $SrFeO_{3-\delta}$ electrocatalysts. *J. Power Sources* **2014**, *267*, 148–154. [CrossRef]

15. Matsuzaki, Y.; Yasuda, I. The poisoning effect of sulfur-containing impurity gas on a SOFC anode: Part I. Dependence on temperature, time, and impurity concentration. *Solid State Ion.* **2000**, *132*, 261–269. [CrossRef]

16. Khan, M.S.; Lee, S.-B.; Song, R.-H.; Lee, J.-W.; Lima, T.-H.; Park, S.-J. Fundamental mechanisms involved in the degradation of nickel–yttria stabilized zirconia (Ni–YSZ) anode during solid oxide fuel cells operation: A review. *Ceram. Int.* **2016**, *42*, 35–48. [CrossRef]

17. Choi, J.-J.; Ryu, J.; Hahn, B.-D.; Ahn, C.-W.; Kim, J.-W.; Yoon, W.-H.; Park, D.-S. Low temperature preparation and characterization of solid oxide fuel cells on FeCr-based alloy support by aerosol deposition. *Int. J. Hydrog. Energy* **2014**, *39*, 12878–12883. [CrossRef]

18. Han, F.; Semerad, R.; Constantin, G.; Dessemond, L.; Costa, R. Beyond the 3rd Generation of Planar SOFC: Development of Metal Foam Supported Cells with Thin Film Electrolyte. *ECS Trans.* **2015**, *68*, 1889–1896. [CrossRef]

19. Costa, R.; Ansar, A. Evolved Materials and Innovative Design for High Performance, Durable and Reliable SOFC Cell and Stack Presentation and Status of the European Project EVOLVE. *ECS Trans.* **2013**, *57*, 533–541. [CrossRef]

20. Marina, O.A.; Canfield, N.L.; Stevenson, J.W. Thermal, electrical, and electrocatalytical properties of lanthanum-doped strontium titanate. *Solid State Ion.* **2002**, *149*, 21–28. [CrossRef]

21. Verbraeken, M.C.; Ramos, T.; Agersted, K.; Ma, Q.; Savaniu, C.D.; Sudireddy, B.R.; Irvine, J.T.S.; Holtappels, P.; Tietz, F. Modified strontium titanates: From defect chemistry to SOFC anodes. *RSC Adv.* **2015**, *5*, 1168–1180. [CrossRef]

22. Canales-Vasquez, J.; Tao, S.W.; Irvine, J.T.S. Electrical properties in $La_2Sr_4Ti_6O_{19-delta}$: A potential anode for high temperature fuel cells. *Solid State Ion.* **2003**, *159*, 159–165. [CrossRef]

23. Savaniu, C.D.; Irvine, J.T.S. La-doped $SrTiO_3$ as anode materials for IT-SOFC. *Solid State Ion.* **2011**, *192*, 491–493. [CrossRef]

24. Burnat, D.; Heel, A.; Holzer, L.; Kata, D.; Lis, J.; Graule, T. Synthesis and Performances of A-site deficient lanthanum-doped strontium titanate by nanoparticle based spray pyrolysis. *J. Power Sources* **2012**, *201*, 26–36. [CrossRef]

25. Miller, D.N.; Irvine, J.T.S. B-site doping of lanthanum strontium titanate for solid oxide fuel cell anodes. *J. Power Sources* **2011**, *196*, 7323–7327. [CrossRef]

26. Zhou, X.; Yan, N.; Chuang, K.T.; Luo, J. Progress in La-doped $SrTiO_3$ (LST)-based anode materials for solid oxide fuel cells. *RSC Adv.* **2014**, *4*, 118–131. [CrossRef]

27. Yurkiv, V.; Constantin, G.; Hornes, A.; Gondolini, A.; Mercadelli, E.; Sanson, A.; Dessemond, L.; Costa, R. Towards understanding surface chemistry and electrochemistry of $La_{0.1}Sr_{0.9}TiO_{3-\alpha}$ based solid oxide fuel cell. *J. Power Sources* **2015**, *287*, 58–67. [CrossRef]

28. Gondolini, A.; Mercadelli, E.; Constantin, G.; Dessemond, L.; Yurkiv, V.; Costa, R.; Sanson, A. On the manufacturing of low temperature activated $Sr_{0.9}La_{0.1}TiO_{3-\delta}$-$Ce_{1-x}Gd_xO_{2-\delta}$ anode for solid oxide fuel cell. *J. Eur. Ceram. Soc.* **2018**, *38*, 153–161. [CrossRef]

29. Chen, G.; Kishimoto, H.; Yamaji, K.; Kuramoto, K.; Horita, T. Electrical performance of La-substituted $SrTiO_3$ anode material with different deficiencies in A-site. *ECS Trans.* **2013**, *50*, 63–71. [CrossRef]

30. Karczewski, J.; Riegel, B.; Molin, S.; Winiarski, A.; Gazda, M.; Jasinski, P.; Murawski, L.; Kusz, B. Electrical properties of $Y_{0.08}Sr_{0.92}Ti_{0.92}Nb_{0.08}O_{3-\delta}$ after reduction in different reducing conditions. *J. Alloys Compd.* **2009**, *473*, 496–499. [CrossRef]

31. Ishihara, T.; Matsuda, H.; Bustam, M.A.B.; Takita, Y. Oxide ion conductivity in doped Ga based perovskite type oxide. *Solid State Ion.* **1996**, *86–88*, 197–201. [CrossRef]

32. Slater, P.R.; Irvine, J.T.S.; Ishihara, T.; Takita, Y. High-Temperature Powder Neutron Diffraction Study of the Oxide Ion Conductor $La_{0.9}Sr_{0.1}Ga_{0.8}Mg_{0.2}O_{2.85}$. *J. Solid State Chem.* **1998**, *139*, 135–143. [CrossRef]

33. Slater, P.R.; Fagg, D.P.; Irvine, J.T.S. Synthesis and electrical characterization of doped perovskite titanates as potential anode materials for solid oxide fuel cells. *J. Mater. Chem.* **1997**, *7*, 2495–2498. [CrossRef]

34. Savaniu, C.; Irvine, J.T.S. Reduction studies and evaluation of surface modified A-site deficient La-doped $SrTiO_3$ as anode material for IT-SOFCs. *J. Mater. Chem.* **2009**, *19*, 8119–8128. [CrossRef]

35. Hui, S.; Petric, A. Evaluation of yttrium-doped $SrTiO_3$ as an anode for solid oxide fuel cells. *J. Eur. Ceram. Soc.* **2002**, *22*, 1673–1681. [CrossRef]

36. Hui, S.; Petric, A. Electrical Properties of Yttrium-Doped Strontium Titanate under Reducing Conditions. *J. Electrochem. Soc.* **2002**, *149*, J1–J10. [CrossRef]

37. Hui, S.; Petric, A. Electrical conductivity of yttrium-doped $SrTiO_3$: Influence of transition metal additives. *Mater. Res. Bull.* **2002**, *37*, 1215–1231. [CrossRef]

38. Kolodiazhnyi, T.; Petric, A. The Applicability of Sr-deficient n-type $SrTiO_3$ for SOFC Anodes. *J. Electroceram.* **2005**, *15*, 5–11. [CrossRef]

39. Blennow, P.; Hansen, K.K.; Wallenberg, L.R.; Mogensen, M. Effects of Sr/Ti-ratio in $SrTiO_3$-based SOFC anodes investigated by the use of cone-shaped electrodes. *Electrochim. Acta* **2006**, *52*, 1651–1661. [CrossRef]

40. Cumming, D.J.; Kharton, V.V.; Yaremchenko, A.A.; Kovalevsky, A.V.; Kilner, J.A. Electrical Properties and Dimensional Stability of Ce-Doped $SrTiOO_{3-\delta}$ for Solid Oxide Fuel Cell Applications. *J. Am. Ceram. Soc.* **2011**, *94*, 2993–3000. [CrossRef]

41. Zulueta, Y.A.; Dawson, J.A.; Mune, P.D.; Froeyen, M.; Nguyen, M.T. Oxygen vacancy generation in rare-earth-doped $SrTiO_3$. *Phys. Status Solidi (b)* **2016**, *253*, 2197–2203. [CrossRef]

crystals

MDPI

Article

Fluorine Translational Anion Dynamics in Nanocrystalline Ceramics: SrF$_2$-YF$_3$ Solid Solutions

Stefan Breuer [1]*, **Bernhard Stanje** [1], **Veronika Pregartner** [1], **Sarah Lunghammer** [1],
Ilie Hanzu [1,2] **and Martin Wilkening** [1,2,*]

1   Christian Doppler Laboratory for Lithium Batteries, and Institute for Chemistry and Technology of
    Materials, Graz University of Technology (NAWI Graz), Stremayrgasse 9, 8010 Graz, Austria;
    bernhard.stanje@gmx.at (B.S.); veronika.pregartner@tugraz.at (V.P.); sarah.lunghammer@tugraz.at (S.L.);
    hanzu@tugraz.at (I.H.)
2   Alistore-ERI European Research Institute, 33 rue Saint Leu, 80039 Amiens, France
*   Correspondence: breuer@tugraz.at (S.B.); wilkening@tugraz.at (M.W.); Tel.: +43-316-873-32330 (M.W.)

Received: 24 January 2018; Accepted: 2 March 2018; Published: 5 March 2018

Abstract: Nanostructured materials have already become an integral part of our daily life. In many applications, ion mobility decisively affects the performance of, e.g., batteries and sensors. Nanocrystalline ceramics often exhibit enhanced transport properties due to their heterogeneous structure showing crystalline (defect-rich) grains and disordered interfacial regions. In particular, anion conductivity in nonstructural binary fluorides easily exceeds that of their coarse-grained counterparts. To further increase ion dynamics, aliovalent substitution is a practical method to influence the number of (i) defect sites and (ii) the charge carrier density. Here, we used high energy-ball milling to incorporate Y^{3+} ions into the cubic structure of SrF$_2$. As compared to pure nanocrystalline SrF$_2$ the ionic conductivity of Sr$_{1-x}$Y$_x$F$_{2+x}$ with $x = 0.3$ increased by 4 orders of magnitude reaching 0.8×10^{-5} S cm^{-1} at 450 K. We discuss the effect of YF$_3$ incorporation on conductivities isotherms determined by both activation energies and Arrhenius pre-factors. The enhancement seen is explained by size mismatch of the cations involved, which are forced to form a cubic crystal structure with extra F anions if x is kept smaller than 0.5.

Keywords: nanocrystalline ceramics; binary fluorides, ionic conductivity; ball milling; cation mixing; aliovalent substitution

---

1. Introduction

Nanostructured materials assume a variety of functions in quite different applications and devices of our daily life [1]. Considering nanocrystalline ionic conductors [2–5], a range of studies report on enhanced anion and cation dynamics [6,7]. Structural disorder and defects [8–10], lattice mismatch [11–13] as well as size effects [14–19], which results in extended space charge regions, are used to explain the properties of nanocrystalline compounds. Especially for nano-engineered systems, which were prepared by bottom-up procedures, such as gas condensation or epitaxial methods [20,21], space charge regions lead to non-trivial effects that may enhance ion transport. This effect has not only been shown for the pioneering prototype system BaF$_2$-CaF$_3$ [20] but also, quite recently, for nanoscopic, grain-boundary engineered SrF$_2$-LaF$_3$ heterolayers [22].

If prepared via a top-down approach such as ball milling [8,23–25], the nanocrystalline material obtained is anticipated to consist of nm-sized crystals surrounded by structurally disordered regions [6,26]. In particular, this structural model helped rationalize ion dynamics in oxides [8]. For fluorides instead, amorphous regions are present to a much lesser extent [27,28]. High-energy ball milling also increases the density of defects in the interior of the nanograins [10,28]. Starting with binary fluorides, such as nanocrystalline BaF$_2$, whose F anion conductivity exceeds that of BaF$_2$ monocrystals by some orders of magnitude [12], iso- and aliovalent substitution of the metal cations drastically improves ion transport [11,29]. As an example, in the nanocrystalline solid solutions $Me_{1-x}Sr_xF_2$ (Me = Ca, Ba) the ionic conductivity passes through a maximum at intermediate values of x while the corresponding activation energy shows a minimum, the same holds for $Ba_{1-x}Ca_xF_2$ [11]. Although the original cubic crystal structure remains untouched for the mixed systems, strain and defects resulting from mixing sensitively affects F anion dynamics [11,12,30]. Metastable $Ba_{1-x}Ca_xF_2$, however, cannot be prepared via solid state synthesis; instead, it is only accessible by high-energy ball milling that forces the cations ions, substantially differing in size, to form a solid solution at atomic scale [12].

Provided the cubic crystal system is retained, aliovalent substitution, e.g., with LaF_3 or YF_3, increases the number density of mobile F anions. This effect has been shown for $Ba_{1-x}La_xF_{2+x}$ [29,31]. Starting with SrF_2 and YF_3 (see Figure 1) a similar behavior is expected. So far, $Sr_{1-x}Y_xF_{2+x}$ solid solutions have been prepared by ceramic synthesis and wet chemical approaches. Here, we report the dynamic parameters of nanocrystalline $Sr_{1-x}Y_xF_{2+x}$ obtained through high-energy ball milling. By using high impact planetary mills we managed to incorporate YF_3 into SrF_2 until the composition $Sr_{0.7}Y_{0.3}F_{2.3}$. Ionic conductivity greatly differs from that of Y-free nanocrystalline SrF_2. Ionic conductivities at elevated temperature might be high enough to realize high-temperature all-solid-state fluorine batteries [32–35]. In addition, as pointed out by Ritter et al. nanoscopic compounds based on SrF_2 as host for rare elements might act as new luminescent materials [36].

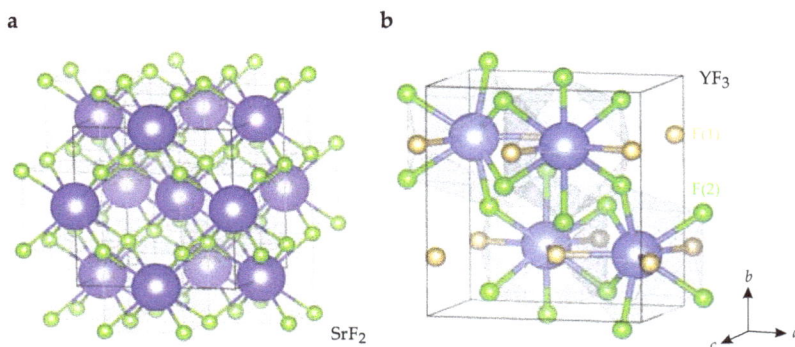

Figure 1. (a) Crystal structure ($Fd\bar{3}m$) of cubic SrF_2 with the Sr^{2+} ions (spheres in light blue, ionic radius 0.99 Å) occupying the $4b$ positions coordinated by eight fluorine anions residing on $8c$; In (b) the crystal structure orthorhombic of YF_3 ($Pnma$) is shown; the trivalent Y^{3+} ions (light blue spheres, ionic radius 0.90 Å) occupy the $4c$ sites, two further crystallographically inequivalent F sites exist which are filled by the F anions F(1) and F(2).

2. Materials and Methods

Nanocrystalline $Sr_{1-x}Y_xF_{2+x}$ samples were prepared in a planetary high-energy ball mill (Fritsch P7, Premium line, Fritsch, Idar-Oberstein, Germany) by mixing SrF_2 and YF_3 in the desired molar ratio. We used a 45 mL ZrO_2 cub with 140 zirconium oxide milling balls (5 mm in diameter). To guarantee complete transformation the mixtures were milled for 10 h at a rotation speed of 600 rpm under dry conditions, i.e., without the addition of any solvents.

The phase purity and crystal system was checked via X-ray diffraction (Bruker D8 Advance, Bragg Brentano geometry, 40 kV, Bruker AXS, Rheinstetten, Germany). With the help of the Scherrer equation [37] we roughly estimated the crystallite size of the samples, see ref. [10,38] for further details on this procedure. Usually, after sufficiently long milling periods the mean crystallite size reaches 10 to 20 nm [39]. ^{19}F magic angle spinning nuclear (MAS) magnetic resonance (NMR) was used to characterize SrF_2 and YF_3 as well as the mixed fluorides at atomic level. We recorded rotor-synchronized Hahn echoes by employing a Bruker 2.5-mm MAS probe placed in a 11.4 T cryomagnet. Spectra were measured with a Bruker Avance 500 spectrometer (Bruker BioSpin, Rheinstetten, Germany). Spinning was carried out with ambient bearing gas pressure. The $\pi/2$ pulse length was 2.1 µs at 50 W power level; the recycle delay was 20 s and 64 transients were accumulated for each spectrum. We took advantage of LiF powder as a secondary reference (-204.3 ppm) [40] to determine chemical shifts δ_{iso}; CFCl$_3$ (0 ppm) served as the primary reference.

To analyze ionic conductivities, the powders were uniaxially cold-pressed to pellets (8 mm in diameter, 0.5 to 1 mm in thickness). After applying electrically conducting Au electrodes (100 to 200 nm) via sputtering, a Novocontrol Concept 80 broadband analyzer (Novocontrol Technologies, Montabaur, Germany) was employed to record conductivity isotherms and complex impedances as a function of temperature. The broadband dielectric analyzer was connected to a QUATRO cryo system (Novocontrol Technologies, Montabaur, Germany) that allows precise settings of temperature and an automatic execution of the measurements. The QUATRO is operated with a stream of freshly evaporated nitrogen gas that passes a heater and enters the sample chamber where the temperature is measured using Pt-100 thermocouples. We used a voltage amplitude of 100 mV and varied the frequency ν from 0.1 Hz to 10 MHz. The isotherms revealed distinct frequency-independent plateaus from which *direct current* DC conductivities σ_{DC} (see above) were read off for each temperature T.

3. Results and Discussion

Structure: X-ray Diffraction and ^{19}F High-Resolution MAS NMR Spectroscopy

Long-range and local structures of Sr$_{1-x}$Y$_x$F$_{2+x}$ ($0 \leq x \leq 1$) were studied by X-ray diffraction and ^{19}F MAS NMR, see Figures 2 and 3.

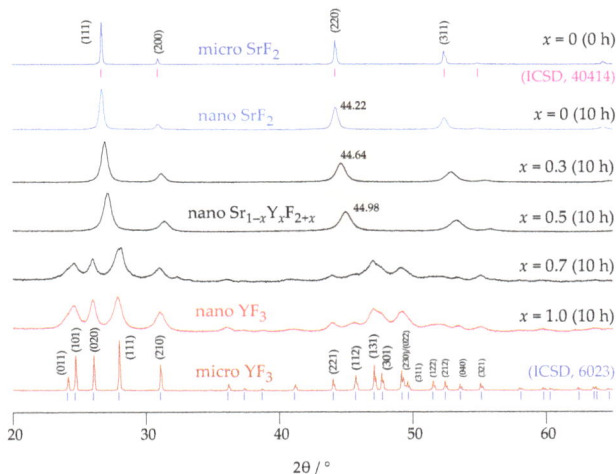

Figure 2. X-ray powder diffractograms recorded (CuK$_{\alpha1}$, 1.54059 Å) to characterize the phase purity of the starting materials as well as that of the mixed phases formed. The periods indicated in brackets show the milling times. The patterns of microcrystalline SrF$_2$ and YF$_3$ perfectly match with those in literature; verticals bars show the average positions of the CuK$_{\alpha1}$ and CuK$_{\alpha2}$ reflections. The broadening of the reflections, which is evident for the milled samples, indicates nm-sized crystallites; lattice strain and a distribution of lattice constants due to cation mixing may contribute to the broadening effect seen as well. Mixing of the cations is anticipated to have a much larger effect on ionic conductivity than the decrease in crystallite size seen when going from $x = 0$ to $x = 0.3$. See text for further explanation.

SrF$_2$ crystallizes in the cubic crystal system and YF$_3$ adopts an orthorhombic structure (see Figure 1). In Figure 2 the X-ray powder patterns of the starting materials as well as those of the milled samples, with the composition Sr$_{1-x}$Y$_x$F$_{2+x}$, are shown.

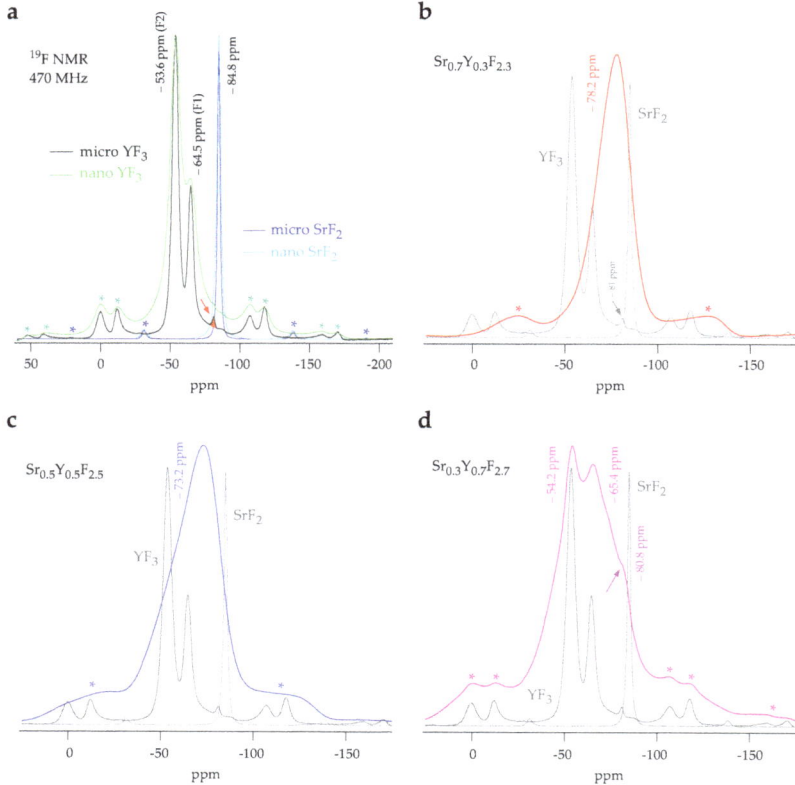

Figure 3. (**a–d**) ^{19}F magic angle spinning nuclear magnetic resonance (MAS NMR) spectra of Sr$_{1-x}$Y$_x$F$_{2+x}$ prepared by ball milling. We used a rotor-synchronized Hahn echo pulse sequence to record the spectra at a spinning speed of 25 kHz; asterisks (∗) denote spinning sidebands. The spectra are referenced to CFCl$_3$ (0 ppm). In (**a**) the spectra of microcrystalline, i.e., unmilled, and nanocrystalline SrF$_2$ and YF$_3$ are shown. The nanocrystalline samples were obtained after milling the starting materials for 10 h under dry conditions. Mechanical treatment leads to broadening of the NMR lines because of the defects and polyhedra distortions introduced. The spectra of YF$_3$ reveal the two magnetically equivalent sites of YF$_3$; their isotropic chemical shifts δ_{iso} are -53.6 ppm and -84.8 ppm. For the Sr-rich and Y-rich samples; see (**b,d**), Sr$_{1-x}$Y$_x$F$_{2+x}$ with $x = 0.3$ and $x = 0.7$ the center of the broad MAS NMR lines shift toward that of SrF$_2$ and YF$_3$, respectively. These isotropic shifts, together with the broadening of the lines, owing to a distribution of chemical shifts caused by atomic disorder, reveals the formation of mixed (Sr,Y) environments the F anions are subjected to. For $x = 0.5$ the extent of mixing is, of course, most effective; see (**c**).

As a result of heavy ball-milling in a planetary mill, the two fluorides form a product with cubic symmetry up to $x = 0.5$. Broadening of the reflections is caused by both lattice strain (ϵ) and nm-sized crystallites. The equation introduced by Scherrer yields 36(2) nm ($\epsilon = 0.0031(5)$), 13(2) nm ($\epsilon = 0.0083(5)$) and 10(2) nm ($\epsilon = 0.0123(10)$) as average crystallite size for the samples with $x = 0$,

$x = 0.3$ and $x = 0.5$, respectively. The values in brackets denote microstrain as estimated using the method proposed by Williamson and Hall [41], see also [42] for details. With increasing extent of cation mixing lattice strain enhances.

Furthermore, we clearly observe a shift of the reflections toward larger diffraction angles, see the positions of the (220) reflections as indicated in Figure 2. The lattice constant *a* gradually changes from 5.8040 Å (SrF_2) to 5.7442 Å ($Sr_{0.7}Y_{0.3}F_{2.3}$). The linear decrease of *a*, as a result of incorporation of the smaller Y cations, indicates Vegard behaviour. At the same time, F interstitials are generated that are expected to boost F^- ionic conductivity, vide infra (Figure 4).

When coming from pure YF_3, Sr ions are incorporated into the trifluoride. Once again, small crystallites significantly broaden the X-ray reflections; the orthorhombic structure of YF_3 is, however, still conserved (see Figure 2). The patterns do not show large amounts of X-ray amorphous material as featureless, broad humps are missing. Hence, we deal with nanocrystalline samples with local disorder.

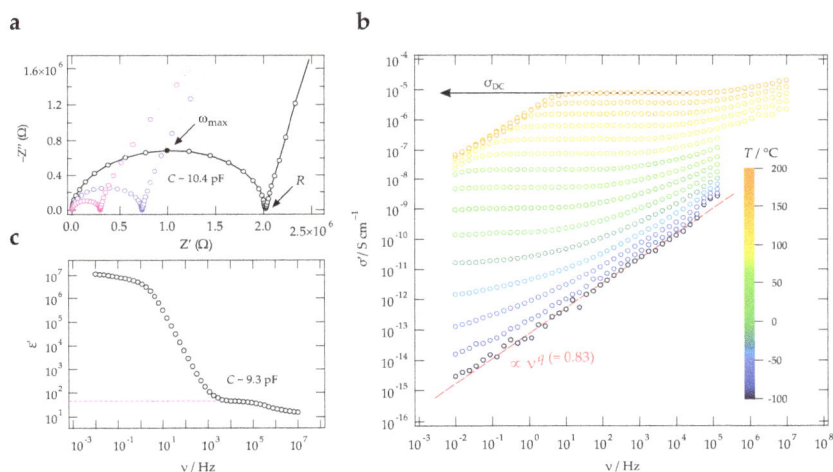

Figure 4. (a) Nyquist plots of the complex impedance of $Sr_{1-x}Y_xF_{2+x}$ with $x = 0.3$ recorded at the temperatures indicated. The solid lines are to guide the eye. The depressed semicircles indicate correlated ion movements in the ternary fluoride; from $\omega_{max} (= \nu_{max} 2\pi)$ and R electrical capacities, $C = 1/(\omega_{max}R)$, in the order of 10 pF were estimated indicating that the overall electrical response is mainly given by bulk properties rather than by ion-blocking grain boundary contributions; (b) Conductivity isotherms of $Sr_{0.7}Y_{0.3}F_{2.3}$ recorded from -100 °C to $+180$ °C in steps of 20 °C. The pronounced plateaus at intermediate frequencies contain the DC conductivity σ_{DC}, as indicated; (c) Real part, ϵ', of the complex primitivity as a function of frequency. Above 10^3 Hz the bulk response is seen; the capacity associated with that response is in the order of 10 pF.

Local disorder is also sensed by ^{19}F MAS NMR spectroscopy. In Figure 3a the spectra of the non-substituted binary fluorides are shown. As expected, coarse grained SrF_2 only shows a single resonance at -84.8 ppm and the spectrum of YF_3 reveals the two crystallographically inequivalent F positions F(1) and F(2). The samples with μm-sized crystallites are composed of sharp lines that perfectly agree with those presented and analyzed in literature [36,40]. Broadening, but no change in isotropic chemical shift, is observed for the samples treated mechanically. YF_3 shows a negligible amount of an impurity phase, see the NMR signal at $\delta_{iso} = -81$ ppm. Upon incorporation of YF_3, the ^{19}F MAS NMR lines drastically broaden because of the introduction of structural disorder, see also [36]. The various Sr-F environments formed as well as the various defects present, including F vacancies and interstitials, result in a broad distribution of ^{19}F NMR chemical shifts. Depending on which cation is the major component in $Sr_{1-x}Y_xF_{2+x}$ the center of the broad lines is either located near the original chemical shifts δ_{iso} of SrF_2 (Figure 3b) or those of YF_3 (Figure 3d). For the equimolar solid-solution with $x = 0.5$, the center of the line is in between that of SrF_2 and YF_3. The shoulder near -50 ppm might indicate Y-rich regions or clusters. For a detailed analysis of ^{19}F MAS spectra we refer to [36]. Indeed, ^{89}Y MAS NMR measurements by Ritter et al. [36] suggested F clustering taking place in samples with more than 10 mol % of Y^{3+}.

In contrast to other solid-solutions, such as $Me_{1-z}Sr_zF_2$ (Me $=$ Ca, Ba; $0 \leq z \leq 1$) and $Ba_{1-z}Ca_zF_2$ [11], line broadening, i.e., the distribution of chemical shifts, is much larger for $Sr_{1-x}Y_xF_{2+x}$. Because of this effect, even under the MAS conditions applied here, we cannot resolve the (Sr,Y) environments as it was possible for the (Ca,Sr) species in cubic $Ca_{1-z}Sr_zF_2$ with a similar range of chemical shifts, see the MAS NMR spectra presented in [11].

In Figure 4a,b Nyquist plots and conductivity isotherms of $Sr_{1-x}Y_xF_{2+x}$ with $x = 0.3$ are displayed; Figure 5 gives an overview of the temperature dependence of the ionic conductivities.

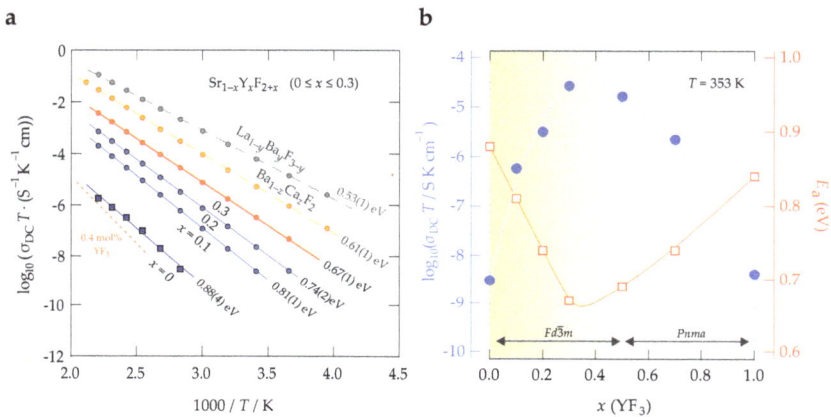

Figure 5. (a) Arrhenius plot of $\sigma_{DC}T$ vs. the $1/T$ of mechanochemically prepared $Sr_{1-x}Y_xF_{2+x}$ with $x \leq 0.3$. Solid and dashed lines represent fits according to an Arrhenius equation, $\sigma_{DC}T \propto \exp(-E_a/(k_BT))$; k_B denotes Boltzmanns constant. The activation energies obtained are indicated; (b) Change of E_a (see right axis) and $\sigma_{DC}T$ of nanocrystalline $Sr_{1-x}Y_xF_{2+x}$ as a function of x. See text for further explanation.

Plotting the real part σ' of the complex conductivity $\hat{\sigma}$ results in curves with (i) distinct DC plateaus at sufficiently low frequencies and (ii) a dispersive part at low T and high frequencies. The collapse of σ' at the lowest frequencies and highest temperature is because of F anion charge accumulation in front of the ion-blocking electrode. We used the frequency-independent plateau to determine specific ionic conductivities as a function of temperature. These values refer to long-range ion transport and characterize successful charge displacements in the fluoride. The same values for σ' can be inferred from analyzing Nyquist plots (see Figure 3a) that show the imaginary part $(-Z'')$ of the complex impedance $\hat{Z}$ as a function of its real part Z'. From the intercept of the semicircle the variable-temperature resistance R can be extracted and converted into σ'. The electrical relaxation process seen in conductivity (or impedance) spectroscopy mainly refers to a bulk process as the corresponding electrical capacity takes values in the pF range, see Figure 4a. Most likely, the broad semicircle also contains contributions from grain boundaries and interfacial regions. These regions do not seem to block ion transport as is known for other fast ion conductors [43].

The shape of the isotherms $\sigma'(\nu)$, cf. Figure 4b, does not change much with temperature indicating that the same relaxation process is probed at elevated T. At sufficiently high T, the data fulfill the so-called time-temperature superposition principle [44,45]. At low T, however, the dispersive part, which is sensitive to ion dynamics on shorter length scales than those seen in the DC regime, dominates the spectra. This regime can be approximated with Jonscher's power law ansatz [46], $\sigma'(\nu) \propto \nu^q$ [44]. The exponent q increases with temperature, most likely reaching $q = 1$ at cryogenic temperatures. At such low temperatures localized ion movements might play a rôle that do not contribute to long-range charge transport any longer [47]. The isotherm of $Sr_{0.7}Y_{0.3}F_{2.3}$ recorded at $T = 100\,°C$ is characterized by $q = 0.83(2)$. q values of approximately 0.8 are frequently seen for disordered ion conductors in this T range [29].

In Figure 5a the product $\sigma_{DC}T$ is plotted vs. the inverse temperature $1/T$ to determine activation energies E_a of the samples prepared and to illustrate the change in ionic conductivity upon Y^{3+} incorporation. For comparison, we included the Arrhenius lines of two other mechanochemically prepared F^- anion conductors *viz.* $Ba_{0.5}Ca_{0.5}F_2$ and $La_{0.9}Ba_{0.1}F_{2.9}$. In addition, ionic conductivities of YF_3-doped SrF_2 crystals is also shown [48]. While pure (nanocrystalline) SrF_2 is a poor ionic conductor with an activation energy as high as 0.88 eV; E_a is significantly decreased to 0.67 eV for $Sr_{0.7}Y_{0.3}F_{2.3}$. The ionic conductivity, as well as the activation energy (0.84 eV) of nanocrystalline YF_3 turned out to be very similar to that of nano-SrF_2. In Figure 5b the change in $\sigma_{DC}T$ and E_a is shown as a function of x. From $x = 0$ to $x = 0.3$ the activation energy almost linearly decrease; $\sigma_{DC}T$ passes through a maximum at ca. $x = 0.3$. Further incorporation of YF_3 causes $\sigma_{DC}T$ to decrease again. Note that the $Sr_{1-x}Y_xF_{2+x}$ system adopts cubic structure up to approximately $x = 0.5$, while the Y-rich samples crystallize with orthorhombic structure. Ultimately, at 80 °C (see Figure 5b) we notice an enhancement in ionic conductivity by 4 orders of magnitude when going from SrF_2 to $Sr_{0.7}Y_{0.3}F_{2.3}$. This opens the way to systematically vary the ionic conductivity of the ternary fluoride over a broad dynamic range.

Interestingly, already at low Y^{3+} contents, that is, at 10 mol%, we see an increase in ionic conductivity by one order of magnitude. For comparison, the ionic conductivity of the YF_3-doped SrF_2 crystals studied by Bollmann et al. [48] in 1970 is still lower than that of nanocrystalline SrF_2 highlighting the importance of size effects on ionic transport even in the cation-mixed system. For single crystals based on the LaF_3 tysonite structure also quite high ion conductivities have, however, been reported [49].

The change in E_a from 0.88 eV to 0.81 eV is accompanied by a significant increase of the pre-exponential factor σ_0 in $\sigma_{DC}T = \sigma_0 \exp(-E_a/(k_B T))$, see also [50]. While $\log(\sigma_0/(S\ cm^{-1}K))$ of SrF_2 amounts to 4.1, it increases to 5.3 for $Sr_{0.9}Y_{0.1}F_{2.1}$. Thus, the new defect structure, most likely involving F clustering as seen by ^{89}Y NMR [36], also influences the attempt frequencies (and/or the number of effective charge carriers) of the underlying jump processes. Presumably, substitution of Sr^{2+} with Y^{3+} will also influence the activation entropy for ionic migration. For $x > 0.1$ no further increase of σ_0 is seen; the additional increase in σ_{DC} is mainly governed by the reduction in activation energy. As the defect concentration is already high in this regime, we expect that predominantly lower migration energies facilitate ion transport further.

4. Conclusions

The mixed system $Sr_{1-x}Y_xF_{2+x}$ represents an attractive system to control ionic conductivities via the concept of cation mismatch. High-energy ball milling is able to force the distinctly sized cations to form a mixed phase with structural disorder at atomic scale. Treating SrF_2 with YF_3 in a high-energy planetary mill causes the original conductivity to increase by some orders of magnitude. Simultaneously, the corresponding activation energy decreases from 0.88 eV to 0.67 eV clearly showing a lowering of the migration barrier the ions have to surmount to take part in long-range transport. Besides this decrease in E_a we also observed the influence of the Arrhenius pre-factor to decisively initiate a further increase in ionic transport at $x \leq 0.1$, i.e., in the compositional range where F clustering starts.

Impedance spectroscopy shows that no grain boundary regions hinder the ions from moving over long distances. Most likely, the cation-mixed interfacial regions also provide fast diffusion pathways similar to those in the bulk structure. As we deal with nm-sized crystallites, space charge regions are expected to also contribute to overall F anion dynamics.

Acknowledgments: We thank M. Gabriel for his help with sample preparation. Moreover, we thank the Deutsche Forschungsgemeinschaft (DFG) for financial support (SPP 1415).

Author Contributions: Stefan Breuer, Bernhard Stanje and Martin Wilkening conceived and designed the experiments; Stefan Breuer, Veronika Pregartner and Sarah Lunghammer performed the experiments; all authors were involved in analyzing and interpreting the data. Stefan Breuer and Martin Wilkening wrote the paper.

References

1. Vollath, D. *Nanomaterials—An Introduction to Synthesis, Properties and Applications*; Wiley-VCH: Weinheim, Germany, 2008; pp. 21–22.

2. Tuller, H.L. Ionic conduction in nanocrystalline materials. *Solid State Ion.* **2000**, *131*, 143–157.

3. Knauth, P. Inorganic solid Li ion conductors: An overview. *Solid State Ion.* **2009**, *180*, 911–916.

4. Kobayashi, S.; Tsurekawa, S.; Watanabe, T. A new approach to grain boundary engineering for nanocrystalline materials. *Beilstein J. Nanotechnol.* **2016**, *7*, 1829.

5. Uitz, M.; Epp, V.; Bottke, P.; Wilkening, M. Ion dynamics in solid electrolytes for lithium batteries. *J. Electroceram.* **2017**, *38*, 142–156.

6. Heitjans, P.; Indris, S. Diffusion and ionic conduction in nanocrystalline ceramics. *J. Phys. Condens. Matter* **2003**, *15*, R1257.

7. Prutsch, D.; Breuer, S.; Uitz, M.; Bottke, P.; Langer, J.; Lunghammer, S.; Philipp, M.; Posch, P.; Pregartner, V.; Stanje, B.; et al. Nanostructured Ceramics: Ionic Transport and Electrochemical Activity. *Z. Phys. Chem.* **2017**, *231*, 1361–1405.

8. Heitjans, P.; Masoud, M.; Feldhoff, A.; Wilkening, M. NMR and impedance studies of nanocrystalline and amorphous ion conductors: Lithium niobate as a model system. *Faraday Discuss.* **2007**, *134*, 67–82.

9. Epp, V.; Wilkening, M. Motion of Li$^+$ in nanoengineered LiBH$_4$ and LiBH$_4$:Al$_2$O$_3$ comparison with the microcrystalline form. *ChemPhysChem* **2013**, *14*, 3706–3713.

10. Wilkening, M.; Epp, V.; Feldhoff, A.; Heitjans, P. Tuning the Li diffusivity of poor ionic conductors by mechanical treatment: High Li conductivity of strongly defective LiTaO$_3$ nanoparticles. *J. Phys. Chem. C* **2008**, *112*, 9291–9300.

11. Düvel, A.; Ruprecht, B.; Heitjans, P.; Wilkening, M. Mixed Alkaline-Earth Effect in the Metastable Anion Conductor Ba$_{1-x}$Ca$_x$F$_2$ ($0 \leq x \leq 1$): Correlating Long-Range Ion Transport with Local Structures Revealed by Ultrafast ^{19}F MAS NMR. *J. Phys. Chem. C* **2011**, *115*, 23784–23789.

12. Ruprecht, B.; Wilkening, M.; Steuernagel, S.; Heitjans, P. Anion diffusivity in highly conductive nanocrystalline BaF$_2$:CaF$_2$ composites prepared by high-energy ball milling. *J. Mater. Chem.* **2008**, *18*, 5412–5416.

13. Sorokin, N.I.; Buchinskaya, I.I.; Fedorov, P.P.; Sobolev, B.P. Electrical conductivity of a CaF$_2$-BaF$_2$ nanocomposite. *Inorg. Mater.* **2008**, *44*, 189–192.

14. Maier, J. Ionic conduction in space charge regions. *Prog. Solid State Chem.* **1995**, *23*, 171–263.

15. Maier, J. Nanoionics: Ion transport and electrochemical storage in confined systems. *Nat. Mater.* **2005**, *4*, 805–815.

16. Maier, J. Nano-Ionics: Trivial and Non-Trivial Size Effects on Ion Conduction in Solids. *Zeitschrift für Physikalische Chemie* **2003**, *217*, 415–436.

17. Maier, J. Nanoionics: Ionic charge carriers in small systems. *Phys. Chem. Chem. Phys.* **2009**, *11*, 3011–3022.

18. Puin, W.; Rodewald, S.; Ramlau, R.; Heitjans, P.; Maier, J. Local and overall ionic conductivity in nanocrystalline CaF$_2$. *Solid State Ion.* **2000**, *131*, 159–164.

19. Blanchard, D.; Nale, A.; Sveinbjörnsson, D.; Eggenhuisen, T.M.; Verkuijlen, M.H.W.; Vegge, T.; Kentgens, A.P.; de Jongh, P.E. Nanoconfined LiBH$_4$ as a fast lithium ion conductor. *Adv. Funct. Mater.* **2015**, *25*, 184–192.

20. Sata, N.; Eberman, K.; Eberl, K.; Maier, J. Mesoscopic fast ion conduction in nanometre-scale planar heterostructures. *Nature* **2000**, *408*, 946–949.

21. Jin-Phillipp, N.Y.; Sata, N.; Maier, J.; Scheu, C.; Hahn, K.; Kelsch, M.; Rühle, M. Structures of BaF_2-CaF_2 heterolayers and their influences on ionic conductivity. *J. Chem. Phys.* **2004**, *120*, 2375–2381.

22. Vergentev, T.; Banshchikov, A.; Filimonov, A.; Koroleva, E.; Sokolov, N.; Wurz, M.C. Longitudinal conductivity of LaF_3/SrF_2 multilayer heterostructures. *Sci. Technol. Adv. Mater.* **2016**, *17*, 799–806.

23. Šepelak, V.; Düvel, A.; Wilkening, M.; Becker, K.D.; Heitjans, P. Mechanochemical reactions and syntheses of oxides. *Chem. Soc. Rev.* **2013**, *42*, 7507–7520.

24. Wilkening, M.; Bork, D.; Indris, S.; Heitjans, P. Diffusion in amorphous $LiNbO_3$ studied by ^7Li NMR—Comparison with the nano-and microcrystalline material. *Phys. Chem. Chem. Phys.* **2002**, *4*, 3246–3251.

25. Wilkening, M.; Indris, S.; Heitjans, P. Heterogeneous lithium diffusion in nanocrystalline Li_2O: Al_2O_3 composites. *Phys. Chem. Chem. Phys.* **2003**, *5*, 2225–2231.

26. Gleiter, H. Nanoglasses: A new kind of noncrystalline materials. *Beilstein J. Nanotechnol.* **2013**, *4*, 517–533.

27. Chadwick, A.V.; Düvel, A.; Heitjans, P.; Pickup, D.M.; Ramos, S.; Sayle, D.C.; Sayle, T.X.T. X-ray absorption spectroscopy and computer modelling study of nanocrystalline binary alkaline earth fluorides. In *IOP Conference Series: Materials Science and Engineering*; IOP Publishing: Bristol, UK, 2015; Volume 80, p. 012005.

28. Preishuber-Pflügl, F.; Wilkening, M. Mechanochemically synthesized fluorides: Local structures and ion transport. *Dalton Trans.* **2016**, *45*, 8675–8687.

29. Preishuber-Pflügl, F.; Bottke, P.; Pregartner, V.; Bitschnau, B.; Wilkening, M. Correlated fluorine diffusion and ionic conduction in the nanocrystalline F^- solid electrolyte $Ba_{0.6}La_{0.4}F_{2.4}$—^{19}F $T_{1(\rho)}$ NMR relaxation vs. conductivity measurements. *Phys. Chem. Chem. Phys.* **2014**, *16*, 9580–9590.

30. Ruprecht, B.; Wilkening, M.; Feldhoff, A.; Steuernagel, S.; Heitjans, P. High anion conductivity in a ternary non-equilibrium phase of BaF_2 and CaF_2 with mixed cations. *Phys. Chem. Chem. Phys.* **2009**, *11*, 3071–3081.

31. Rongeat, C.; Reddy, M.A.; Witter, R.; Fichtner, M. Nanostructured fluorite-type fluorides as electrolytes for fluoride ion batteries. *J. Phys. Chem. C* **2013**, *117*, 4943–4950.

32. Anji Reddy, M.; Fichtner, M. Batteries based on fluoride shuttle. *J. Mater. Chem.* **2011**, *21*, 17059–17062.

33. Nowroozi, M.A.; Wissel, K.; Rohrer, J.; Munnangi, A.R.; Clemens, O. $LaSrMnO_4$: Reversible Electrochemical Intercalation of Fluoride Ions in the Context of Fluoride Ion Batteries. *Chem. Mater.* **2017**, *29*, 3441–3453.

34. Grenier, A.; Porras-Gutierrez, A.G.; Groult, H.; Beyer, K.A.; Borkiewicz, O.J.; Chapman, K.W.; Dambournet, D. Electrochemical reactions in fluoride-ion batteries: Mechanistic insights from pair distribution function analysis. *J. Mater. Chem. A* **2017**, *5*, 15700–15705.

35. Grenier, A.; Porras-Gutierrez, A.G.; Body, M.; Legein, C.; Chrétien, F.; Raymundo-Piñero, E.; Dollé, M.; Groult, H.; Dambournet, D. Solid Fluoride Electrolytes and Their Composite with Carbon: Issues and Challenges for Rechargeable Solid State Fluoride-Ion Batteries. *J. Phys. Chem. C* **2017**, *121*, 24962–24970.

36. Ritter, B.; Krahl, T.; Scholz, G.; Kemnitz, E. Local Structures of Solid Solutions $Sr_{1-x}Y_xF_{2+x}$ ($x = 0 \ldots 0.5$) with Fluorite Structure Prepared by Sol-Gel and Mechanochemical Syntheses. *J. Phys. Chem. C* **2016**, *120*, 8992–8999.

37. Scherrer, P. Bestimmung der Größe und der inneren Struktur von Kolloidteilchen mittels Röntgenstrahlen. *Nachrichten Ges. Wiss. Göttingen* **1918**, *2*, 98.

38. Düvel, A.; Wegner, S.; Efimov, K.; Feldhoff, A.; Heitjans, P.; Wilkening, M. Access to metastable complex ion conductors via mechanosynthesis: Preparation, microstructure and conductivity of (Ba, Sr) LiF_3 with inverse perovskite structure. *J. Mater. Chem.* **2011**, *21*, 6238–6250.

39. Indris, S.; Bork, D.; Heitjans, P. Nanocrystalline oxide ceramics prepared by high-energy ball milling. *J. Mater. Synth. Proc.* **2000**, *8*, 245–250.

40. Sadoc, A.; Body, M.; Legein, C.; Biswal, M.; Fayon, F.; Rocquefelte, X.; Boucher, F. NMR parameters in alkali, alkaline earth and rare earth fluorides from first principle calculations. *Phys. Chem. Chem. Phys.* **2011**, *13*, 18539–18550.

41. Williamson, G.; Hall, W. X-ray line broadening from filed aluminium and wolfram. *Acta Metall.* **1953**, *1*, 22–31.

42. Düvel, A.; Wilkening, M.; Uecker, R.; Wegner, S.; Šepelak, V.; Heitjans, P. Mechanosynthesized nanocrystalline $BaLiF_3$: The impact of grain boundaries and structural disorder on ionic transport. *Phys. Chem. Chem. Phys.* **2010**, *12*, 11251–11262.

43. Breuer, S.; Prutsch, D.; Ma, Q.; Epp, V.; Preishuber-Pflügl, F.; Tietz, F.; Wilkening, M. Separating bulk from grain boundary Li ion conductivity in the sol-gel prepared solid electrolyte $Li_{1.5}Al_{0.5}Ti_{1.5}(PO_4)_3$. *J. Mater. Chem. A* **2015**, *3*, 21343–21350.

44. Dyre, J.C.; Maass, P.; Roling, B.; Sidebottom, D.L. Fundamental questions relating to ion conduction in disordered solids. *Rep. Prog. Phys.* **2009**, *72*, 046501.

45. Sakellis, I. On the origin of time-temperature superposition in disordered solids. *Appl. Phys. Lett.* **2011**, *98*, 072904.

46. Jonscher, A. The 'universal' dielectric response. *Nature* **1977**, *267*, 673–679 .

47. Funke, K.; Cramer, C.; Wilmer, D. *Diffusion in Condensed Matter*, 2nd ed.; Heitjans, P., Kärger, J., Eds.; Springer: Berlin, Germany, 2005; Chapter 21, pp. 857–893.

48. Bollmann, W.; Görlich, P.; Hauk, W.; Mothes, H. Ionic conduction of pure and doped CaF_2 and SrF_2 crystals. *Phys. Status Solidi (A)* **1970**, *2*, 157–170.

49. Sorokin, N.I.; Sobolev, B.P.; Krivandina, E.A.; Zhmurova, Z.I. Optimization of single crystals of solid electrolytes with tysonite-type structure (LaF_3) for conductivity at 293 K: 2. Nonstoichiometric phases $R_{1-y}M_yF_{3-y}$ (R = La-Lu, Y; M = Sr, Ba). *Cryst. Rep.* **2015**, *60*, 123.

50. Breuer, S.; Wilkening, M. Mismatch in cation size causes rapid anion dynamics in solid electrolytes: The role of the Arrhenius pre-factor. *Dalton Trans.* **2018**, in press, doi:10.1039/C7DT04487A.

crystals

MDPI

Article

Effect of MnO_2 Concentration on the Conductivity of $Ce_{0.9}Gd_{0.1}Mn_xO_{2-\delta}$

Kerstin Neuhaus [1,*] , Stefan Baumann [2] , Raimund Dolle [3] and Hans-Dieter Wiemhöfer [1,3]

[1] Institute for Inorganic and Analytical Chemistry, University of Münster, Corrensstr. 28/30, 48149 Münster, Germany; hdw@uni-muenster.de

[2] Forschungszentrum Jülich GmbH, Institute of Energy and Climate Research, Materials Synthesis and Processing (IEK-1), 52425 Jülich, Germany; s.baumann@fz-juelich.de

[3] Helmholtz Institute Münster (HI MS), IEK-12, Forschungszentrum Jülich GmbH, Corrensstraße 46, 48149 Münster, Germany; r.dolle@fz-juelich.de

* Correspondence: kerstin.neuhaus@uni-muenster.de; Tel.: +49-251-83-36095

Received: 15 November 2017; Accepted: 15 January 2018; Published: 17 January 2018

Abstract: Samples with the composition $Ce_{0.9}Gd_{0.1}Mn_xO_{2-\delta}$ with x = 0.01, 0.02, and 0.05 Mn-addition were prepared by mixed oxide route from $Ce_{0.9}Gd_{0.1}O_{2-\delta}$ and MnO_2 and sintered at 1300 °C. The electronic conductivity was measured using a modified Hebb-Wagner technique, the electrical conductivity was investigated by impedance spectroscopy, and oxygen permeation was measured for the sample with x = 0.05. An increase of the electronic partial conductivity with increasing Mn addition was observed, which can be attributed to an additional Mn 3d-related state between the top of the valence band and the bottom of the Ce 4f band. The grain boundary conductivity was found to be suppressed for low Mn contents, but enhanced for the sample with x = 0.05.

Keywords: ceria; Hebb-Wagner measurements; electronic conductivity

1. Introduction

Materials based on doped ceria have been investigated mainly for high temperature applications in solid oxide fuel cells and as sensor materials. In the last ten years, the scientific focus changed from the high temperature regime to application temperatures in the range of room temperature to 400 °C, as ceria has become more and more important in the area of catalytic applications [1–3] (e.g., for the car exhaust catalyst), and also in the pharmaceutical area [4,5]. Manganese doping of ceria has until now been mainly investigated for catalytic applications [1,6] and as additive to improve the sinterability of ceria [7,8]. It has been found additionally that manganese ions are able to diffuse into ceria when it is in contact to Mn containing materials at high temperatures, e.g., during the co-sintering of ceria-perovskite composites for solid oxide fuel cell components. This also caused a certain interest in the effect of Mn doping [7].

So far, the effect of Mn addition has mostly been investigated using Mn as a co-dopant or rather sintering additive [7–10]. Mainly Gd_2O_3 [7,10,11], but in some cases also other oxides [1], were used as main dopants to fix the oxygen vacancy concentration of the material. The solubility limit of Mn in pure ceria has been found to be in the range between 5–10 mol % [7,8,12], or <1 mol % in another study [13]. Also, the solubility was found to be lower in combination with Gd as a co-dopant ($\leq$5 mol %) [12].

The improvement of the sinterability is lower than found for Fe and Co [10,14], but much better than found for Cr addition, which actually inhibits the densification of ceria [15]. Excess Mn, which is not soluble in the ceria matrix, was reported to form the perovskite phase $GdMnO_3$ in combination with Gd [12].

Pikalova et al. [10] and Park et al. [13] reported a decrease in the electrical conductivity for Mn- and Gd-codoped ceria. Apart from this, a change of redox state of Mn^{3+} to Mn^{2+} was found [9,16,17]

for Mn doping in ceria, as well as in yttria stabilized zirconia (YSZ). Sasaki and Maier [17] confirmed a depression of the chemical diffusivity by Mn doping of YSZ because of this redox process, which takes places at roughly $pO_2 = 10^{-5}$ bar at 800 °C. The sintering temperature in literature for Mn doped ceria materials is mainly in the range between 1200 °C and 1300 °C. Previous investigations found that after sintering at temperatures above 1200 °C, manganese ions exist in the divalent state (Mn^{2+}) within the ceria matrix [9]. Therefore, as sintering was performed at $\geq$1300 °C for the samples presented here, Mn^{3+} and Mn^{4+} should only be present at oxygen partial pressures above the pO_2 of air, which are not addressed in our study.

Theoretical work to corroborate the positive effect on catalytic performance of MnO_x-CeO_2 oxides showed that Mn doping in ceria induces an additional Mn 3d-related gap state between the top of the valence band and the bottom of the Ce 4f band, which facilitates oxygen vacancy formation [16]. This has not yet been verified experimentally. The second aim of this study was to verify whether small amounts of Mn cations work as acceptor dopant or are more or less immiscible within the ceria matrix. Indications for an acceptor doping effect were already found for Fe addition [14,18] and a possible substitution of Ce was also postulated for Mn addition [19], while other studies suggested a low miscibility of Mn within the ceria matrix [13,20]. In case of a low miscibility, Mn-rich segregates could either be found as single defects at the grain boundaries [20] or forming a secondary phase [7,21], which would both influence the grain boundary charge transport characteristics.

Samples with the composition $Ce_{0.9}Gd_{0.1}Mn_xO_{2-\delta}$ with x = 0.01, 0.02, and 0.05 were prepared by the mixed oxide route from $Ce_{0.9}Gd_{0.1}O_{2-\delta}$ and MnO_2. Both, the Mn 3d-related gap states, as well as Mn working as an acceptor dopant in low concentrations, can be verified by the evaluation of the electronic conductivity of Mn doped samples over a broad continuous temperature and oxygen partial pressure range and a comparison of these values to the electrical conductivity. The effect on the grain boundary and bulk conductivity was therefore investigated using impedance spectroscopy in air and the variation of the electronic partial conductivity with varying Mn content was measured using a modified Hebb-Wagner technique. This measurement method permits the temperature dependent detection of the electronic partial conductivity of a sample over a continuous oxygen partial pressure range of roughly 0.21–10^{-20} bar. Additionally, for the sample with x = 0.05, the oxygen permeation was measured to validate the Hebb-Wagner measurements.

2. Materials and Methods

Powders of the composition $Ce_{0.9}Gd_{0.1}Mn_xO_{2-\delta}$ with x = 0.01, 0.02, and 0.05 were synthesized from cerium nitrate hexahydrate (Ce(NO$_3$)$_3$·6H$_2$O by abcr, 99.9%), gadolinium nitrate hexahydrate (Gd(NO$_3$)$_3$·6H$_2$O by abcr, 99.9%), and manganese dioxide (MnO_2 by Acros Organics, 99%).

Cerium and gadolinium nitrate were dissolved in water, NH_4OH was added until pH = 9 and subsequently H_2O_2 was added dropwise to oxidize Ce^{3+} to Ce^{4+}. The brownish gel was filtered, washed with distilled water, and then dried for 24 h at 70 °C. The resulting powder was then calcined for 4 h at 500 °C in air.

The appropriate amounts of MnO_2 were added. After ball milling for 4 h in ethanol, the powders were dried and then uniaxially dry pressed into pellets. These were sintered subsequently at 1300 °C for 2 h (Mn-containing materials) or at 1600 °C for 4 h (Mn-free reference sample). The samples with 5 mol % Mn, which were applied in the permeation experiments were sintered at 1350 °C for 2 h to achieve sufficient gas-tightness. The grain size of the samples was determined by scanning electronic microscopy using the lineal intercept method [22]. All of the SEM images were obtained in secondary electronic imaging mode with an acceleration voltage of 7 kV and 8 mm working distance.

Impedance measurements in air were carried out as a function of temperature in ambient air using a Novotherm HT 1200 (NovoControl GmbH, Montabaur, Germany). Pt resinate paste (RP 070107, Heraeus GmbH, Hanau, Germany) was applied directly on the sample pellets and was sintered at 850 °C before measurement to minimize any contact resistance. The paste electrodes were contacted by a Pt-sheet. An AC peak-to-peak amplitude of 40 mV was chosen. All the measurements were carried out in ambient air and in a temperature range from 200–800 °C. The experiments were executed three times at each temperature to control for hysteresis of the conductivity.

The experimental frequency dependence of the impedance was fitted using the software package ZView 2 (Scribner Associates, Inc., Southern Pines, NC, USA) with regard to the medium and high frequency range. For fitting, an equivalent circuit with two resistors in series, which each have a constant phase element (CPE) in parallel, was used [23]. One resistor/constant phase element can be assigned to bulk characteristics and one to grain boundary characteristics, respectively. Figure 1 shows an example of the measured data for pure CGO in the temperature range between 500–200 °C and the equivalent circuit used for evaluation. (σ_b) and grain boundary conductivities (σ_{gb}) have the following connection:

$$R_{total} = R_b + R_{gb} = \frac{n}{A} \cdot \left(\frac{1}{\sigma_b} \cdot L_b + \frac{1}{\sigma_{gb}} \cdot L_{gb} \right) = \frac{1}{\sigma_t} \tag{1}$$

Here, n is the number of grains perpendicular to the electric field, A is the contact area, L_g is the average grain size (cf.), L_{gb} is the average size of the grain boundary, which was in our case assumed to be 1 nm [24,25]. The grain boundary conductivity was corrected with the factor $\frac{L_{gb}}{L_g}$ to calculate the corrected grain boundary conductivity σ^*_{gb}, according to the brick layer model [24,26]. Activation energies were calculated from Arrhenius diagrams of the respective conductivities.

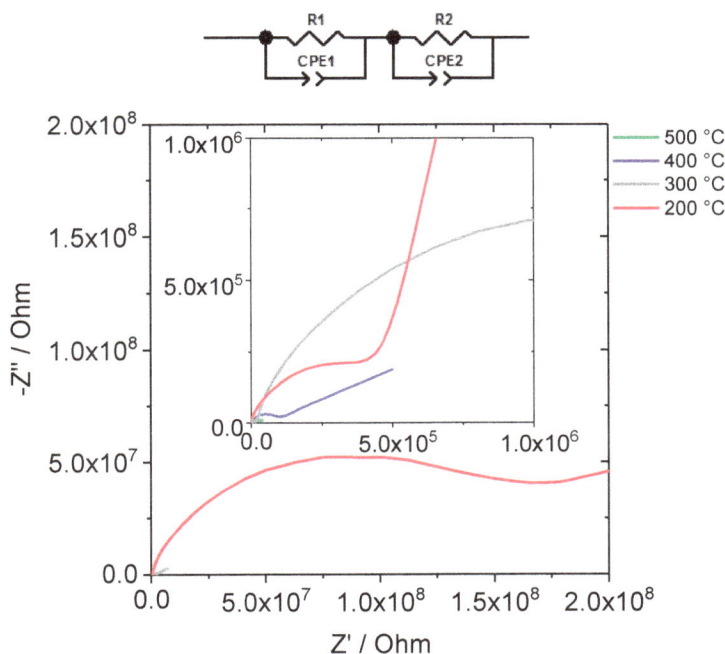

Figure 1. Exemplary impedance spectra for pure CGO at 200–500 °C. Higher temperatures are shown in the inlay. The equivalent circuit used for data evaluation is shown above the graph.

The electronic partial conductivity of the samples was investigated using a home-made modified Hebb-Wagner setup (Sourcemeter: Keithley 2601A, Keithley Instruments, Solon, OH, USA). In this experimental setup, an encapsulated platinum micro contact is applied as working electrode and a Pt sheet as counter electrode. The sample was fixed to the Pt sheet with Pt resinate paste

(RP 070107, Heraeus GmbH). Measurements were carried out at temperatures between 500–800 °C. Each measurement was performed three times and the average value from all three measurements was taken for further calculations.

Permeation experiments were performed on a uniaxially pressed disc 15 mm in diameter sintered at 1350 °C for 2 h in order to achieve sufficient gas-tightness. The thickness of the disc was 0.91 mm after grinding both surfaces with P1200 emery paper. Thin (approx. 5–10 μm) porous surface activation layers of $La_{0.6}Sr_{0.4}Co_{0.2}Fe_{0.8}O_{3-\delta}$ (LSCF) were screen printed on both sides of the disc and calcined at 1050 °C, as described elsewhere [27]. The disc was sealed with two gold rings in an asymmetric glass tube setup. The membrane was heated to 1000 °C, while a spring force was applied to the gold ring sealing. Permeation measurements were carried out between 1000 °C and 400 °C in steps of approx. 50 K with air and argon as feed and sweep gas at a flow rate of 250 and 50 mL/min, respectively. The permeate gas composition was monitored using mass spectrometry (Omnistar, Pfeiffer Vacuum, Asslar, Germany). The nitrogen content in the permeate gas was used to monitor leakage. The nitrogen content in the permeate gas was constantly low (approx. 150 ppm). However, also the concentration of permeated oxygen is very low (<500 ppm), and, hence, the permeation rate was corrected by subtracting 25% of the nitrogen flow rate in the permeate gas from the oxygen flux, reflecting the feed gas composition of approx. 80:20% N_2/O_2. The measurements below 850 °C showed too high leakages, and, hence, were excluded from the investigation.

3. Results

3.1. Microstructure and XRD Measurements

As can be derived from Figure 2, all of the samples were found to be single-phase fluorite materials within limits of detection of the XRD method (limit of detection of crystalline secondary phases is in the area of 3 mol %). The lattice constant of the different samples does not follow a clear trend: 1 and 2 mol % Mn addition lead to an increase of the lattice constant, but measurements for the sample with 5 mol % Mn addition show a decreased lattice constant with respect to the pristine Mn-free material.

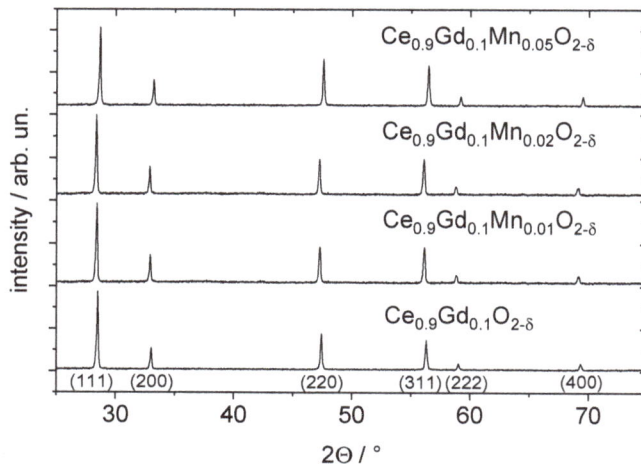

Figure 2. Powder XRD measurements of the four different samples.

As the sintering temperatures and sintering times were different for the Mn-free sample and the three different Mn-containing samples (see Section 2, Materials and Methods), it was already expected that the grain size for the pure CGO10 should be larger than for $Ce_{0.9}Gd_{0.1}Mn_xO_{2-\delta}$ (cf. Table 1 and Figure 3). It can be seen that the average grain size does not follow a clear trend with increasing

Mn addition. The density of the samples with Mn addition is also higher than for the pure CGO10 (cf. Table 1), confirming the findings from literature that addition of small amounts of Mn leads to an improved sinterability.

Table 1. Relative densities calculated from Archimedean balance results and the lattice constants, average grain sizes from microscopic evaluation and measured lattice constants from XRD measurements.

Composition	Density	Average Grain Size	Lattice Constant/Å
x = 0.00	87%	2.6 ± 1.2 μm	5.416
x = 0.01	92%	1.2 ± 0.5 μm	5.432
x = 0.02	88%	1.7 ± 0.7 μm	5.432
x = 0.05	88%	1.4 ± 0.5 μm	5.386

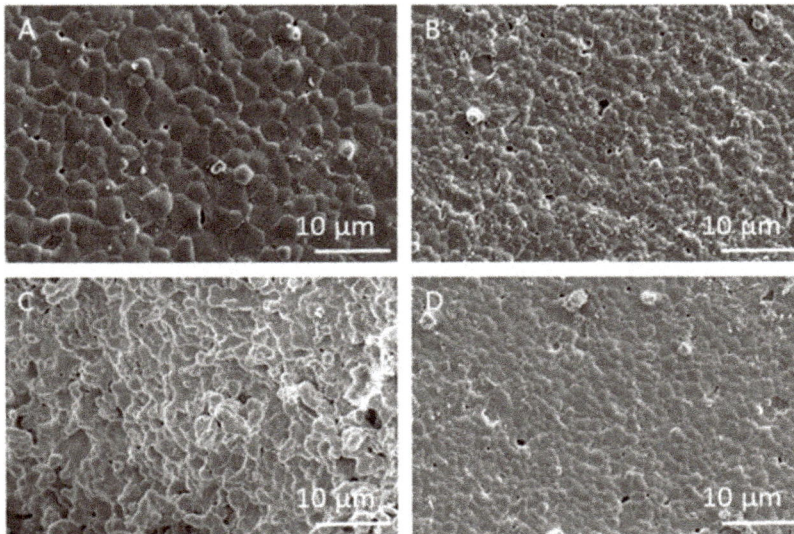

Figure 3. SEM images of the four different samples. (**A**) x = 0.00; (**B**) x = 0.01; (**C**) x = 0.02; and (**D**) x = 0.05.

3.2. Impedance Spectroscopy

It can be observed by comparison of the temperature dependent conductivities presented in Figure 4 that the Mn addition only has a small influence on the electrical conductivity of the samples, which remains in a similar range for all four measured samples, although a slight increase can be observed with increasing Mn addition. This is largely consistent with findings by other groups [10,13].

In contrast, the effect on the grain boundary conductivity is comparably large: a significant drop of $\sigma_{gb}{}^*$ can be observed for small additions of Mn (x = 0.01 and 0.02) while an addition of 5 mol % Mn leads to a slight increase of the grain boundary conductivity when compared to the pristine CGO10.

Activation energies for the respective transport pathways were calculated from Arrhenius diagrams and for the temperature range of 400–800 °C, as at lower temperatures, a slight decrease of the temperature dependence of the conductivity is visible. The activation energies for grain boundary and total charge transport ($E_{A,gb}$ and $E_{A,tot}$ cf. Table 2) remain in the same range for all of the compositions, while there is a strong variation in the activation energy for bulk transport: for x = 0.01 a pronounced decrease of this values can be found, while for all other compositions, the values remain in a similar range.

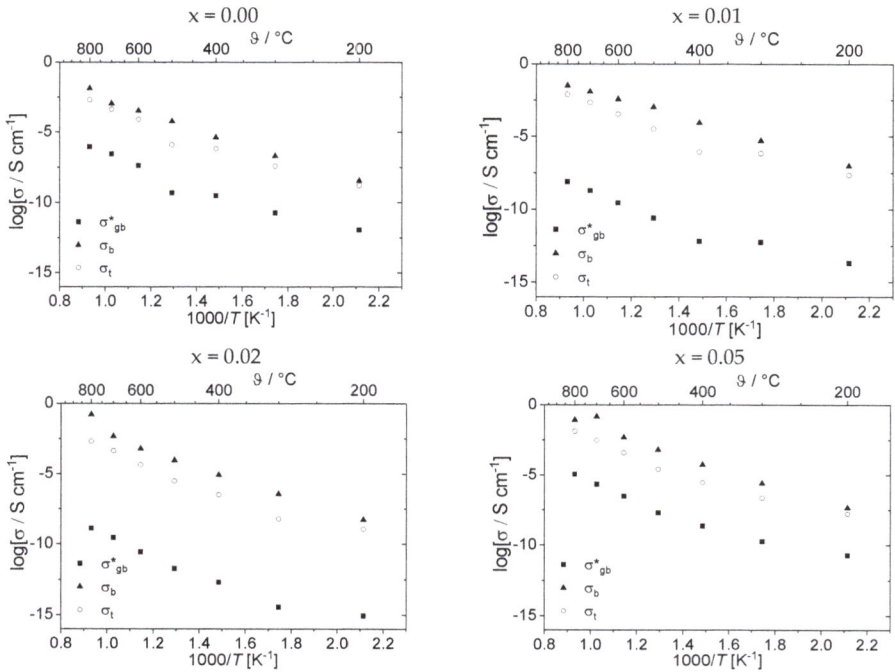

Figure 4. Grain boundary, grain, and electrical conductivity from impedance spectroscopy measurements in air. The grain boundary conductivity was corrected by the factor L_{gb}/L_b.

Table 2. Activation energies for bulk, grain boundary and total transport from impedance spectroscopy measurements in air at 400–800 °C. Standard deviation shows the goodness of the linear fit for calculating the slope of the Arrhenius diagram. The activation energies for hole or electronic transport were calculated from the Hebb-Wagner measurements in chapter 2.3 at $pO_2 = 0.21$ bar ($E_{A,h}$) or at $pO_2 = 10^{-10}$ bar ($E_{A,e}$), respectively. Values for $Ce_{0.79}Mn_{0.01}Gd_{0.2}O_{2-\delta}$ by Zhang et al. [7] and for $Ce_{0.985}Mn_{0.005}Gd_{0.01}O_{2-\delta}$ by Avila-Paredes et al. [20] are listed for comparison.

Composition	$E_{A,gb}$/eV	$E_{A,bulk}$/eV	$E_{A,tot}$/eV	$E_{A,h}$/eV	$E_{A,e}$/eV
x = 0.00	1.45 ± 0.23	1.26 ± 0.10	1.42 ± 0.21	1.20 ± 0.04	2.24 ± 0.08
x = 0.01	1.51 ± 0.26	0.97 ± 0.04	1.49 ± 0.05	1.00 ± 0.05	2.06 ± 0.09
x = 0.02	1.47 ± 0.08	1.51 ± 0.20	1.47 ± 0.08	1.02 ± 0.03	1.98 ± 0.08
x = 0.05	1.43 = 0.08	1.34 ± 0.17	1.41 ± 0.07	0.85 ± 0.06	2.15 ± 0.09
$Ce_{0.79}Mn_{0.01}Gd_{0.2}O_{2-\delta}$ [7]	1.03	0.86	1.01	-	-
$Ce_{0.985}Mn_{0.005}Gd_{0.01}O_{2-\delta}$ [20]	1.35 ± 0.03	0.65	-	-	-

The bulk and grain boundary resistances can be utilized to determine the barrier function of the grain boundaries with respect to ionic diffusion, assuming a phase-pure material. Acceptor doped ceria grains were previously described by a space charge model, where the grain boundaries present a Mott-Schottky type contact [28]. Hence, the electrostatic (or space charge) potential barrier $\Delta\varphi_s$, which is equivalent to the Schottky barrier height, can be calculated from the impedance spectroscopy measurements by solving the following equation numerically [20]:

$$\frac{\rho_{gb}}{\rho_{bulk}} = \frac{\exp\left(\frac{2e\Delta\varphi_s}{kT}\right)}{\frac{4e\Delta\varphi_s}{kT}} \tag{2}$$

Here, ρ_{gb} is the grain boundary resistivity, ρ_{bulk} is the bulk resistivity, e is the elementary charge, k is the Boltzmann constant in J/K, and T is the absolute temperature in K [20]. The space charge potential difference can be equated to the grain boundary core potential:

$$\Delta\varphi_s = \varphi_s - \varphi_{bulk} \qquad (3)$$

Avila-Paredes et al. [20] calculated the grain boundary core potential for the composition $Ce_{0.985}TM_{0.005}Gd_{0.01}O_{2-\delta}$ (TM = Co, Fe, Cu, and Mn) and found potentials in the area of $\Delta\varphi_s = 0.4$ to 0.55 V. This is roughly two times higher than the values found in our study (cf. Figure 5), but the space charge potential barrier is heavily dependent on grain size [29] and dopant distribution.

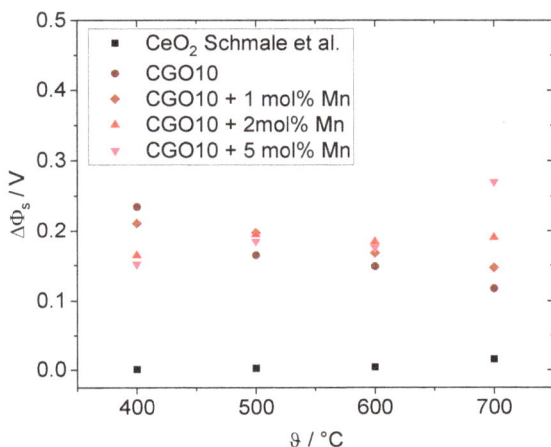

Figure 5. Temperature dependence of the space charge potential of $Ce_{0.9-x}Gd_{0.1}MnxO_{2-\delta}$ in comparison to nominally pure ceria (99.9% purity) [23] calculated by Equation (2).

When comparing the values for pure CeO_2 by [23] to the values for $Ce_{0.9}Gd_{0.1}Mn_xO_{2-\delta}$, it can be seen that Gd doping leads to an increase of the space charge potential barrier at the grain boundaries of ceria. The effect is more intense for low temperatures and decreases with increasing temperature. The influence of Mn differs: at high temperatures, Mn addition leads to an increase of the barrier height, while at low temperatures, a decreasing effect can be observed as compared to pristine CGO10.

3.3. Electronic Conductivity

By applying an electrical potential difference (potential V_{exp} in Figure 6) to the cell, the oxygen partial pressure at the microcontact can be controlled under steady-state conditions. The oxygen partial pressure at the microcontact (denoted by $pO_{2,cont}$) is given with respect to the reference partial pressure of air at the large planar back contact by the Nernst Equation:

$$V_{exp} = \frac{RT}{4F}ln\left[\frac{pO_{2,\,cont}}{pO_{2,\,air}}\right] + V_{cont} \qquad (4)$$

The sign of the applied potential V_{exp} refers to the polarity of the microcontact. Only negative potentials were applied, i.e., $pO_{2,cont} < pO_{2,air}$. An oxygen partial pressure that significantly exceeds the pressure of air leads to a breakdown of the glass encapsulation due to formation of gas bubbles. Oxygen partial pressures above 0.21 bar were therefore not addressed in this study.

The non-negligible ohmic contact resistance R_{cont} of the interface between sample and microcontact is considered by addition of the voltage V_{cont} in (4). As V_{cont} is variable with temperature and applied voltage, it has not been corrected in these measurements. This leads to an underestimation of the measured electronic conductivities presented in this study.

The oxygen partial pressure at the microcontact is given by

$$pO_{2,\,\text{cont}} = 0.21\,\text{bar} \cdot \exp\left[\frac{4F(V_{\text{exp}} - V_{\text{cont}})}{RT}\right] \tag{5}$$

with R = gas constant, T = temperature in K, and F = Faraday constant. In accordance with the Hebb-Wagner model, the local electronic conductivity (σ_e) of the ceria sample at the ion-blocking microelectrode can be calculated by multiplying the slope of the steady-state potential curve (see exemplary experimental I-V curve in Figure 6) with a geometric factor considering the radius of the tip. For a flat, disk shaped platinum tip, σ_e as a function of the local oxygen partial pressure at the microcontact is obtained by:

$$\sigma_e(pO_{2,\,\text{cont}}) = \left(\frac{\delta I}{\delta V}\right)_{\text{st}} \cdot \frac{1}{2\pi a} \tag{6}$$

The index 'st' of the derivative denotes that values are obtained from the steady state I-V curve. For each data point, the respective voltage was held for 30 min in the temperature range between 600–800 °C or 45 min for all measurements below 600 °C, before the average value of the steady state current was determined. As "abort criterion" to minimize the measurement time, the polarization was considered complete (and the steady state reached) if the value of the measured current did not change more than 0.5% over a time of 180 s (240 s at temperatures below 600 °C). All of the electronic conductivity values were corrected for the microcontact diameter, which was determined by measuring the contact print in the glass encapsulation after the end of the measurement. The diameter of the contact area was in the range between 220–330 μm for the various experiments.

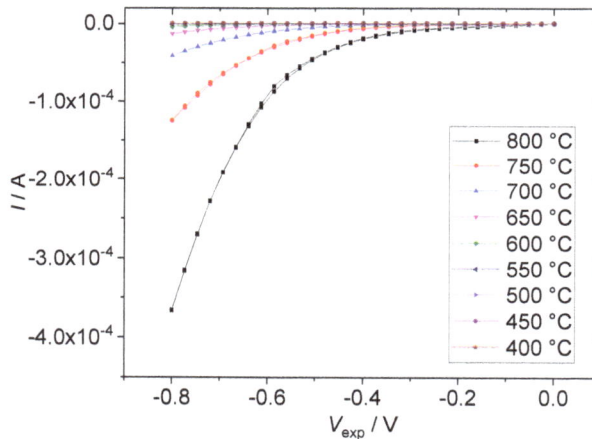

Figure 6. Experimental I-V curves for one Hebb-Wagner experiment for the sample with the composition $Ce_{0.9}Gd_{0.1}Mn_{0.01}O_{2-\delta}$.

As can be seen from Figure 7, a significant increase of the electronic conductivity was found for Mn-doped materials in comparison with pristine CGO10. At the same time, the position of the minimum of electronic conductivity remains at the same oxygen partial pressure.

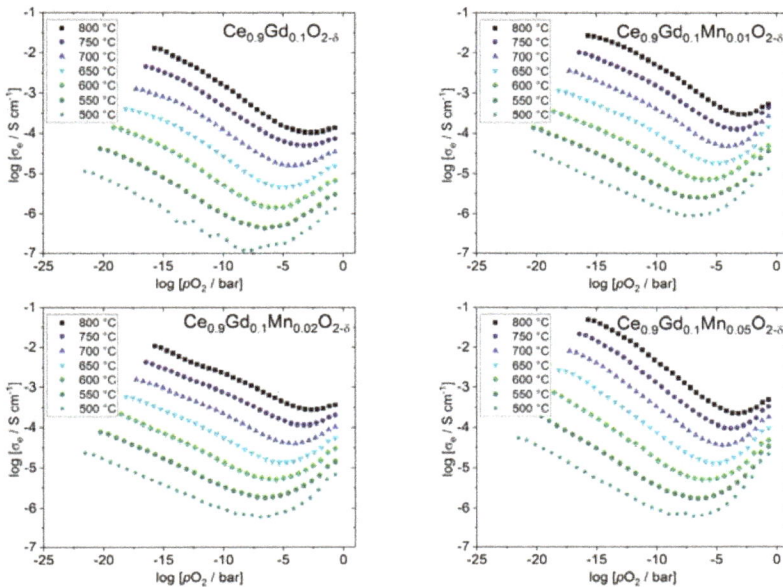

Figure 7. Oxygen partial pressure dependent conductivity for the four different samples. Average error for each point is $\leq 4\%$. The values were corrected for the microcontact diameter, but not for contact resistance. Indication for $\pm 1/4$ and $\pm 1/6$ slopes are given in each graphic.

In Figure 8, the electronic conductivities for the respective samples at $pO_2 = 0.21$ bar and $pO_2 = 10^{-10}$ bar are plotted. It can be observed that the p-type conduction at 0.21 bar is significantly enhanced, and also the activation energy for charge transfer ($E_{A,h}$), which was calculated for this partial pressure range (cf. Table 2) shows a significant decrease with increasing Mn addition.

For the lower oxygen partial pressure, this is not the case: here, the activation energies ($E_{A,e}$ in Table 2) are significantly lowered for the addition of 1 and 2 mol % Mn but get back to a similar value for 5 mol % Mn addition. Also, the activation energies for p-type charge transport ($E_{A,h}$) are by a factor of 2 lower than for n-type charge transport ($E_{A,e}$) for all samples in this study.

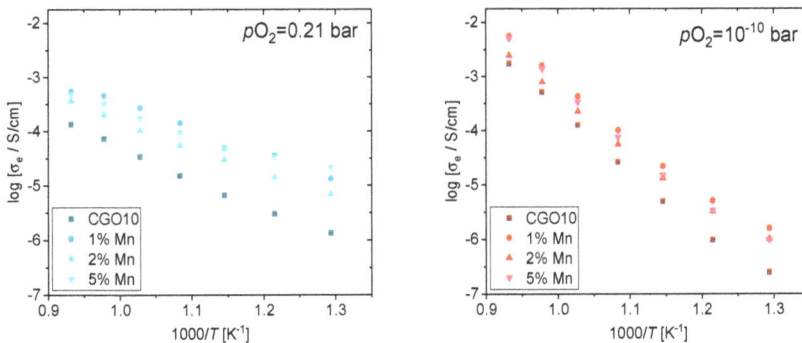

Figure 8. Electronic conductivity at $pO_2 = 0.21$ bar (blue symbols) and at $pO_2 = 10^{-10}$ bar (red symbols) for the respective compositions. A list of activation energies for electronic transport at these partial pressures can be found in Table 2.

3.4. Oxygen Permeation

The oxygen permeation rate j_{O_2} can be described using the simplified Wagner equation (Equation (4)), assuming that the ambipolar conductivity is independent of oxygen partial pressure within the gradient between feed and permeate side.

$$j_{O_2} = \frac{R}{16F^2} \cdot \frac{1}{L} \cdot \sigma_{amb} T \cdot \ln \frac{p'_{O_2}}{p''_{O_2}} \tag{7}$$

R is the gas constant, F is the Faraday constant, L is the membrane thickness, σ_{amb} is the ambipolar conductivity, T is the absolute temperature, and p'_{O2} and p''_{O2} are the oxygen partial pressures at the feed and permeate side, respectively. σ_{amb} is composed of the ion conductivity σ_i and the electronic conductivity σ_e and can also be expressed by the ionic end electronic transference numbers t_i and t_e, respectively (Equation (8)).

$$\sigma_{amb} = \frac{\sigma_i \cdot \sigma_e}{\sigma_i + \sigma_e} = t_i t_e \sigma_t \tag{8}$$

In contrast to other mixed ionic electronic conducting materials, such as $La_{0.6}Sr_{0.4}Co_{0.2}Fe_{0.8}O_{3-\delta}$ (LSCF), where the electronic conductivity is orders of magnitude larger compared to the ion conductivity [30], Mn co-doped CGO has the opposite properties, i.e., $\sigma_e << \sigma_i$. With this boundary condition, σ_e can be calculated and plotted versus the electrical conductivity, as shown in Figure 9. The slope of the linear regression gives an average of the electronic transference of 0.009. The calculated electronic conductivities at high temperature are in very good agreement to those extrapolated from values of the Hebb-Wagner-method at lower temperatures, as shown in Figure 10. The oxygen permeation measurement also prove, that the Hebb-Wagner polarization experiments yield consistent data, as the density of the samples (see Table 1) was relatively low and Hebb-Wagner experiments should normally only be performed for samples with a high density ($\geq$95%).

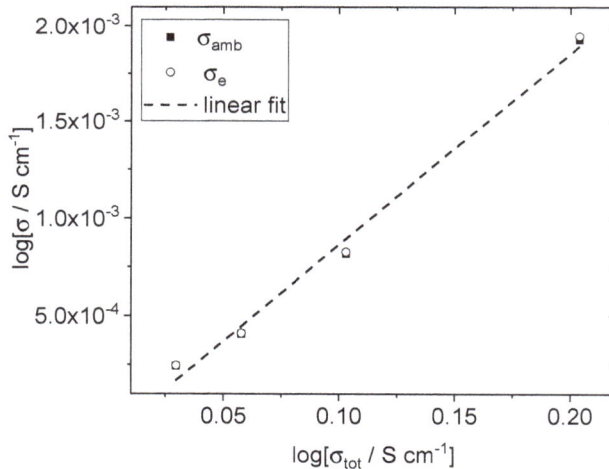

Figure 9. Electronic conductivity calculated via Equations (6) and (7) as function of electrical conductivity determined by impedance spectroscopy.

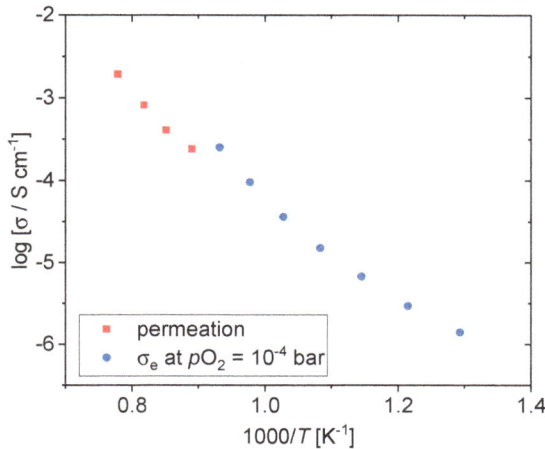

Figure 10. Electronic conductivities determined by oxygen permeation (red) and Hebb-Wagner measurements (blue).

4. Discussion

The lattice constant of CGO with Mn-addition presented in Section 3.1 did not show a clear trend. A lattice expansion with Mn addition could be taken as hint for Mn in interstitial positions in the ceria structure at least for 1 and 2 mol % Mn addition. Mn^{2+} has a similar ionic radius (91 pm) like Ce^{4+}, but substitution of Ce^{4+} therefore would not lead to a lattice expansion. Still, this scenario does not explain the decrease of the lattice constant for x = 0.05. A possible explanation could be, that after reaching a certain Mn threshold concentration, a secondary phase is formed (formation of $GdMnO_3$ has already been observed for Mn-addition to highly Gd-doped ceria [12]), which incorporates most of the Mn, but this cannot be verified with the data measured in this study.

In our electrochemical measurements, a strong effect of Mn addition on the grain boundary conductivity has been observed, with 1 and 2 mol % Mn addition, leading to a significant decrease of the grain boundary conductivity. In contrast, a substantial increase was found for 5 mol % Mn addition when compared to the pristine CGO10. This further substantiates the idea of the segregation of Mn as a second phase at the grain boundaries.

The observed strong increase of the electronic conductivity without any change of the position of the conductivity minimum in dependence on oxygen partial pressure confirms the assumption that Mn doping leads to an additional Mn 3d-related state between the top of the valence band and the bottom of the Ce 4f band [16]. In addition, in the low temperature regime a lowered potential barrier $\Delta\varphi_s$ between grains and grain boundaries, which was found with increasing Mn concentration, could also improve the electronic transport between grains.

The constant position of the minimum of the electronic conductivity can however only be taken as hint that Mn does neither work as a donor or acceptor dopant within the ceria matrix, because the defect chemistry of the material is already strongly influenced by the Gd co-dopant. As no additional maximum of the electronic conductivity was found in the oxygen partial pressure range between 10^{-4} and 10^{-7} bar [17] (which would be the case if predominantly trivalent or tetravalent Mn was existing in the samples and would be reduced during Hebb-Wagner measurements), Mn exists predominantly in a Mn^{2+} state, as initially assumed [9]. Reduction of Mn^{3+} or Mn^{4+} would hypothetically lead to a small polaron hopping process, which could be comparable to the effect of the reduction process of $Pr^{3+/4+}$ [31], but this mechanism was not observed.

By comparing electronic and electrical conductivity in the high temperature regime (cf. Table 3), we can ascertain that Mn addition has only a minor effect for 700–800 °C, while for lower temperatures, the electronic transport in slightly Mn doped materials becomes more prominent. The strongest effect can be observed for 1 mol% Mn addition. For 2 and 5 mol%, a behavior that is comparable to pristine CGO10 can be found. Permeation measurements confirmed the electronic conductivity data from Hebb-Wagner measurements for 5 mol% Mn addition.

Table 3. Electronic transference numbers t_e for the different samples at chosen temperatures. Data for 900 and 1000 °C were determined from permeation measurements.

$Ce_{0.9}Gd_{0.1}Mn_xO_{2-\delta}$	600 °C	700 °C	800 °C	900 °C	1000 °C
x = 0	0.41	0.17	0.07	-	-
x = 0.01	0.78	0.11	0.07	-	-
x = 0.02	0.39	0.14	0.10	-	-
x = 0.05	0.45	0.11	0.04	0.007	0.010

An enhanced influence of Mn at lower temperature can also be found regarding the potential barrier at the grain boundaries: at high temperatures, Mn addition increases the grain boundary core potential by roughly a factor of 2.5 with respect to the bulk potential, while at low temperatures a decreasing effect (about 40% lower) can be observed when compared to pristine CGO10. This behavior is especially pronounced for 5 mol% Mn addition.

Two possible options can be the cause for this effect: on the one hand, the contrarious temperature dependence of the grain boundary potential barrier with respect to pure CGO could be caused by a Mn-rich second phase at the grain boundaries. Such an effect was already confirmed for secondary phases formed by Fe addition to ceria [18], and secondary phase formation (GdMnO$_3$) was already observed for Mn addition to heavily Gd-doped ceria [12]. Such a secondary phase does not necessarily appear in the XRD measurements, as it could either have a concentration below the detection limit of XRD or it could be amorphous. Alternatively, the effect on the grain boundary potential barrier could also be triggered by a local increase of the Mn^{2+} concentration and a parallel depletion of segregated Gd^{3+} at the grain boundaries. Both of the effects could potentially occur at the same time.

In conclusion, an impact on the electronic partial conductivity, even of small amounts of Mn was detected. This can on the one hand be mainly attributed to Mn introducing additional states in the ceria band gap, which facilitates electronic transport. On the other hand, a reduced potential barrier at the grain boundaries, which was found to decrease for increasing Mn content for temperatures below 500 °C, can also facilitate electronic transport. A clear indication for a donor (Mn on the interstitial lattice sites as proposed by XRD measurements) or acceptor doping effect (Mn substituting for Ce as initially thought) was not discernible from the electrochemical measurements or the XRD data. This is consistent with findings by Park et al. [13].

Altogether, this means, that for example co-sintering of ceria and Mn-containing materials will lead to slightly changed characteristics of the ceria phase itself. A detailed TEM-based analysis of the grain boundaries could additionally show, whether the lowered potential barrier at low temperatures for the sample with 5 mol % Mn content can be ascribed to secondary phase formation or to a lowered Gd^{3+} segregation at the grain boundaries. Both of the effects could be exploited to tailor the grain boundary characteristics of ceria-based composite materials for particular low temperature applications by carefully manipulating the Mn concentration within the material.

Acknowledgments: Many thanks to S. Heinz for sample preparation and execution of the oxygen permeation tests. Investigations were performed during preliminary studies for a collaborative project funded by the German Research Foundation—project number 387282673.

Author Contributions: K.N. and S.B. conceived and designed the experiments; K.N. and S.B. performed the experiments; K.N. and S.B. analyzed the data; R.D. and H.D.W. contributed to the analysis and discussion; K.N. wrote the paper.

Conflicts of Interest: The authors declare no conflict of interest. The founding sponsors had no role in the design of the study; in the collection, analyses, or interpretation of data; in the writing of the manuscript, and in the decision to publish the results.

References

1. Terribile, D.; Trovarelli, A.; de Leitenburg, C.; Primavera, A.; Dolcetti, G. Catalytic combustion of hydrocarbons with Mn and Cu-doped ceria–zirconia solid solutions. *Catal. Today* **1999**, *47*, 133–140. [CrossRef]
2. Yao, H.C.; Yao, Y.F.Y. Ceria in automotive exhaust catalysts: I. Oxygen storage. *J. Catal.* **1984**, *86*, 254–265. [CrossRef]
3. Trovarelli, A. Catalytic properties of ceria and CeO_2-containing materials. *Catal. Rev.* **1996**, *38*, 439–520. [CrossRef]
4. Tarnuzzer, R.W.; Colon, J.; Patil, S.; Seal, S. Vacancy engineered ceria nanostructures for protection from radiation-induced cellular damage. *Nano Lett.* **2005**, *5*, 2573–2577. [CrossRef] [PubMed]
5. Wason, M.S.; Zhao, J. Cerium oxide nanoparticles: Potential applications for cancer and other diseases. *Am. J. Transl. Res.* **2013**, *5*, 126–131. [PubMed]
6. Mandal, S.; Santra, C.; Bando, K.K.; James, O.O.; Maity, S.; Mehta, D.; Chowdhury, B. Aerobic oxidation of benzyl alcohol over mesoporous Mn-doped ceria supported au nanoparticle catalyst. *J. Mol. Catal. A* **2013**, *378*, 47–56. [CrossRef]
7. Zhang, T.; Kong, L.; Zeng, Z.; Huang, H.; Hing, P.; Xia, Z.; Kilner, J. Sintering behavior and ionic conductivity of $Ce_{0.8}Gd_{0.2}O_{1.9}$ with a small amount of MnO_2 doping. *J. Sol. State Electrochem.* **2003**, *7*, 348–354. [CrossRef]
8. Zhang, T.; Hing, P.; Huang, H.; Kilner, J. Sintering study on commercial CeO_2 powder with small amount of MnO_2 doping. *Mater. Lett.* **2002**, *57*, 507–512. [CrossRef]
9. Tianshu, Z.; Hing, P.; Huang, H.; Kilner, J. Sintering and densification behavior of Mn-doped CeO_2. *Mater. Sci. Eng. B* **2001**, *83*, 235–241. [CrossRef]
10. Pikalova, E.Y.; Demina, A.N.; Demin, A.K.; Murashkina, A.A.; Sopernikov, V.E.; Esina, N.O. Effect of doping with Co_2O_3, TiO_2, Fe_2O_3, and Mn_2O_3 on the properties of $Ce_{0.8}Gd_{0.2}O_{2-\delta}$. *Inorg. Mater.* **2007**, *43*, 735–742. [CrossRef]
11. Kondakindi, R.R.; Karan, K. Characterization of Fe- and Mn-doped GDC for low-temperature processing of solid oxide fuel cells. *Mater. Chem. Phys.* **2009**, *115*, 728–734. [CrossRef]
12. Kang, C.Y.; Kusaba, H.; Yahiro, H.; Sasaki, K.; Teraoka, Y. Preparation, characterization and electrical property of Mn-doped ceria-based oxides. *Solid State Ion.* **2006**, *177*, 1799–1802. [CrossRef]
13. Park, S.-H.; Yoo, H.-I. Defect-chemical role of Mn in Gd-doped CeO_2. *Solid State Ion.* **2005**, *176*, 1485–1490. [CrossRef]
14. Fagg, D.P.; Kharton, V.V.; Frade, J.R. P-type electronic transport in $Ce_{0.8}Gd_{0.2}O_{2-\delta}$: The effect of transition metal oxide sintering aids. *J. Electroceram.* **2002**, *9*, 199–207. [CrossRef]
15. Taub, S.; Neuhaus, K.; Wiemhöfer, H.-D.; Ni, N.; Kilner, J.A.; Atkinson, A. The effects of Co and Cr on the electrical conductivity of cerium gadolinium oxide. *Solid State Ion.* **2015**, *282*, 54–62. [CrossRef]
16. Cen, W.; Liu, Y.; Wu, Z.; Wang, H.; Weng, X. A theoretic insight into the catalytic activity promotion of CeO_2 surfaces by Mn doping. *Phys. Chem. Chem. Phys.* **2012**, *14*, 5769–5777. [CrossRef] [PubMed]
17. Sasaki, K.; Maier, J. In situ EPR studies of chemical diffusion in oxides. *Phys. Chem. Chem. Phys.* **2000**, *2*, 3055–3061. [CrossRef]
18. Tsipis, E.V.; Waerenborgh, J.C.; Kharton, V.V. Grain-boundary states in solid oxide electrolyte ceramics processed using iron oxide sintering aids: A Mössbauer spectroscopy study. *J. Solid State Electrochem.* **2017**, *21*, 2965–2974. [CrossRef]
19. Pereira, G.J.; Castro, R.H.R.; de Florio, D.Z.; Muccillo, E.N.S.; Gouvêa, D. Densification and electrical conductivity of fast fired manganese-doped ceria ceramics. *Mater. Lett.* **2005**, *59*, 1195–1199. [CrossRef]
20. Avila-Paredes, H.J.; Kim, S. The effect of segregated transition metal ions on the grain boundary resistivity of gadolinium doped ceria: Alteration of the space charge potential. *Solid State Ion.* **2006**, *177*, 3075–3080. [CrossRef]
21. Foschini, C.R.; Souza, D.P.F.; Paulin Filho, P.I.; Varela, J.A. Ac impedance study of Ni, Fe, Cu, Mn doped ceria stabilized zirconia ceramics. *J. Eur. Ceram. Soc.* **2001**, *21*, 1143–1150. [CrossRef]

22. Wurst, J.C.; Nelson, J.A. Lineal intercept technique for measuring grain size in two-phase polycrystalline ceramics. *J. Am. Ceram. Soc.* **1972**, *55*, 109. [CrossRef]

23. Schmale, K.; Daniels, M.; Buchheit, A.; Grünebaum, M.; Haase, L.; Koops, S.; Wiemhöfer, H.-D. Influence of zinc oxide on the conductivity of ceria. *J. Electrochem. Soc.* **2013**, *160*, F1081–F1087. [CrossRef]

24. Jasper, A.; Kilner, J.A.; McComb, D.W. TEM and impedance spectroscopy of doped ceria electrolytes. *Solid State Ion.* **2008**, *179*, 904–908. [CrossRef]

25. Hwang, J.-H.; McLachlan, D.S.; Mason, T.O. Brick layer model analysis of nanoscale-to-microscale cerium dioxide. *J. Electroceram.* **1999**, *3*, 7–16. [CrossRef]

26. Christie, G.M.; Van Berkel, F.P.F. Microstructure—Ionic conductivity relationships in ceria-gadolinia electrolytes. *Solid State Ion.* **1996**, *83*, 17–27. [CrossRef]

27. Ramasamy, M.; Persoon, E.S.; Baumann, S.; Schroeder, M.; Schulze-Küppers, F.; Görtz, D.; Bhave, R.; Bram, M.; Meulenberg, W.A. Structural and chemical stability of high performance $Ce_{0.8}Gd_{0.2}O_{2-\delta}$-$FeCo_2O_4$ dual phase oxygen transport membranes. *J. Membr. Sci.* **2017**, *544*, 278–286. [CrossRef]

28. Kim, S.; Maier, J. On the conductivity mechanism of nanocrystalline ceria. *J. Electrochem. Soc.* **2002**, *149*, J73–J83. [CrossRef]

29. Wang, B.; Lin, Z. A schottky barrier based model for the grain size effect on oxygen ion conductivity of acceptor-doped ZrO_2 and CeO_2. *Int. J. Hydrog. Energy* **2014**, *39*, 14334–14341. [CrossRef]

30. Gryaznov, D.; Baumann, S.; Kotomin, E.A.; Merkle, R. Comparison of permeation measurements and hybrid density-functional calculations on oxygen vacancy transport in complex perovskite oxides. *J. Phys. Chem. C* **2014**, *118*, 29542–29553. [CrossRef]

31. Neuhaus, K.; Eickholt, S.; Maheshwari, A.; Schulze-Küppers, F.; Baumann, S.; Wiemhöfer, H.-D. Analysis of charge transport in $Ce_{0.8}Gd_{0.2-x}Pr_xO_{2-\delta}$ at t $\leq$ 600 °C. *J. Electrochem. Soc.* **2017**, *164*, H491–H496. [CrossRef]

crystals

MDPI

Review

Thermal and Chemical Expansion in Proton Ceramic Electrolytes and Compatible Electrodes

Andreas Løken [1,2], Sandrine Ricote [3] and Sebastian Wachowski [4,*]

1   Centre for Earth Evolution and Dynamics, University of Oslo, N-0315 Oslo, Norway;
    andreas.loken@geo.uio.no
2   Jotun Performance Coatings, Jotun A/S, N-3202 Sandefjord, Norway
3   Department of Mechanical Engineering, Colorado School of Mines, Golden, CO 80401, USA;
    sricote@mines.edu
4   Department of Solid State Physics, Faculty of Applied Physics and Mathematics, Gdańsk University of
    Technology, 80233 Gdańsk, Poland
*   Correspondence: sebastian.wachowski@pg.edu.pl; Tel.: +48-58-348-66-12

Received: 18 August 2018; Accepted: 6 September 2018; Published: 14 September 2018

Abstract: This review paper focuses on the phenomenon of thermochemical expansion of two specific categories of conducting ceramics: Proton Conducting Ceramics (PCC) and Mixed Ionic-Electronic Conductors (MIEC). The theory of thermal expansion of ceramics is underlined from microscopic to macroscopic points of view while the chemical expansion is explained based on crystallography and defect chemistry. Modelling methods are used to predict the thermochemical expansion of PCCs and MIECs with two examples: hydration of barium zirconate ($BaZr_{1-x}Y_xO_{3-\delta}$) and oxidation/reduction of $La_{1-x}Sr_xCo_{0.2}Fe_{0.8}O_{3-\delta}$. While it is unusual for a review paper, we conducted experiments to evaluate the influence of the heating rate in determining expansion coefficients experimentally. This was motivated by the discrepancy of some values in literature. The conclusions are that the heating rate has little to no effect on the obtained values. Models for the expansion coefficients of a composite material are presented and include the effect of porosity. A set of data comprising thermal and chemical expansion coefficients has been gathered from the literature and presented here divided into two groups: protonic electrolytes and mixed ionic-electronic conductors. Finally, the methods of mitigation of the thermal mismatch problem are discussed.

Keywords: thermal expansion; chemical expansion; protonic conductors; proton ceramic fuel cells; TEC; CTE; high temperature proton conductors

1. Introduction

The expansion of a solid upon the exposure to heat is a phenomenon known to mankind for centuries. This process is called thermal expansion and examples of scientists studying thermal expansion can be dated back as far as 1730, when Petrus van Musschenbroek [1] measured the expansion of metals used in pendulum clocks. This posed a significant problem for time measurements at the time, as the length of the pendulum would change with small temperature variations, inducing a change in the period of the oscillating pendulum of up to $\pm 0.05\%$ [2]. With passing decades, the search for reliable pendulum clocks became less relevant for the scientific community, mostly due to the arrival of cheaper synchronous electric clocks in the 1930s. Nevertheless, changes in the size and shape of solids upon heating still remain a large obstacle in applied science and engineering today, often being an underlying cause for device failures. In this review, we will specifically address and discuss expansion processes that are relevant for applications involving Proton Conducting Ceramics (PCCs).

Proton Conducting Ceramics are a group of materials exhibiting protonic conductivity at elevated temperatures (400–700 °C). The group can be divided in two classes; materials which are predominantly protonic with low transport numbers for other charged species, and mixed protonic conductors where the conductivity is dominated by protons and at least one other type of charge carrier contributing to the overall total conductivity. The former are often referred to as Proton Ceramic Electrolytes (PCEs) or High Temperature Proton Conductors (HTPCs) [3,4], where typical examples are acceptor doped barium zirconates [5] and cerates [6]. The mixed-conducting class can be further divided into two subclasses: mixed protonic-electronic conductors, conducting protons and electrons or electron holes, and the so-called Triple Conducting Oxides (TCO) [7]—a group of oxides exhibiting high levels of conductivity of three charge carriers; protons, oxygen ions and electron holes (or electrons).

In recent years, PCCs have become increasingly more popular, mostly stemming from the large number of potential applications [7–17], of which gas sensors [13–21], hydrogen separation membranes [22–25], fuel cells [7,9,11,14–16,26] and electrolysers [13,17,27] are typical examples. However a constant evolution of the field has also led to new exciting developments such as electrochemical synthesis of ammonia [28,29], conversion of methane into aromatics in a membrane reactor [8] or thermo-electrochemical production of hydrogen from methane [10]. Especially, the two latter applications seem to be particularly interesting—the first enables efficient production of elementary petrochemicals from methane, while the second may be a new cost-efficient source of compressed hydrogen, useful for instance for hydrogen-fueled vehicles, exhibiting superior energy efficiency and lower greenhouse gas emissions compared to battery electric vehicles [30]. Recent efforts in the development of PCCs have also demonstrated that fuel cells based on a PCC, Proton Ceramic Fuel Cells (PCFCs), can operate with high efficiencies, while remaining cost-competitive in the medium temperature range of 300–600 °C compared to traditional Solid Oxide Fuel Cells (SOFCs) operating at 800–1000 °C [7,31,32]. The advances made in the field in recent years clearly show that PCCs deserve a broader attention from the scientific community to enable the implementation of these in future applications.

At this point it is important to note that each of the previously described devices is in fact a sort of electrochemical cell, which typically has a sandwich-type construction consisting of at least three layers; an anode, a cathode and an electrolyte. Additional layers that could be present but are not as often considered in lab-scale tests, are interconnects, sealing, current collectors or other functional layers. These devices typically operate at elevated temperatures, for example, 300–800 °C, making thermal mismatch between the layers a challenge. Differences in the thermal expansion coefficients between the layers will cause thermal stresses in the materials, leading to cracks, delamination and—in extreme cases—a total failure of the device. Moreover, chemical expansion—a result of defect formation caused by chemical reactions between the material and its surroundings—may also lead to similar degradation processes in the cells. In the case of Proton Conducting Ceramics, a chemical expansion is caused predominantly by hydration—a reaction in which protonic defects are formed in the oxide. Although a full assessment of the stability and durability of an electrochemical device requires a number of different input parameters, the thermal and chemical expansion coefficients provide a very strong indicator whether individual electrochemical cell components will be compatible. Thus, a prerequisite for any electrochemical device is to ensure that each component exhibits similar expansion coefficients under the operating conditions of interest.

Unfortunately, many of the works studying thermal mismatch in electrochemical cells are done for SOFCs in which the electrolytes conduct oxygen ions [33–37]. In the field of Proton Conducting Ceramics, the availability of necessary data is limited. Moreover, the literature presents thermal expansion coefficients determined by different measurement/simulation techniques and at different conditions, resulting in data sets with different physical meaning, which should not necessarily be compared. In addition, effects of chemical expansion can also obscure the picture. For the electrolytes, it is specifically a chemical expansion due to the hydration reaction, given in (1), causing an additional expansion of the crystal lattice upon the incorporation of water vapor. This can in turn lead to additional mismatch between the individual material components in an electrochemical device. However, chemical expansion is not restricted to the hydration reaction, as any chemical reaction can also cause an expansion or contraction of the lattice. For instance, for electrode materials often containing high concentrations of transition metals, chemical expansion upon oxidation/reduction should also be accounted for.

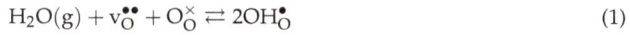

$$H_2O(g) + v_O^{\bullet\bullet} + O_O^{\times} \rightleftarrows 2OH_O^{\bullet} \tag{1}$$

This work aims to gather the existing data of thermal and chemical expansion of Proton Conducting Ceramics. The underlying theory of the phenomena will be explained and discussed to give a proper basis for analysis and comparison of expansion coefficients determined in various ways and conditions. Additionally, the essential guidelines for selecting matching materials and managing thermal mismatch will be provided.

2. Theory of Expansion of Solids

2.1. Basic Principles of Thermal Expansion

Thermal expansion of solids is a known and well-described phenomenon and its theoretical description can be found in many textbooks [2,38–43]. The most important parameter for this phenomenon is the thermal expansion coefficient, denoted as TEC, or alternatively as the coefficient of thermal expansion, CTE. In this work, to avoid misconceptions, we are consistently using α as the symbol for thermal expansion coefficients.

As there are many ways of determining α, distinct differences may arise in the evaluated parameters unless proper re-calculations are applied. For instance, if the solid expands anisotropically, or if measurements are conducted under conditions in which the lattice is contracting or expanding due to a chemical reaction, the determined values of α may differ significantly. Moreover, solid matter is, in itself, a very broad category and the expansion of different types of solids (e.g., glasses, single crystals and polycrystals) should be approached in a different way. Especially in polycrystalline ceramics, which are typically used in the context intended in this review, many phenomena can overlap leading to distorted values, overshadowing the true meaning of the obtained data. Therefore, extra caution should be taken when comparing different data sets—both self-measured and obtained from the literature. A thorough and complete understanding of the underlying theory is required and this will be outlined and discussed in the current section.

Thermal expansion can be considered from two separate perspectives; macroscopically and microscopically. The former is the expansion of a bulk material, being useful for technical applications, whereas the latter reflects the expansion of the crystal lattice due to atomic vibrations. This difference in perspective is an important distinction, and although they are highly correlated, they carry essentially different information of the material as we shall see in the following sections, starting with thermal expansion of bulk materials.

2.1.1. Thermal Expansion of Bulk Materials

A macroscopic bulk material will generally expand upon exposure to heat and along one selected direction, the mean coefficient of linear thermal expansion is phenomenologically defined as:

$$\langle \alpha_{L_{\text{mat}}} \rangle = \frac{L_2 - L_1}{L_1(T_2 - T_1)} = \frac{\Delta L}{L_1 \Delta T} \tag{2}$$

where L_1 and L_2 are material lengths at temperatures, T_1 and T_2, respectively. This relation can also be expressed in differential form:

$$\alpha_{L_{\text{mat}}} = \frac{dL}{L dT} \tag{3}$$

In this case the coefficient is defined for a given temperature and is thus no longer a mean value. For that reason, this parameter is referred to as the true coefficient, or simply the coefficient of thermal expansion [2]. Similar considerations can be done for two or three dimensions, for example, the volumetric expansion coefficient, $\langle \alpha_{V_{mat}} \rangle$, can be given in its mean form:

$$\langle \alpha_{V_{mat}} \rangle = \frac{V_2 - V_1}{V_1(T_2 - T_1)} = \frac{\Delta V}{V_1 \Delta T} \tag{4}$$

where V_1 and V_2 now refer to the material volume at T_1 and T_2, respectively. In differential form, we arrive at the true thermal coefficient of volumetric expansion:

$$\alpha_{V_{mat}} = \frac{dV_{mat}}{V_{mat}dT} \tag{5}$$

If the bulk material is isotropic then the relation between the true coefficients in (3) and (5) can be given as:

$$\alpha_{V_{mat}} = 3\alpha_{L_{mat}} \tag{6}$$

For anisotropic materials, where the bulk material expands differently in each direction, the volumetric expansion coefficient is expressed by the sum of the true linear expansion coefficients measured along three orthogonal directions, $\alpha_{1,mat}$, $\alpha_{2,mat}$ and $\alpha_{3,mat}$ [2]:

$$\alpha_{V_{mat}} = \alpha_{1,mat} + \alpha_{2,mat} + \alpha_{3,mat} \tag{7}$$

2.1.2. Crystal Lattice Thermal Expansion

Moving away from the macroscopic perspective, we now consider the material at an atomic level consisting of a periodic three-dimensional array of species (atoms, ions or molecules) making up the crystal lattice. At a finite temperature, each species is vibrating around its equilibrium position in a potential well. The shape of this well is given by the interatomic interactions and in the simplest approximation (harmonic), it is expressed by a simple parabolic function. While the harmonic approximation successfully predicts the heat capacity at constant volume of real solids at finite temperatures, it cannot account for the existence of thermal expansion, which is an anharmonic effect. In textbooks [2,40–42], the anharmonic well is typically given by the potential:

$$U(x) = cx^2 - gx^3 \tag{8}$$

where c and g are non-negative constants. Then, the time-averaged position, $\langle x \rangle$, is expressed by [2,40]:

$$\langle x \rangle = \alpha_{latt} T \tag{9}$$

The average position increases proportionally with respect to temperature T, constituting an expression for thermal expansion from a simple lattice model, where the proportionality constant, α_{latt}, is the thermal expansion coefficient. Using this basic thermodynamic model, one may extend these considerations to a crystal lattice of a given symmetry, where we now consider the expansion of the unit cell parameters, a, b and c, as a function of temperature:

$$\alpha_a = \frac{da}{adT}; \ \alpha_b = \frac{db}{bdT}; \ \alpha_c = \frac{dc}{cdT} \tag{10}$$

Additionally, the true volumetric thermal expansion can be defined, respectively, as follows:

$$\alpha_V = \frac{dV}{VdT} \tag{11}$$

where V is the unit cell volume. For a cubic crystal, where $a = b = c$ and the crystal properties are isotropic, only one thermal expansion coefficient is sufficient to describe the thermal expansion of the entire crystal lattice, analogously to an isotropic bulk material given in (6):

$$\alpha_V = 3\alpha_a \tag{12}$$

For crystals possessing lower symmetries with orthogonal principal axes, for example, orthorhombic or tetragonal, the relation instead becomes:

$$\alpha_V = \alpha_a + \alpha_b + \alpha_c \tag{13}$$

For anisotropic crystals exhibiting non-orthogonal principle axes, for example, monoclinic or triclinic, relations between the linear and volumetric thermal expansion coefficients can unfortunately not be expressed in such a simple manner, also requiring temperature dependencies of the unit cell angles [2,44,45].

2.1.3. Significance and Relation between Bulk and Lattice Expansion

In the two previous subsections, we have briefly described thermal expansion from a macroscopic and microscopic perspective. Although thermal expansion coefficients of a bulk polycrystalline sample (macroscopic) and a crystal lattice unit cell (microscopic) are often very similar, the values are not necessarily interchangeable, underlining the importance of keeping this distinction.

While bulk coefficients are more important with respect to device fabrication and applications, the thermal expansion of a unit cell provides fundamental characteristics of the crystal itself. The thermal expansion of a crystal depends on bond strength and collective lattice vibrations (phonons), linking it to many other physical properties of the material. For instance, the coefficient of thermal expansion along a specific direction can be defined phenomenologically in terms of uniaxial strain ε [46,47]:

$$\varepsilon = \frac{\Delta L}{L} = \alpha \Delta T \tag{14}$$

where L and T represent the length and temperature of the crystal, respectively. Other factors, such as heat capacity at constant pressure [2,48], Debye temperature [2,49], Grüneisen parameter [2] and anharmonic terms of lattice vibrations [50–52], are also correlated to thermal expansion. Thus, the thermal expansion of a crystal lattice provides a set of fundamental properties of a given system.

A problem emerges when one would like to extrapolate crystal lattice expansion parameters to bulk material properties. Although the values are correlated and sometimes similar, the lattice thermal expansion coefficients can only be extrapolated accurately for a macroscopic body in the case of an isotropic single crystal. However, Proton Conducting Ceramics (PCCs) are predominantly polycrystalline materials, exhibiting different microstructures and thermal expansion coefficients can for instance be affected by the grain size [53,54] and texturing effects [55,56]. It becomes even more complicated for a composite material, consisting of two or more different material phases.

Several theoretical models have been proposed to predict the bulk thermal expansion coefficient of a composite material (some of which are described in detail in Section 4.2). This is useful to model the thermal expansion of certain electrode materials, requiring high ionic and electronic conductivity. For instance, for mixed protonic-electronic conduction, a cermet consisting of Ni and $BaCe_{0.9-x}Zr_xY_{0.1}O_{3-\delta}$—providing electronic and protonic conductivity, respectively—can typically be used in a proton ceramic electrochemical cell [7,8,10]. These models can also be applied to polycrystalline single-phase materials composed of anisotropic grains with random crystal orientations and thermal expansion coefficients. Similarly for textured ceramics, where one or more crystal directions are preferred, such models can be implemented to predict the thermal expansion coefficient. However, extrapolating values from thermal expansion coefficients of a crystal lattice is not necessarily trivial and some caution must be taken upon choosing the appropriate model and assumptions.

2.2. Chemical Expansion in Proton Conducting Oxides

While thermal expansion in materials is related to a change in their inherent vibrational properties, chemical expansion arises from a change in the materials' chemical composition. We can divide chemical expansion into *stoichiometric* and *phase change* expansion processes, where the former reflects a continuous change in the lattice parameter with composition, whereas a phase change expansion typically induces an abrupt change in the lattice parameter due to a phase change or phase separation. An example of a stoichiometric expansion include the gradual expansion of $CeO_{2-\delta}$ with increasing oxygen nonstoichiometry, δ [57–60], while the oxidation of Ni to NiO [61,62] and the phase transition from the monoclinic to tetragonal polymorph of $LaNbO_4$ [63–65], constitute phase change expansions. We can envisage both processes for the cubic proton conducting oxide, $BaZr_{1-x}Y_xO_{3-\delta}$, where the lower valent Y^{3+}-cation is substitutionally replacing Zr^{4+}. Starting from x = 0, the volume will first increase with increasing Y-content in correspondence with Vegard's law [66], as more and more of the larger Y^{3+} cations replace the host (Zr^{4+}). If this concentration is increased further, we will eventually reach what is known as the solubility limit, where the volume of the system will abruptly change, as it becomes energetically more favorable for the system to separate into Y_2O_3 and Y-substituted $BaZrO_3$. This volume change corresponds to the start of a miscibility gap in the $BaZrO_3$-Y_2O_3 phase diagram, being an example of a phase change expansion. Further increasing the yttria content will then only result in a larger proportion of Y_2O_3 at the expense of the amount of Y-substituted $BaZrO_3$. The Y_2O_3 solubility limit for $BaZrO_3$ has typically been estimated to be around 30 mol% [67], such that all chemical expansion processes below this limit will be that of a stoichiometric expansion. As most commercial developments in the usage of PCCs typically use oxides with up to 10–20 mol% acceptor dopants, we will for simplicity primarily focus on stoichiometric expansion for the remaining part of the paper, unless specified otherwise.

Pure proton conductors, such as Y-doped $BaZrO_3$, will typically only chemically expand upon hydration, whereas mixed protonic electronic conductors, often consisting of one or more transition metal cations, may also expand due to reduction at higher temperatures. As such, both expansion processes are relevant for this review and we will start by considering chemical expansion due to hydration.

2.2.1. Chemical Expansion upon Hydration

A stoichiometric volumetric chemical expansion coefficient can for any defect, *i*, be expressed by

$$\beta_{i,V} = \frac{1}{\delta_i}\frac{(V-V_0)}{V_0} \tag{15}$$

where δ_i constitutes the defect concentration in mole fractions, whereas V and V_0 are the final and initial volume of the material, respectively. Thus, for the formation of a proton, $OH_O^{\bullet}$, the volumetric chemical expansion will be

$$\beta_{OH_O^{\bullet},V} = \frac{1}{[OH_O^{\bullet}]}\frac{(V-V_0)}{V_0} \tag{16}$$

For proton conductors such as acceptor doped $BaZrO_3$ and $BaCeO_3$, the concentration of protons, $[OH_O^{\bullet}]$, will typically be fixed by the acceptor concentration, $[Acc']$, under moist conditions at lower temperatures, i.e., $[OH_O^{\bullet}] = [Acc']$. The chemical expansion upon acceptor doping per mol acceptor (volume or linear) can then be expressed by

$$\beta_{doping,wet} = \beta_{Acc'} + \beta_{OH_O^{\bullet}} \tag{17}$$

where $\beta_{Acc'}$ represents the chemical expansion coefficient upon the introduction of an acceptor and $\beta_{OH_O^{\bullet}}$ is given by (16). While the formation of a hydroxide ion generally results in a volume contraction, due to its smaller size (ionic radius of 1.37 Å compared to 1.4 Å for O^{2-}) [68], the sign and magnitude of $\beta_{\text{Acc}'}$ depends on the relative size difference between the acceptor and the host ion. Reverting back to our example of acceptor doped BaZrO$_3$, $\beta_{\text{Acc}'}$ will be positive, i.e., the lattice expands, for larger trivalent cations such as Y^{3+} or Gd^{3+} with ionic radii of 0.9 and 0.938 Å (coordination VI), respectively, whereas the host cation Zr^{4+} has a radius of 0.72 Å [68]. On the other hand, $\beta_{\text{Acc}'}$ will be much smaller in magnitude, being close to zero or even negative for smaller cations such as Sc^{3+} (0.745 Å). In fact, we can, for the smaller cations, often set $\beta_{\text{Acc}'} \approx -\beta_{\text{OH}_O^\bullet}$, resulting in $\beta_{\text{doping,wet}} \approx 0$, such that the resulting lattice parameter (or volume) upon acceptor doping is very close to that of an undoped specimen. This is the case for Sc-doped BaZrO$_3$, where the lattice parameter changes minimally with Sc-content. While undoped BaZrO$_3$ has a reported lattice parameter of 4.193–4.194 Å [69–71], remarkably similar lattice parameters have been determined for Sc-doped BaZrO$_3$, being 4.194–4.195 Å with 6 mol% Sc [72] and 4.193–4.197 and 4.191 Å for 10 mol% and 19 mol%, respectively [73–75]. There is a remarkable difference in the lattice parameters for the same samples if we instead expose them to dry conditions or high temperatures. Under such conditions, the acceptors will instead be charge compensated by oxygen vacancies, i.e., $2[v_O^{\bullet\bullet}] = [\text{Acc}']$. The change in the lattice parameter (or volume) per mol acceptor is then expressed by

$$\beta_{\text{doping,dry}} = \beta_{\text{Acc}'} + \frac{1}{2}\beta_{v_O^{\bullet\bullet}} \tag{18}$$

where $\beta_{v_O^{\bullet\bullet}}$ is the chemical expansion upon the removal of an oxide ion, which induces a crystal lattice contraction due to the smaller size of $v_O^{\bullet\bullet}$ (1.16–1.18 Å estimated for CeO$_{2-\delta}$, BaZrO$_3$ and BaCeO$_3$ [59,76,77]) compared to O^{2-} (1.4 Å [68]), that is, $\beta_{v_O^{\bullet\bullet}} < 0$. The two expansion processes in (18) are generally competing processes, where the volume contraction from an oxygen vacancy overshadows the expansion caused by the acceptor substitution, i.e., $\beta_{v_O^{\bullet\bullet}} \ll \beta_{\text{Acc}'}$. Note that the factor $\frac{1}{2}$ in (18) simply stems from the imposed electroneutrality condition, where each acceptor is charge compensated by $\frac{1}{2} v_O^{\bullet\bullet}$.

We have now described the chemical expansion processes upon acceptor doping under moist and dry conditions, in (17) and (18), respectively. By subtraction of these two expressions, we arrive at the chemical expansion upon hydration of oxygen vacancies per mol water:

$$\beta_{\text{hydr}} = 2\beta_{\text{OH}_O^\bullet} - \beta_{v_O^{\bullet\bullet}} \tag{19}$$

Although, both defects induce crystal lattice contractions, the difference in their magnitudes causes the lattice to expand upon hydration, i.e., the lattice contraction of a single oxygen vacancy is larger than the contraction of two protons, $\beta_{v_O^{\bullet\bullet}} < 2\beta_{\text{OH}_O^\bullet}$. For most proton conducting oxides, β_{hydr} has been determined in the region of 0.05–0.2 [75,77–86]. This corresponds to a volume increase of +0.25–1.0% upon hydration for a proton concentration of 10 mol%.

Note that we have explicitly not accounted for any defect interactions, making our treatment strictly only applicable in a dilute limit. The interactions of defects can in principle also introduce changes to the volume, which may subsequently alter the chemical expansion coefficients. We can consider the following defect associations for an acceptor (Acc$'$) doped oxide:

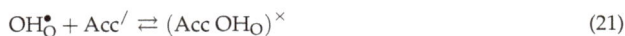

$$v_O^{\bullet\bullet} + \text{Acc}' \rightleftarrows (\text{Acc }v_O)^\bullet \tag{20}$$

$$\text{OH}_O^\bullet + \text{Acc}' \rightleftarrows (\text{Acc OH}_O)^\times \tag{21}$$

For the sake of simplicity, we will neglect larger defect clusters consisting of three or more defects, such as two acceptors and an oxygen vacancy, $(\text{Acc }v_O \text{ Acc})^\times$, although studies have indicated that such configurations may be present in In-doped BaCeO$_3$ and BaZrO$_3$ [87–89]. However, the concentration of such clusters will generally be lower than the corresponding concentrations of the defect associate pairs, $(\text{Acc }v_O)^\bullet$ and $(\text{Acc OH}_O)^\times$ [89], thus minimizing their effect on chemical expansion.

If we consider $(\text{Acc v}_O)^{\bullet}$ and $(\text{Acc OH}_O)^{\times}$ to dominate at lower temperatures, i.e., no defects are unassociated, then the chemical expansion upon hydration becomes

$$\beta_{\text{hydr,assoc}} = 2\beta_{(\text{Acc OH}_O)^{\times}} - \beta_{(\text{Acc v}_O)^{\bullet}} \tag{22}$$

where $\beta_{(\text{Acc OH}_O)^{\times}}$ and $\beta_{(\text{Acc v}_O)^{\bullet}}$ represent the chemical expansion coefficients for the defect associates $(\text{Acc OH}_O)^{\times}$ and $(\text{Acc v}_O)^{\bullet}$, respectively. Such defect associations will only affect the chemical expansion upon hydration if $\beta_{(\text{Acc OH}_O)^{\times}}$ and/or $\beta_{(\text{Acc v}_O)^{\bullet}}$ differ significantly from the corresponding chemical expansion coefficients for the isolated proton and oxygen vacancy, $\beta_{\text{OH}_O^{\bullet}}$ and $\beta_{\text{v}_O^{\bullet\bullet}}$, respectively. Although there is little data available specifically addressing the effects of different dopants on β_{hydr}, there appears to be a general tendency for the chemical expansion coefficient upon hydration to increase with increasing dopant size [83,85]. Furthermore, such defect associations may also impose a temperature dependence on β_{hydr}, as experiments may be conducted under conditions where the oxygen vacancies and/or protons are partially associated to the acceptors. In such a case, a sample would exhibit an apparent chemical expansion coefficient upon hydration, varying from $\beta_{\text{hydr,assoc}}$ at lower temperatures, where all defects are associated, to β_{hydr} at higher temperatures, where the protons and/or oxygen vacancies are completely unassociated. This also means that if β_{hydr} and $\beta_{\text{hydr,assoc}}$ are distinctly different, then measurements conducted with different $p_{\text{H}_2\text{O}}$ on the same sample will exhibit different volumetric expansions upon hydration. This can in other words be a source for small discrepancies in measured volume changes upon hydration.

2.2.2. Chemical Expansion upon Reduction

Although the chemical expansion in Proton Conducting Ceramics (PCCs) is generally due to hydration, compositions displaying mixed protonic electronic conductivity may also expand due to reduction at higher temperatures. This is mostly encountered for electrode type materials consisting of one or more transition metal cations, which display different oxidation states depending on the conditions that they are exposed to. To illustrate the chemical expansion upon reduction, we will consider the reduction of $\text{CeO}_{2-\delta}$, which has already been addressed in great detail in the literature [59,60,90–92]:

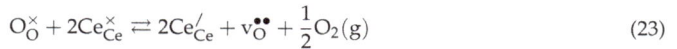

$$\text{O}_O^{\times} + 2\text{Ce}_{\text{Ce}}^{\times} \rightleftarrows 2\text{Ce}_{\text{Ce}}' + \text{v}_O^{\bullet\bullet} + \frac{1}{2}\text{O}_2(\text{g}) \tag{23}$$

This equilibrium clearly demonstrates that with decreasing p_{O_2} and/or increasing temperature, we form more oxygen vacancies, $\text{v}_O^{\bullet\bullet}$, charge compensated by reduced cerium, Ce_{Ce}'. The associated chemical expansion coefficient per mol $\text{v}_O^{\bullet\bullet}$ for (24) is given by:

$$\beta_{\text{red}} = \beta_{\text{v}_O^{\bullet\bullet}} + 2\beta_{\text{Ce}_{\text{Ce}}'} \tag{24}$$

where $\beta_{\text{v}_O^{\bullet\bullet}}$ and $\beta_{\text{Ce}_{\text{Ce}}'}$ represent the chemical expansion coefficients for the formation of an oxygen vacancy and the reduction of Ce^{4+} to Ce^{3+}, respectively. As discussed in the preceding Section 2.2.1, the formation of an oxygen vacancy results in a lattice contraction, that is, $\beta_{\text{v}_O^{\bullet\bullet}}$ is negative. $\beta_{\text{Ce}_{\text{Ce}}'}$, on the other hand, is positive, stemming from the increase in the cation radius going from Ce^{4+} to Ce^{3+}. In total, the expansion due to the reduction of cerium is larger in magnitude than the contraction upon forming an oxygen vacancy, i.e., $2\beta_{\text{Ce}_{\text{Ce}}'} > \left|\beta_{\text{v}_O^{\bullet\bullet}}\right|$, resulting in a net expansion of the crystal lattice upon reduction.

We can also extend the chemical expansion due to reduction to more complicated examples, such as $\text{La}_{1-x}\text{Sr}_x\text{Co}_y\text{Fe}_{1-y}\text{O}_{3-\delta}$ (LSCF), which has been used as a cathode in SOFCs [93–95] and PCFCs [96,97]. Although Fe and Co are both redox active elements, previous work has shown that they both can be treated as an indistinguishable elements, B, in the defect chemical analysis [98–100]. Both elements can be present in three oxidation states; B^{2+}, B^{3+} or B^{4+}. Assigning B^{3+} as the reference state, $\text{B}_B^{\times}$, the oxidation states +2 and +4 become effectively negative, B_B' and positive, $\text{B}_B^{\bullet}$, respectively, whereas Sr and La are consistently +2 and +3, respectively. The reduction of LSCF can then be considered to involve the reduction of B^{4+} to B^{3+}, accompanied by the formation of oxygen vacancies:

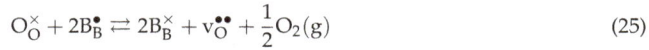

$$O_O^\times + 2B_B^\bullet \rightleftharpoons 2B_B^\times + v_O^{\bullet\bullet} + \frac{1}{2}O_2(g) \tag{25}$$

Further reduction of B^{3+} to B^{2+} is also possible, forming even more oxygen vacancies:

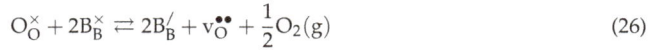

$$O_O^\times + 2B_B^\times \rightleftharpoons 2B_B' + v_O^{\bullet\bullet} + \frac{1}{2}O_2(g) \tag{26}$$

Note that by subtracting (26) with (25), we obtain a disproportionation reaction for B:

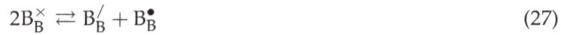

$$2B_B^\times \rightleftharpoons B_B' + B_B^\bullet \tag{27}$$

The complete chemical expansion coefficient upon reduction of LCSF per mol $v_O^{\bullet\bullet}$ then becomes:

$$\beta_{red,LSCF} = \beta_{v_O^{\bullet\bullet}} + 2\beta_{B_B'} - 2\beta_{B_B^\bullet} \tag{28}$$

where $\beta_{v_O^{\bullet\bullet}}$, $\beta_{B_B'}$ and $\beta_{B_B^\bullet}$ are the chemical expansion coefficients for the formation of $v_O^{\bullet\bullet}$, B_B' and $B_B^\bullet$, respectively. Thus, to completely describe the chemical expansion upon reduction of LCSF, we need the defect concentrations of all three species: $v_O^{\bullet\bullet}$, B_B' and $B_B^\bullet$. This requires a defect chemical model, where the thermodynamics of two of the equilibria, (25)–(27), are known. An electroneutrality condition is imposed:

$$[Sr_{La}'] + [B_B'] = [B_B^\bullet] + 2[v_O^{\bullet\bullet}] \tag{29}$$

along with a site restriction to conserve the total number of regular B and oxide ion sites

$$[B_B'] + [B_B^\bullet] + [B_B^\times] = 1 \tag{30}$$

$$[O_O^\times] + [v_O^{\bullet\bullet}] = 3 \tag{31}$$

Note that $[Sr_{La}']$ is considered to be constant in this model. The volume upon reduction of LCSF is thereby expressed by a combination of the previous set of equations, along with two of the equilibrium constants for reactions (25)–(27).

3. Methods of Determining Expansion Coefficients

We will treat the methods of determining expansion of materials by again considering the material from a *macroscopic* and *microscopic* perspective, adhering with Sections 2.2.1 and 2.2.2, respectively.

3.1. Bulk Thermal Expansion Coefficient Measurement

There are several methods available to measure the thermal expansion coefficient of a bulk specimen [2,101]. These can be divided into methods measuring either absolute or relative thermal expansion coefficients. While the absolute methods directly measure the thermal expansion coefficient, relative methods rely on data of other material samples, using them as a reference during the measurement. Examples of the former include interferometry, twin-telemicroscopy or scissors-type dilatometry, while push-rod dilatometry is an example of the latter and is by far the most popular technique amongst materials scientists and will also be the focus of this review [101]. Push-rod dilatometry has several advantages, being both a simple and reliable technique but it is also easily automated, which has resulted in a number of commercial measurement systems ready to be used in laboratories across the globe. In a push-rod dilatometer, the push rod is in direct contact with the sample, protruding a displacement resulting from the expansion of the specimen. As this is a relative measurement, it requires the use of a reference sample, such that the expansion of the reference is subtracted from that of the sample. This is typically done in succession, measuring the expansion of the specimen and the reference in separate steps under the same conditions, but some dilatometers can also do this simultaneously by measuring the sample and reference using two separate push-rods.

3.2. Measuring Thermal Expansion of Crystal Lattice

To measure the thermal expansion of a crystal lattice, any method determining the unit cell parameters as a function of temperature can be used. Although this can be achieved by several different techniques, common diffraction methods, such as powder X-ray Diffraction (XRD) [63,64,77,102] or Neutron Diffraction (ND) [103], are typically used. Resulting diffraction patterns are refined by the Rietveld [104] or Le Bail [105] method yielding unit cell parameters. Plotting their evolution against temperature and determining the slope or differentiating the unit cell parameter change with respect to temperature leads to the determination of the mean or true thermal expansion coefficients, respectively.

3.3. Chemical Expansion Coefficient Measurement

While thermal expansion coefficients can be obtained directly from dilatometry or XRD, the determination of chemical expansion coefficients requires the input of several different data sets measured under different conditions, along with detailed knowledge about the chemical composition of the sample. For instance, the expansion of a Proton Conducting Ceramic (PCC) can be measured under wet and dry conditions, where the volume difference between the two stems from chemical expansion upon hydration. By relating this difference in volume with the specific change in the proton concentration, the chemical expansion coefficient can be determined. The proton concentration is typically measured by thermogravimetric analysis (TG), where relative mass changes are evaluated at different temperatures and water vapor partial pressures.

For chemical expansion coefficients upon reduction, the procedure is very similar (weight or lattice changes as a function of p_{O_2}), although it may require a more complicated defect chemical model due to the presence of one or more transition metals, which can be present in several oxidation states (see Section 2.2.2, where LSCF is treated as an example). The volume changes due to reduction are then related to changes in the oxygen nonstoichiometry, which can be determined by TG, coloumetric [106,107] or iodometric titration [107–109].

3.4. Chemical Expansion Coefficients from Computational Modelling

Unlike experiments, where chemical expansion coefficients require the combination of different data sets and measurement techniques, computational approaches, such as density functional theory (DFT), have the advantage of having the chemical composition controlled entirely by the user. This means that volume changes with respect to small changes in the stoichiometry of the material can easily be obtained. To demonstrate the benefit of this, we will briefly discuss the chemical expansion upon reduction in $CeO_{2-\delta}$, where computational modelling was recently used to provide valuable insight to the underlying expansion mechanism [59]. The volume expansion upon reduction of $CeO_{2-\delta}$ was for a very long time debated in the literature [57,90,92]. It was hypothesized whether the expansion stemmed from an increase in the volume due to repulsive interactions between the oxygen vacancies or cerium cations, or from the increase in the cationic radius of cerium upon the reduction from Ce^{4+} to Ce^{3+}. By using first principles calculations, Marrocchelli et al. [59] were able to distinguish these processes, and for the first time systematically demonstrate that the chemical expansion upon reduction arises primarily from the volume expansion from reducing Ce^{4+} to Ce^{3+}, while the formation of oxygen vacancies actually results in a volume contraction. The superposition of these two effects results in an overall expansion of the lattice as the magnitude of volume expansion from cerium reduction overshadows the volume contraction induced by the oxygen vacancy formation.

3.5. Thermal Expansion Coefficients from Computational Modelling

Determining thermal expansion coefficients computationally can be achieved by calculating harmonic phonon spectra for a set of different volumes (lattice parameters), through the so-called quasi-harmonic approximation (QHA). This approximation is based on the assumption that the harmonic approximation (HA) holds at each and every volume and each atom/ion is considered to be a non-interacting harmonic oscillator. Phonon frequencies at each volume can then be determined by perturbation theory [110] or the finite displacements technique [111], providing vibrational thermodynamics, such as the Helmholtz free energy, F_{vib}

$$F_{\text{vib}} = \sum_i^{3N} \left[\frac{\hbar\omega(q,v)}{2} + k_B T \ln\left(1 - \exp\left(-\frac{\hbar\omega(q,v)}{k_B T}\right)\right) \right] \tag{32}$$

where $3N$ represents the number of degrees of freedom (3 directions × number of atoms, N), while q and v are the wave vector and band index, respectively. $\omega(q,v)$ is the phonon frequency at the specific q and v, while k_B is the Boltzmann constant.

By repetition for multiple volumes, the Gibbs free energy, $G(T,p)$, is determined by the unique minimum of the sum of the total electronic energy, $E^{\text{el}}(V)$, the Helmholtz vibrational energy, F_{vib} in (33) and pV work with respect to volume given by:

$$G(T,p) = \min_V [E^{\text{el}}(V) + F_{\text{vib}}(T;V) + pV] \tag{33}$$

By fitting the thermodynamic functions on the right hand side of (33) to an equation of state (EoS) for solids, such as the Rose-Vinet EoS [112] or the third order Birch-Murnaghan EoS [113], all thermodynamic parameters are obtained. More importantly, the fitting procedure also provides equilibrium volumes at finite temperatures, which subsequently yields volumetric thermal expansion coefficients.

The overall procedure to calculate thermal expansion coefficients computationally can be quite strenuous and time-consuming, requiring strict convergence criteria to obtain reliable forces. As an example, we will briefly indicate the computational cost involved in the QHA to calculate the thermal expansion coefficient for orthorhombic *Pnma* BaCeO$_3$. In the finite displacement method, which is generally the method of choice in the literature, ions in the crystal system are sequentially shifted slightly from their equilibrium positions (by ~0.01 Å). The resulting forces from such displacements are then used to calculate the phonon spectra at each volume. By taking advantage of crystal symmetry, the number of displacements can be effectively reduced and for a 2 × 2 × 2 BaCeO$_3$ supercell (160 atoms), only 17 displacements are required for each volume. The QHA typically requires at least 10 volume points for the EoS fitting to be scientifically sound, such that almost 200 static calculations are necessary to obtain the thermal expansion coefficient. While this is not too costly in itself, the situation worsens for lower symmetries, which can arise by the introduction of defects. For a single $OH_O^{\bullet}$ or $v_O^{\bullet\bullet}$, the symmetry reduces to the monoclinic space group *Pm*, requiring more than 500 displacements for each volume. Thus, the computational cost rises by a factor of 30 to calculate the thermal expansion coefficient for a defective supercell compared to that of the pristine bulk system. For a completely non-symmetric system, the force matrix is obtained by displacing every ion in all 6 directions, which amounts to a total of 10,000 displacements to calculate the thermal expansion coefficient for a supercell of 160 atoms. For this reason, there are very few computational studies using the QHA to calculate thermal expansion of low symmetry systems.

An alternative computational method to calculate thermal expansion is by running Molecular Dynamics (MD) simulations, where the trajectories of a set of interacting ions/atoms are calculated as a function of time by integrating their Newtonian equations of motion. The system is then heated up to the desired temperature by employing a computational thermostat, for example, the Nosé-Hoover algorithm [114–116], providing constant temperature conditions. The pressure can similarly be controlled by, for example, a Berendsen barostat [117]. Although the principles of MD simulations are relatively simple, they typically require long simulation times (several picoseconds) to ensure that the system is equilibrated and that there is enough statistics to determine specific properties, such as the volume. Thermal expansion coefficients are then determined by running a series of MD simulations at a range of desired temperatures and/or pressures. Because MD simulations are more or less unaffected by crystal symmetry, the method is particularly advantageous for low symmetry systems, where the QHA is very computationally expensive.

3.6. Assessment of the Influence of Heating Rate on the Thermal Expansion Coefficient—Experimental Details

While this work is intended to be a review, experiments have been conducted to assess the influence of the heating/cooling rates on the resulting expansion coefficients, as a large variety of heating rates are typically employed in the literature. To do so bulk samples of $BaZr_{0.7}Ce_{0.2}Y_{0.1}O_{3-\delta}$ (BZCY72) and ZrO_2 with 8 mol% Y_2O_3 (YSZ) have been studied by the means of dilatometry. The former is a typical example of a High Temperature Proton Conductor (HTPC), while YSZ is used as a reference material as it does not undergo chemical expansion due to hydration or reduction.

BZCY72 extruded rods were prepared by solid state reactive sintering using precursor oxides (ZrO_2 (Neo Performance Materials (AMR Ltd.)), CeO_2 (Neo Performance Materials (AMR Ltd.), $\geq$99.5%), Y_2O_3 (HJD Intl. 99.99+%) and barium carbonate (Alfa Aesar, tech grade, 99.6%)). The appropriate amounts of precursors were mixed with 2 wt.% NiO (Inco Grade F) and ball milled in deionized water for 24 h. The mixture was subsequently dried in air and sieved through a 40-mesh screen. The resulting powder was blended with water-soluble acrylic and a cellulosic-ether plasticizer and pelletized for wet-extrusion processing. Green rods were extruded through and encapsulated die set, dried and fired in air at 1600 °C for 6 h. The final sample had a cylindrical shape with a length of 25 mm and a diameter of 3 mm.

YSZ bars were prepared by sintering commercial YSZ powder (Tosoh). First the powders were pressed into pellets under a pressure of 12 MPa. The green bodies were sintered at 1400 °C for 2 h. The bars were cut from pellets using a diamond wire saw and were 22 mm long, 4 mm wide and had a thickness of 1 mm.

The linear expansion of the bulk samples was measured as a function of temperature using a Netzsch DIL 402 PC/4 dilatometer. To eliminate the influence of chemical expansion upon hydration, all measurements were performed in a flow of dry Ar. All measurements were carried out by pre-drying the samples at 1000 °C for 8 h, followed by cooling to 100 °C. A total of five heating/cooling rates were used for each sample: 1, 3, 5, 10 and 20 K min^{-1}.

4. Challenges in Assessing Measurement Data—Effects of Thermochemical Expansion, Non-Constant Thermal Expansion, Heating Rates and Porosity

In this section, we will address some of the challenges associated with measuring expansion data experimentally. First, we investigate and discuss the influence of chemical expansion on the determination of thermal expansion coefficients, before analyzing how to appropriately predict the thermal expansion coefficient of a composite material and lastly assessing how factors such as heating rate and porosity can affect thermal expansion.

4.1. Decoupling Thermal and Chemical Expansion Coefficients—Impossible?

As we have addressed in the preceding sections, a material can expand *thermally* and *chemically* and because these expansion processes typically occur simultaneously during a measurement, analyzing measurement data of expansion can often be non-trivial. It will typically result in a non-linearity in the volume (or lattice parameter) with respect to temperature, such that authors tend to report apparent expansion coefficients for different temperature intervals. To illustrate the difficulties in decoupling chemical and thermal expansion coefficients from a single data set, we will consider the thermochemical expansion of two typically used compositions; Y-doped $BaZrO_3$ (BZY) and $La_{1-x}Sr_xCo_{0.2}Fe_{0.8}O_{3-\delta}$

(LSCF). While BZY expands upon hydration under wet conditions at lower temperatures, LSCF expands upon reduction at high temperatures and reducing conditions (see Sections 2.2.1 and 2.2.2 for more details). Lastly, we assess and discuss the effects of having a non-constant thermal expansion coefficient, which has been demonstrated to affect the interpretation of chemical expansion due to hydration in BaCeO$_3$ [86].

4.1.1. Thermochemical Expansion upon Hydration—Case of Y-Doped BaZrO$_3$ (BZY)

To model the thermochemical expansion upon hydration for BZY, a constant linear thermal expansion coefficient of 8×10^{-6} K^{-1} is used, which is in correspondence with the majority of values reported for acceptor doped BaZrO$_3$ [77,80,118,119]. To account for the chemical expansion upon hydration, the proton and oxygen vacancy concentration is modelled based on standard hydration thermodynamics determined by Kreuer et al. on BaZr$_{0.9}$Y$_{0.1}$O$_{3-\delta}$; $\Delta_{hydr}H^\circ = -79.5$ kJ mol^{-1} and $\Delta_{hydr}S^\circ = -88.9$ J K^{-1} mol^{-1} [73], and these parameters are used for all BZY compositions. Note that $[Y'_{Zr}]$ is assumed to be constant and is only charge-compensated by protons and oxygen vacancies for this model, i.e., all other charge carriers are considered to have a negligible concentration. A constant linear chemical expansion coefficient upon hydration, β_{hydr}, of 0.042 (volumetric coefficient of 0.125 divided by 3) is used based on work by Bjørheim et al. [83]. This value is generally consistent with other chemical expansion coefficients reported for BZY [77,78,80,82,84].

Figure 1 shows the linear expansion of BZY as a function of temperature under wet conditions ($p_{H_2O} = 0.03$ atm) with 10 mol% and 20 mol% yttrium in (a), while (b) gives the corresponding thermal, chemical and the combined thermochemical expansion coefficients as a function of temperature. The expansion of BZY under dry conditions (no chemical expansion) is given for reference. As Figure 1a demonstrates, the linear expansion is the same for both compositions at higher temperatures (above 1000 °C), where the change in the lattice parameter with respect to temperature is only due to thermal expansion. Upon cooling from ~1000 °C, the lattice starts to chemically expand due to hydration until the acceptors are fully charge compensated by protons, which is achieved around 300 °C. The lattice expansion due to chemical expansion increases in magnitude with increasing dopant concentration, x. The resulting expansion coefficients, given in Figure 1b, show how the apparent expansion of the lattice is affected by chemical expansion. We see that the thermochemical expansion coefficient equals that of the thermal expansion coefficient (8×10^{-6} K^{-1}) at very low and high temperatures, while it becomes effectively diminished between 250 and 1200 °C. This suggests that extracting thermal expansion coefficients from an experimental data set conducted under humid conditions will very easily result in a lower apparent expansion coefficient. Although this becomes extreme in the case of large proton concentrations such as 20 mol%, it may be overlooked for samples with less protons or for measurements conducted under non-equilibrium conditions, where hydration is kinetically limited in certain temperature intervals. Overall, this demonstrates the need for additional measurements in order to completely decouple chemical and thermal expansion. This can be achieved by either measuring the expansion under dry conditions, or by relating the expansion to the level of hydration as a function of temperature by for example, thermogravimetry. Note that the latter is specifically needed to accurately determine the chemical expansion coefficient upon hydration.

Further, it is important to be aware that the overall thermochemical expansion of a PCC might also depend on the sample history due to kinetics. This is illustrated with HT-XRD data collected on a BZY20 sample [120]. Lattice changes were recorded in dry O$_2$, on one sample kept in a desiccator prior to the measurements and on the same sample previously hydrated. The results show a significant change in the dehydration temperature between the two sets of experiments, which can be attributed to slow hydration (or dehydration) kinetics. This aspect becomes even more important during cell fabrication when different layers of the electrochemical device are co-sintered.

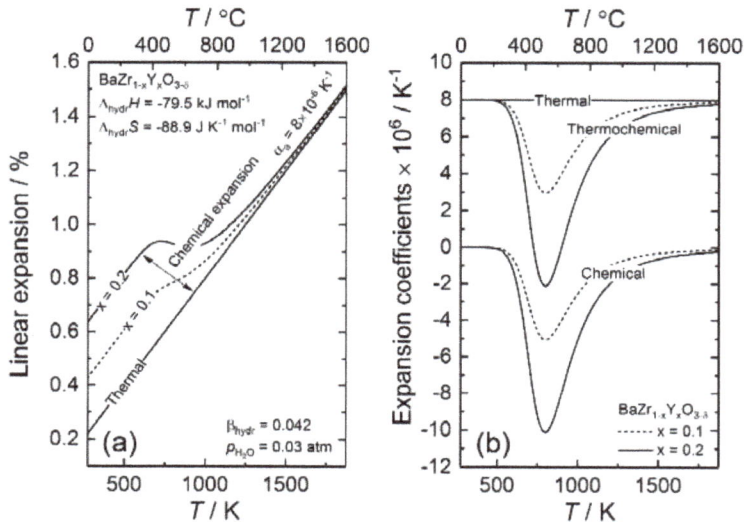

Figure 1. Linear expansion of $BaZr_{1-x}Y_xO_{3-\delta}$ (BZY) as a function of temperature under humid conditions (p_{H_2O} = 0.03 atm) for 10–20 mol% yttrium and β_{hydr} = 0.042 in (a) and the corresponding thermal, chemical and thermochemical expansion coefficients in (b). The expansion of BZY under dry conditions (no chemical expansion) is given for reference. The linear thermal expansion coefficient is consistently 8×10^{-6} K^{-1}.

4.1.2. Thermochemical Expansion upon Reduction—Case of $La_{1-x}Sr_xCo_{0.2}Fe_{0.8}O_{3-\delta}$ (LSCF)

Unlike the thermochemical expansion upon hydration, where the lattice expansion is effectively lowered, reduction causes a non-linearity in the lattice expansion with increasing temperature yielding a higher apparent expansion coefficient. This will be demonstrated in the case of LSCF, where a constant thermal expansion coefficient of 14×10^{-6} K^{-1} is used based on Bishop et al. [99]. This is generally consistent with the coefficients determined by other groups for the same and similar compositions [100,121–125]. The chemical expansion of the lattice is determined using the defect chemical model outlined in Section 2.2.2, along with oxidation (opposite of reduction) and disproportionation thermodynamics for the reactions (25) and (27) determined by Bishop et al. [98] on $La_{0.6}Sr_{0.4}Co_{0.2}Fe_{0.8}O_{3-\delta}$; ΔH_{ox} = −111 kJ mol^{-1}, ΔS_{ox} = −64.5 J K^{-1} mol^{-1}, ΔH_{dis} = +19.3 kJ mol^{-1} and ΔS_{dis} = −37.3 J K^{-1} mol^{-1}. The same thermodynamic parameters are also used to model the defect concentrations for x = 0.2. The linear chemical expansion coefficient upon reduction is consistently set to 0.031 [99,122].

Figure 2 shows the linear expansion of LSCF with increasing temperature in air (p_{O_2} = 0.21 atm) with 20 mol% and 40 mol% strontium in (a), while (b) gives the corresponding thermal, chemical and the combined thermochemical expansion coefficients as a function of temperature. Thermal expansion of LSCF (no chemical expansion) is given for reference. As the figure demonstrates, at lower temperatures (below 600 °C), the lattice expansion is primarily due to thermal expansion. As the temperature increases further, the lattice starts to chemically expand upon reduction, yielding a higher thermochemical expansion coefficient, which increases with increasing Sr-content. Chemical expansion constitutes around 20% and 30% of the total thermochemical expansion for 20 mol% and 40 mol% Sr, respectively, such that the linear expansion coefficient is ~19 × 10^{-6} and ~21 × 10^{-6} K^{-1} in the temperature region 800–1200 °C.

To decouple thermal and chemical expansion for LSCF, it is first and foremost clearly necessary to measure the lattice expansion at lower temperatures (below 600 °C) under oxidizing conditions, for example, air or O_2, to extract the thermal expansion coefficient. The deviation from linearity at higher temperatures can then be attributed to chemical expansion. To accurately determine the chemical expansion coefficient, a defect chemical model is required, where the oxygen nonstoichiometry is expressed as a function of temperature and p_{O_2}. As this requires at least two sets of thermodynamic parameters and some underlying assumptions (see Section 2.2.2 for details), extensive work involving coloumetric titration or thermogravimetry is often needed. Although, it may be tempting to rely on previous studies for the defect thermodynamics, small differences in the chemical composition and/or conditions can very easily result in discrepancies. It is therefore recommended that the defect chemical model be tested on a separate defect concentration data set to ensure its validity before curve-fitting the thermochemical expansion upon reduction.

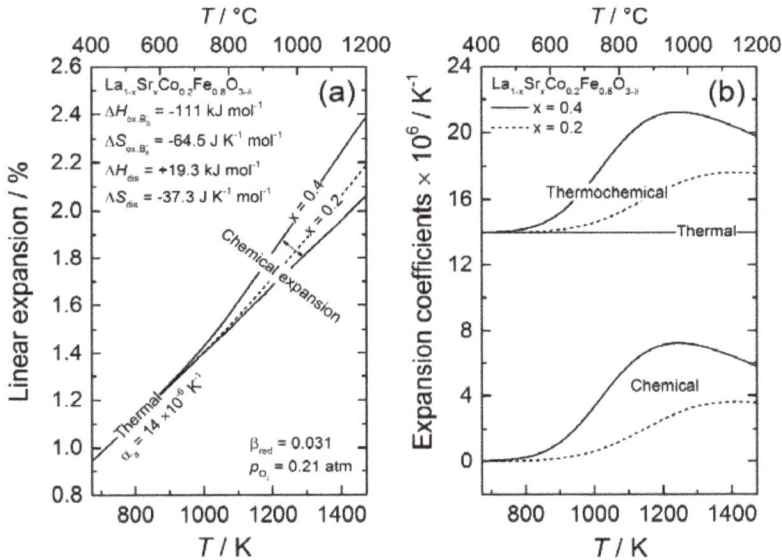

Figure 2. Linear expansion of $La_{1-x}Sr_xCo_{0.2}Fe_{0.8}O_{3-\delta}$ (LSCF) as a function of temperature in air ($p_{O_2} = 0.21$ atm) for 20 mol% and 40 mol% strontium and $\beta_{red} = 0.031$ in (**a**) and the corresponding thermal, chemical and thermochemical expansion coefficients in (**b**). Pure thermal expansion of LSCF (no chemical expansion) is given for reference. The linear thermal expansion coefficient is consistently set to 14×10^{-6} K^{-1}.

4.1.3. Non-Constant Thermal Expansion Coefficient

Thermal expansion coefficients are generally assumed to be constant, such that changes in the slope of the linear expansion with respect to temperature are considered to only reflect chemical expansion (see Figures 1 and 2). However, recent first principles calculations [86] have shown that this may not necessarily be the case and the thermal expansion coefficient may in fact change upon hydration. In the case of $BaCeO_3$, the thermal expansion coefficient was shown to increase from 10.6×10^{-6} to 12.2×10^{-6} K^{-1} for a dry and fully protonated specimen, respectfully [86]. This has also been seen experimentally for $BaCe_{0.8}Y_{0.2}O_{3-\delta}$ and $BaZr_{0.7}Ce_{0.2}Y_{0.1}O_{3-\delta}$, where the thermal expansion coefficients increase upon hydration from 9.9×10^{-6} and 12.7×10^{-6} K^{-1} to 11.1×10^{-6} and 15.0×10^{-6} K^{-1}, respectively [77,126]. This means that a sample will contract and expand to a greater extent with respect to temperature when exposed to wet conditions compared to that of dry conditions. The change in volume upon hydration thus stems from a superposition of two effects; chemical expansion and an increase in the thermal expansion coefficient. By assuming that the thermal expansion coefficient remains unchanged upon cooling, where the value of α is taken from the change in volume (or lattice parameter) at high temperatures, the chemical expansion coefficient will be underestimated. Such an underestimation will increase with decreasing proton concentration, yielding apparent expansion coefficients that are 27% and 55% lower for 20 mol% and 10 mol% protons in BaCeO₃ [86], respectively. A way of circumventing this underestimation can be to measure the chemical expansion isothermally, thus eliminating any effects of thermal expansion. Nonetheless, having a non-constant thermal expansion coefficient will definitely make it more complicated to separate the lattice expansion into thermal and chemical expansion contributions. For that reason, it can, for the sake of simplicity, be easier to assume a constant thermal expansion coefficient, although this can also lead to some discrepancies between different studies. This will not be focused on any further in this review due to the large number of other uncertainties that need to be accounted for, but the reader should be aware that a non-constant thermal expansion coefficient can be a potential source of error, although often neglected.

4.2. Thermal Expansion of a Composite

Up until now, all expansion phenomena have only been described for materials of a single phase. However, most electrochemical devices will more often than not employ a composite, consisting of two or more phases, for at least one of its electrodes. In such cases, it is important to understand how to describe the thermal expansion of the composite material and in this section, we will go through some of the basic models that exist in the literature. We will often consider a typical SOFC anode material NiO/YSZ and its PCFC analogue—NiO/BZY as examples.

It is important to note that all models presented here assume a composite sphere assemblage (CSA), where all particles of the composite are considered to be spherical and that any action on a particle is transmitted through a spherical interphase shell. Using this as a basis, the combined thermal expansion coefficient of a composite material, α_{comp}, can be predicted from the coefficients values of the individual phases of the composite. An example of this is through Kerner's model [127] with a random distribution of spherical grains of one phase (phase 1) in a continuous matrix of phase 2:

$$\alpha_{comp} = \alpha_1 v_1 + \alpha_2 v_2 + v_1 v_2 (\alpha_1 - \alpha_2) \frac{(K_1 - K_2)}{v_2 K_2 + v_1 K_1 + (3 K_1 K_2 / \mu_2)} \tag{34}$$

where α, v, μ and K are the average volumetric thermal expansion coefficient, volume fraction, shear modulus and bulk modulus, respectively. The equation can often be simplified, as in the case of NiO/YSZ and BZY/NiO, where the differences in bulk moduli are negligible, such that the second term becomes zero [128]:

$$\alpha_{comp} = \alpha_1 v_1 + \alpha_2 v_2 \tag{35}$$

Thus, for a sintered composite BZY/NiO, the expression for the thermal expansion becomes:

$$\alpha_{comp} = \alpha_{NiO} v_{NiO} + \alpha_{BZY} v_{BZY} \tag{36}$$

Upon reducing the sintered composite BZY/NiO, porosity is introduced into the cermet and the resulting amount of porosity will depend on the initial volume fraction of NiO used [128].

An important question to address before continuing is how and to what extent porosity affects thermal expansion. Although this has not been considered in great detail in the literature, some studies have attempted to answer this question. Coble and Kingery [129] assessed the influence of porosity on the physical properties of alumina using samples with 5% and 50% porosity (isolated pores). Their results demonstrated that the thermal expansion coefficient was unaffected by the amount of porosity, although increasing porosity decreased the strength and elastic modulus of the material. Similar results have also been reported on cordierite [130]. For porous ceramic composites, there are no systematic studies but we can use data from Mori et al. [131] on NiO/YSZ and Ni/YSZ composites to confirm that expansion coefficients are independent of the amount of porosity. NiO/YSZ composites prepared with 40 vol.% and 60 vol.% of NiO end up with 20% and 33% porosity upon reduction, respectively. Using (36) for NiO/YSZ, along with the following thermal expansion coefficients; $\alpha_{YSZ} = 10.3 \times 10^{-6}$ K^{-1} and $\alpha_{Ni} = 13 \times 10^{-6}$ K^{-1}, α_{comp} after reduction is 11.4 and 11.9 for the 40 vol.% and 60 vol.% of NiO composite, respectively, without explicitly accounting for porosity. This is in good agreement with the values determined experimentally by Mori et al. (11.3 and 12.2, respectively). Now, if we were to attempt to account for the porosity in (36) by adding an additional term for the volume fraction of the porosity and assuming $\alpha_{pore} = 0$, the calculated α_{comp} becomes 9.1 and 8.0, respectively, i.e., much lower than the measured values. Based on this, porosity can be considered to have a negligible effect on the thermal expansion, and also probably the chemical expansion, of a material.

A slightly different approach to Kerner's model was formulated by Schapery, where the effective thermal expansion coefficient of an isotropic composite employing extremum principles of thermoelasticity [132]:

$$\alpha_{comp} = \alpha_1 + (\alpha_2 - \alpha_1) \frac{\frac{1}{K_c} - \frac{1}{K_1}}{\frac{1}{K_2} - \frac{1}{K_1}} \tag{37}$$

Note that this equation will generally only provide upper and lower bounds of α_{comp}, as the bulk modulus of the composite, K_c, will generally be varied in the range of K_1 to K_2. In the case of $K_c = K_1$, then α_{comp} reduces to α_1, while $K_c = K_2$ similarly simplifies (37) to $\alpha_{comp} = \alpha_2$.

Another popular model for composites is Turner's model [132]:

$$\alpha_{comp} = \frac{\sum \alpha_i K_i v_i}{\sum K_i v_i} \tag{38}$$

Note that, upon applying the assumption that the difference between the bulk moduli of the different phases is negligible, (38) reduces to the same as (35).

Elomari et al. [132] compared all three models in their study of the composite Al/SiC/SiO$_2$ and found that their experimentally measured thermal expansion coefficient agreed well with the values predicted by the Schapery model at low temperature (<150 °C), while Kerner's model was better suited at higher temperatures (>400 °C). In general, all models will often provide a satisfactory way of describing the thermal expansion behavior of a composite, although knowing the bulk moduli of the individual phases is necessary before using the simplified expressions. Other models have also been proposed and are described in Reference [133].

One could argue that the geometry used for the above-mentioned models (spheres in a matrix) does not represent typical geometries of actual ceramic-ceramic or ceramic-metal composites, or that none of the models account for particle size. Despite these limitations, Kerner's model has been shown to successfully predict the thermal expansion of several composites, such as the YSZ/NiO and YSZ/Ni [131].

4.3. The Effect of Heating Rate on the Bulk Expansion Coefficients

This section summarizes and discusses the results of the measurements conducted in this work, where the influence of the heating rate on the thermal expansion coefficients has been explored (see Section 3.6 for details regarding the measurement procedure). To ensure that each measurement had identical starting conditions, we have only used the data obtained upon cooling.

The linear expansion as a function of temperature and the corresponding derivative, for YSZ is presented in Figure 3. Upon first inspection, the linear expansion curves appear to be similar, being almost completely parallel for all heating rates used, thus indicating an absence of changes in the expansion coefficients. This is similarly observed in the derivative, given in (b), where the true thermal expansion coefficient changes a little with respect to temperature but is unaffected by the change in heating rate. Overall, the thermal expansion coefficient is found to change continuously with respect to temperature, being in the range of 9–12 $\times$ 10^{-6} K^{-1}, which is in good correspondence with previous work [134].

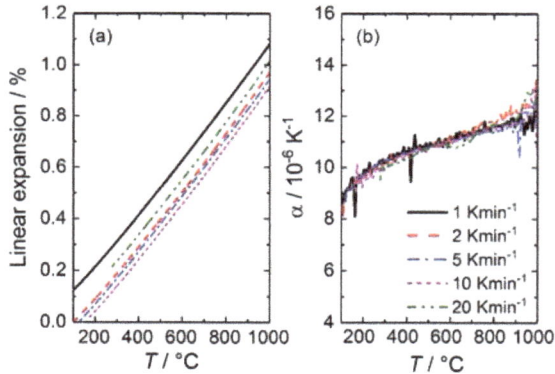

Figure 3. Linear expansion (**a**) and its corresponding first order derivative in (**b**), plotted as a function of temperature for YSZ for all cooling rates (1–20 K min^{-1}), dry Ar.

The results for BZCY72 are presented in a similar way to that of YSZ (Figure 3), and given in Figure 4. Again, we find that the linear expansion is similar for all heating rates. Note that 20 K min^{-1} has been excluded here, as it deviated considerably from the other heating rates used, which can possibly be attributed to incomplete drying at 1000 °C prior to cooling. Unlike YSZ, the true thermal expansion coefficient of BZCY72 does not change considerably with increasing temperature, although it exhibits two distinct temperature intervals with a constant TEC, above and below ~750 °C. At temperatures close to 1000 °C, a rapid increase of thermal expansion coefficient is observed. The temperature threshold of this phenomenon shifts to lower temperatures with increasing heating rate. This is only an experimental artifact, stemming from the transition from the isotherm (1000 °C) to the cooling segment during the measurement. An additional experiment (not shown) was also conducted with a drying segment at 1100 °C, whereupon the experimental artefact was in fact shifted to 1100 °C.

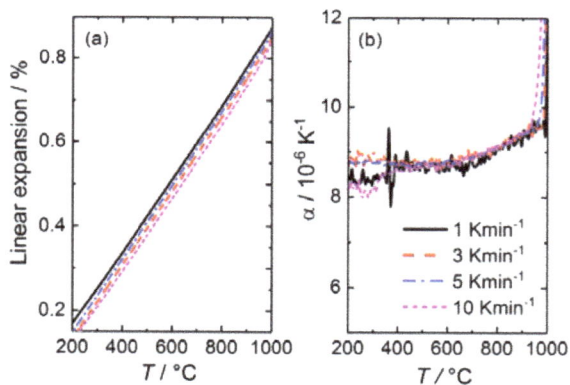

Figure 4. Linear expansion (**a**) and its corresponding first order derivative (**b**) plotted as a function of temperature for BZCY72 for all heating rates except 20 K min^{-1} (1–10 K min^{-1}), dry Ar.

Average thermal expansion coefficients of BZCY72 for the two respective temperature intervals, being above and below ~750 °C, are presented in Table 1 for all heating rates employed. It increases from 8.5–8.8 × 10^{-6} to around 9.3 × 10^{-6} K^{-1}. The distinct difference in the coefficients could be due to partial hydration of the sample at lower temperatures, stemming from small gas leakages allowing water vapor from ambient atmosphere to enter the measurement cell. A slower cooling rate will then give the sample more time to react (and possibly even equilibrate), such that it hydrates to a greater extent, ultimately affecting the measurement (see Section 2.2.1 for more details of thermochemical expansion upon hydration). This shows how important it is to perform thermal expansion measurements in controlled conditions when measuring samples that are prone to undergo chemical expansion.

Table 1. The mean thermal expansion coefficients, $<\alpha>$, determined for BZCY72 in dry Ar.

Cooling Rate K·min^{-1}	Temperature Range °C	$<\alpha>$ 10^{-6} K^{-1}
1	750–1000	9.3(2)
3	750–980	9.3(2)
5	750–960	9.3(2)
10	750–935	9.2(2)
1	100–750	8.5(3)
3	110–750	8.8(3)
5	120–750	8.7(2)
10	170–750	8.5(2)

5. Thermal and Chemical Expansion of Proton Conducting Ceramics (PCCs)

The preceding sections have provided the theoretical basis for thermal and chemical expansion, described experimental and computational methods to determine the corresponding coefficients, while Section 4 has gone through and discussed the challenges and influencing factors when measuring thermal and chemical expansion coefficients. In this section, we present and review literature values for the thermal and chemical expansion coefficients for a series of PCCs in a systematic manner. Following the constituents of an electrochemical cell, the coefficients will first be presented for electrolyte type materials, where protons are the dominating charge carriers, before moving onto materials employed as electrodes, which display both electronic and ionic conductivity.

5.1. Proton Conducting Solid Electrolytes

The proton conducting electrolyte type materials can be separated into a few major groups of oxides: perovskites and perovskite-related compounds, ABO_4 oxides and pyrochlore- and fluorite-related materials [3,22,80,135–138]. Although, this is not a complete list by any means, these materials cover the majority of research activity within the field of solid state high temperature proton conductors. We would like to remind the reader that only chemical expansion due to hydration is presented in this section as PCC materials will generally not exhibit p_{O_2}-induced expansion [139,140].

5.1.1. ABO_3 Perovskites and Perovskite-Related Materials

5.1.1.1. Barium Cerate and Barium Zirconate Based Perovskites

Out of all High Temperature Proton Conductors (HTPCs) that have been worked on over the last four decades, none have been studied in more detail than acceptor doped $BaZrO_3$ or $BaCeO_3$, or even a solid solution of the two. This is mostly due to their superior proton conductivity, reaching levels of ~0.02 S cm^{-1} at intermediate temperatures under humid conditions (400–600 °C) [73,137,141–144]. Due to their popularity, significant work has also been devoted to describing their thermal and chemical expansion behavior. Table 2 lists thermal expansion coefficients of barium zirconate and barium cerate based perovskites, along with some comments regarding the measurement technique and conditions. To avoid effects of chemical expansion upon hydration (see Section 2.2.1 for details), we primarily give coefficients determined under dry conditions and/or high temperatures, especially from studies reporting several coefficients for different temperature intervals and/or conditions. Although there are discrepancies between the values reported in the literature, Table 2 demonstrates a general trend, in which $<\alpha>$ increases with increasing yttrium content, x, for $BaZr_{1-x}Y_xO_{3-\delta}$ (BZY), varying from around ~8×10^{-6} K^{-1} in $BaZrO_3$ to around ~10×10^{-6} K^{-1} when acceptor doped by 20–30 mol% yttrium. For $BaCeO_3$, it is less clear due to the smaller data set available but the thermal expansion coefficients are generally higher in magnitude than BZY, typically being about 11×10^{-6} K^{-1} and appear to be independent of the amount of yttrium. The solid solutions of Y-doped BaZrO$_3$-BaCeO$_3$ similarly reflect this changeover in thermal expansion coefficients, with <α> decreasing systematically with increasing Zr-content. Overall, these trends offer a potential way of tailoring the thermal expansion coefficient of the electrolyte to avoid thermal mismatch with the other components in the electrochemical device. <α> can for instance be increased by either increasing the proportion of Ce in the BaZrO$_3$-BaCeO$_3$ solid solution, or by increasing the amount of yttrium dopant.

Table 2. Thermal expansion coefficients of barium zirconate and barium cerate based perovskites.

Compounds	<α> (10^{-6} K^{-1})	Comments and References
Undoped and Y-doped BaZrO$_3$		
BaZrO$_3$	4.7 [1] 6.87 [2] 7.13 [3] 7.5(2) [4] 8.02 [5] 7.89 [6] 8.65 [7]	[1] Neutron scattering, 100–300 K, gas composition not specified [145] [2] HT-XRD, estimated by $\alpha_V/3$ ($\alpha_V = 20.6 \times 10^{-6}$ K^{-1}), RT–600 °C, gas composition not specified [118] [3] Dilatometry, RT–1000 °C, reducing conditions—not specified [146] [4] HT-XRD, 298–1675 K, air [147] [5] HT-XRD, 30–1000 °C, dry Ar [78] [6] Dilatometry, temperature range used for <α> not clear, sample made by citrate method [148] [7] Dilatometry, temperature range used for <α> not clear, sample made by solid state synthesis [148]
BaZr$_{0.98}$Y$_{0.02}$O$_{3-\delta}$	8.47 [1]	[1] HT-XRD, 30–1000 °C, dry Ar [78]
BaZr$_{0.95}$Y$_{0.05}$O$_{3-\delta}$	8.62 [1] 8.75 [2]	[1] HT-XRD, 30–1000 °C, dry Ar [78] [2] HT-XRD, 373–1123 K, vacuum conditions (0.1 mbar) [77]
BaZr$_{0.9}$Y$_{0.1}$O$_{3-\delta}$	6.19 [1] 8.45 [2] 8.78 [3] 8.80 [4] 13 [5]	[1] HT-ND, 300–773 K, dry N$_2$ [149] [2] HT-XRD, 650–1000 °C, dry 4% H$_2$ in Ar [119] [3] HT-XRD, 30–1000 °C, dry Ar [78] [4] HT-XRD, 373–1123 K, vacuum conditions (0.1 mbar) [77] [5] Dilatometry, 30–1000 °C, air [150]
BaZr$_{0.85}$Y$_{0.15}$O$_{3-\delta}$	8 [1] 8.7 [2] 9.25 [3]	[1] Dilatometry, dry and wet (p_{H_2O} = 0.023 bar) give same result [80] [2] Dilatometry, Estimate made from thermal expansion LT (RT–320 °C), air [120] [3] HT-XRD, 30–1000 °C, dry Ar [78]
BaZr$_{0.8}$Y$_{0.2}$O$_{3-\delta}$	8.2 [1] 9.65 [2] 10.1 [3]	[1] HT-XRD, 100–900 °C, air [151] [2] HT-XRD, 373–1123 K, vacuum conditions (0.1 mbar) [77] [3] HT-XRD, 30–1000 °C, dry Ar [78]
BaZr$_{0.75}$Y$_{0.25}$O$_{3-\delta}$	10.2 [1]	[1] HT-XRD, 30–1000 °C, dry Ar [78]
BaZr$_{0.7}$Y$_{0.3}$O$_{3-\delta}$	10.0 [1]	[1] HT-XRD, 30–1000 °C, dry Ar [78]
Acceptor doped BaZrO$_3$ (other elements than Y)		
BaZr$_{0.8}$Sc$_{0.2}$O$_{3-\delta}$	9.4 [1]	[1] HT-XRD, Estimate made for lattice expansion at HT (600–1000 °C) under dry conditions [75]
BaZr$_{0.8}$Sm$_{0.2}$O$_{3-\delta}$	9.2 [1]	[1] HT-XRD, Estimate made for lattice expansion at HT (600–1000 °C) under dry conditions [75]
BaZr$_{0.8}$Eu$_{0.2}$O$_{3-\delta}$	9.9 [1]	[1] HT-XRD, Estimate made for lattice expansion at HT (600–1000 °C) under dry conditions [75]
BaZr$_{0.8}$Dy$_{0.2}$O$_{3-\delta}$	10.5 [1]	[1] HT-XRD, Estimate made for lattice expansion at HT (600–1000 °C) under dry conditions [75]
Undoped and acceptor doped BaCeO$_3$		
BaCeO$_3$	11.2 [1] 12.2 [2]	[1] Dilatometry, RT–1000 °C, reducing conditions—not specified [146] [2] DFT, QHA, 300 K, conditions not applicable [56]
BaCe$_{0.8}$Y$_{0.2}$O$_{3-\delta}$	11.6 [1] 9.94 [2] 10.5 [3] 11.6 [4]	[1] HT-XRD, 100–900 °C, air [151] [2] HT-XRD, 373–1123 K, vacuum conditions (0.1 mbar) [77] [3] HT-XRD, RT–1000 °C, dry O$_2$, estimate made based on changes in pseudo-cubic volume [152] [4] HT-ND, 30–800 °C, air, estimate made based on changes in pseudo-cubic volume [153]
Y:BaZrO$_3$-BaCeO$_3$ solid solutions (in order of increasing Zr-content)		
BaZr$_{0.1}$Ce$_{0.7}$Y$_{0.2}$O$_{3-\delta}$	11.3 [1] 12.1 [2]	[1] HT-XRD, 600–900 °C, air [151] [2] Dilatometry, 50–900 °C, air, <α> extracted for 50–600 °C where the expansion is linear [154]
BaZr$_{0.1}$Ce$_{0.7}$Y$_{0.1}$Yb$_{0.1}$O$_{3-\delta}$	9.07 [1] 11.60 [2]	[1] Dilatometry, 720–1100 °C, Ar, p_{H_2O} not indicated [155] [2] Dilatometry, 800–1100 °C, air, p_{H_2O} not indicated [155]
BaZr$_{0.2}$Ce$_{0.6}$Y$_{0.2}$O$_{3-\delta}$	11.3 [1]	[1] HT-XRD, 600–900 °C, air [151]
BaZr$_{0.3}$Ce$_{0.5}$Y$_{0.2}$O$_{3-\delta}$	10.8 [1]	[1] HT-XRD, 600–900 °C, air [151]
BaZr$_{0.4}$Ce$_{0.4}$Y$_{0.2}$O$_{3-\delta}$	10.9 [1]	[1] HT-XRD, 600–900 °C, air [151]
BaZr$_{0.5}$Ce$_{0.3}$Y$_{0.2}$O$_{3-\delta}$	9.3 [1]	[1] HT-XRD, 600–900 °C, air [151]
BaZr$_{0.6}$Ce$_{0.2}$Y$_{0.2}$O$_{3-\delta}$	9.1 [1]	[1] HT-XRD, 600–900 °C, air [151]
BaZr$_{0.7}$Ce$_{0.1}$Y$_{0.2}$O$_{3-\delta}$	8.4 [1]	[1] HT-XRD, 600–900 °C, air [151]
BaZr$_{0.7}$Ce$_{0.2}$Y$_{0.1}$O$_{3-\delta}$	12.7 [1] 10.2 [2] 9.3 [3]	[1] HT-ND, pre-hydrated sample, p_{H_2O} estimated to be 0.001 atm based on thermodynamic modelling [126] [2] HT-XRD, 650–1000 °C, dry 4% H$_2$ in Ar [119] [3] Dilatometry, this work, dry Ar, see Section 3.6 for more details
BaZr$_{0.8075}$Ce$_{0.0425}$Y$_{0.15}$O$_{3-\delta}$	8 [1]	[1] Dilatometry, dry and wet (p_{H_2O} = 0.023 bar) give same result [80]

Due to the large concentrations of protons in acceptor doped $BaZrO_3$ and $BaCeO_3$, typically being around 10–20 mol% when fully saturated at lower temperatures, chemical expansion upon hydration poses a serious challenge for device fabrication. A detailed description of the effect of chemical expansion on the linear expansion of Y-doped $BaZrO_3$ (BZY) has already been provided in Section 2.2.1, assuming a constant volumetric chemical expansion coefficient, β_{hydr}, of 0.125. However, this coefficient only represents a mean value of what can already be found in the literature and we will here attempt to collectively describe the changes in chemical expansion upon hydration in a more systematic manner, by first going through available data on BZY. To reduce uncertainties, only studies where the proton concentrations have been determined will be included, i.e., work assuming the proton concentration to be equal to the nominal doping concentration are neglected.

Figure 5 presents experimentally determined volumetric chemical expansion coefficients upon hydration, β_{hydr}, as a function of the proton concentration for BZY (top), along with the corresponding linear expansion of the lattice (bottom). The dashed lines represent a completely linear regime, where β_{hydr} is constant, for reference. For the data by Han et al. [78], volume changes and proton concentrations have been determined at two separate temperatures; 300 and 600 °C, thus giving two values for β_{hydr}. These data are represented by full and open squares, respectively. Although there is significant scattering in the values presented, the top panel of the figure clearly demonstrates a trend in β_{hydr}, which decreases with increasing proton concentration. At a proton concentration of ~0.07 mole fractions, β_{hydr} appears to reach a plateau, whereupon β_{hydr} stays constant (0.136) with increasing $[OH_O^\bullet]$, that is, the lattice expands linearly with increasing $[OH_O^\bullet]$ above 7 mol%. At lower proton concentrations, β_{hydr} is significantly higher, being 0.272 at ~0.02 mole fractions, i.e. double of the constant value at higher $[OH_O^\bullet]$, 0.136. This results in a non-linear lattice expansion below 7 mol% $OH_O^\bullet$, as is demonstrated in the bottom panel of the figure. In other words, the incorporation of the first few protons induces a larger chemical expansion of the lattice, compared to the protons formed when BZY is close to its level of saturation, where $[Y_{Zr}'] = [OH_O^\bullet]$. Although we have limited knowledge as to why there is a change in the chemical expansion coefficient upon hydration, we wish to propose a possible explanation for this change.

For BZY, protons can be formed by protonating and filling oxide ions, $O_O^\times$ and oxygen vacancies, $v_O^{\bullet\bullet}$, or by protonating and filling their associated counterparts, $(Y_{Zr}O_O)'$ and $(Y_{Zr}v_O)^\bullet$. These hydration mechanisms will in turn result in unassociated or associated protons, $OH_O^\bullet$ and $(Y_{Zr}OH_O)^\times$, respectively. Thus, we have to consider a total of six defects, $v_O^{\bullet\bullet}$, $O_O^\times$, $OH_O^\bullet$, $(Y_{Zr}v_O)^\bullet$, $(Y_{Zr}O_O)'$ and $(Y_{Zr}OH_O)^\times$, each of which may chemically expand or contract the lattice differently. At lower temperatures (<400 °C), where full hydration generally is achieved, all defects can be considered to be associated, such that hydration only proceeds by:

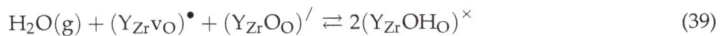

$$H_2O(g) + (Y_{Zr}v_O)^\bullet + (Y_{Zr}O_O)' \rightleftarrows 2(Y_{Zr}OH_O)^\times \tag{39}$$

As the temperature is raised, BZY will start to dehydrate but this will simultaneously be accompanied by the destabilization of the associated defects, $(Y_{Zr}v_O)^\bullet$ and $(Y_{Zr}OH_O)^\times$, causing them to break up into their constituent defects; $v_O^{\bullet\bullet}$, $OH_O^\bullet$ and Y_{Zr}'. The overall hydration mechanism at higher temperatures will thus be a combination of several defect equilibria. This change in hydration mechanisms could therefore be the reason for the change in β_{hydr} as the proton concentration increases.

Although we have now presented an apparent trend for β_{hydr} for BZY, there is far less available data to extend this analysis to other materials. For $BaCeO_3$, the reported values for β_{hydr} vary dramatically—between 0.030 and 0.151 [77,79], making it difficult to make any direct comparison to the work on BZY. In such cases, first principles calculations can be particularly valuable and Bjørheim et al. [84] were recently able to clearly demonstrate that β_{hydr} decreases in the order $BaZrO_3 \rightarrow BaSnO_3 \rightarrow BaCeO_3 \rightarrow SrZrO_3$, suggesting that orthorhombic perovskites ($BaCeO_3$ and $SrZrO_3$) will typically display less chemical expansion than their cubic counterparts ($BaZrO_3$ and $BaSnO_3$). Based on this, addition of cerium in the solid solution $BaZrO_3$—$BaCeO_3$ can be a useful way of reducing the chemical expansion upon hydration. Interestingly, this is in good correspondence with work by Kreuer [80], who showed that the chemical expansion of $BaZr_{0.8075}Ce_{0.0425}Y_{0.15}O_{3-\delta}$ is significantly smaller than that of $BaZr_{0.85}Y_{0.15}O_{3-\delta}$.

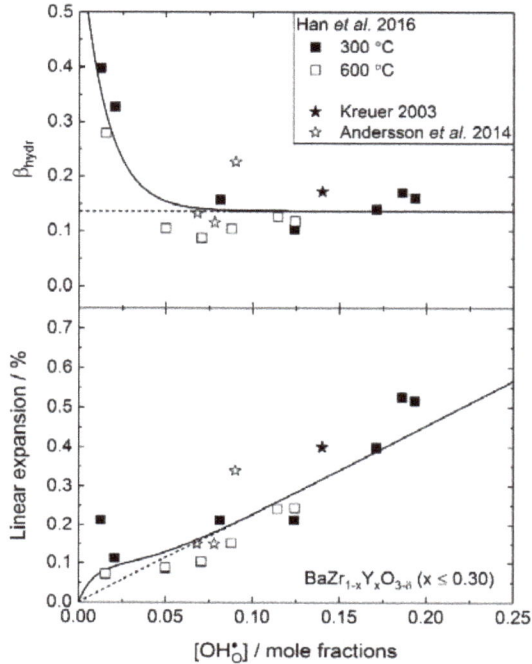

Figure 5. Experimentally determined chemical expansion coefficients upon hydration per mol H_2O, β_{hydr}, for $BaZr_{1-x}Y_xO_{3-\delta}$ as a function of the proton concentration, $[OH_O^\bullet]$ in the top panel, along with the corresponding linear expansion of the lattice in the bottom panel [77,78,80]. The solid lines are based on an exponential decay function fitted to all values of β_{hydr} (top panel) and the same function is used to express the linear expansion (bottom panel). The dashed lines represent the chemical expansion when β_{hydr} is constant (0.136).

5.1.1.2. Other Perovskites and Perovskite-Related Compounds

Although barium zirconates and cerates have received considerable interest in the literature over the last few decades, a number of other promising proton conducting perovskites and perovskite-related compounds have also been studied and these will be the focus of the remaining part of this sub-section.

A common denominator for all perovskites in general is that they are typically acceptor doped by a few to tens of mol%, which is required for them to exhibit protonic conductivity under humid conditions. This means that their lattice will chemically expand upon hydration at lower temperatures. However, due to the lack of literature data on chemical expansion due to hydration in general, being limited to Y-doped $BaZrO_3$ and $BaCeO_3$, the main emphasis will be to describe and review the thermal expansion coefficients of other perovskites. Table 3 lists thermal expansion coefficients of a number of different proton conducting perovskite oxides, along with total conductivities under moist conditions at 600 °C as an indication of their overall performance. Some additional perovskites have also been included for reference, for example, $BaUO_3$, although they have not been shown to conduct protons.

Table 3. Thermal expansion and total conductivity under wet conditions (600 °C) in proton conducting perovskites other than barium cerates and zirconates. The coefficients of some other perovskites are included for completeness, although these have not explicitly been shown to conduct protons. Their conductivities are therefore denoted by N.A.

Compounds	Conductivity at 600 °C (10^{-3} S cm^{-1})	$\langle\alpha\rangle$ (10^{-6} K^{-1})	Comments and References
Ca- and Sr-based perovskites			
CaZrO$_3$	0.3 in air [1]	$8.02 + 0.005 \times T$ [2] $8.92 + 0.003 \times T$ [3] T: temperature	[1] for 3 mol% Sc-doped specimen [156] [2] Dilatometry, dry (p_{H_2O} = 0.0004 atm) air, nonlinear thermal expansion [156] [3] Dilatometry, moist (p_{H_2O} = 0.025 atm) air, nonlinear thermal expansion [156]
SrZrO$_3$	0.8 in H$_2$ [1]	9.69 [2]	[1] for 5 mol% Y-doped specimen [157,158] [2] Dilatometry, gas conditions not specified, undoped sample [159]
SrCeO$_3$	2 in H$_2$ [1]	11.1 [2]	[1] for 5 mol% Yb-doped specimen [27,160] [2] Dilatometry and HT-XRD under reducing conditions and helium, respectively, undoped sample [161]
SrMoO$_3$	N.A.	7.98 [1]	[1] Dilatometry and HT-XRD under reducing conditions and helium, respectively, undoped sample [161]
SrHfO$_3$	N.A.	11.3 [1]	[1] Dilatometry and HT-XRD under reducing conditions and helium, respectively, undoped sample [161]
SrRuO$_3$	N.A.	10.3 [1]	[1] Dilatometry and HT-XRD under reducing conditions and helium, respectively, undoped sample [161]
Ba-based perovskites			
BaSnO$_3$	0.05 in air [1]	9.3 [2]	[1] $t_{OH} \approx 0.75$, doped with 10 mol% Lu, measured at 450 °C [162] [2] Dilatometry, gas composition not specified, undoped sample [163]
BaHfO$_3$	0.05 in air [1]	6.93 [2]	[1] t_{OH} unknown, doped with 10 mol% Y, measured at 450 °C [164] [2] Dilatometry, gas composition not specified, undoped sample [165]
BaThO$_3$	6 in Ar [1]	11.09 [2]	[1] $t_{OH} \approx 0.55$, doped with 5 mol% Nd [166] [2] Dilatometry, gas composition not specified, undoped sample [167]
BaTbO$_3$	6 in Ar [1]	9 [2]	[1] t_{OH} not specified, doped with 5 mol% Yb [166] [2] Estimated from HT-XRD data for temperature range 280–800 K, undoped sample [168]
BaMoO$_3$	N.A.	9.46 [1]	[1] Dilatometry and HT-XRD under reducing conditions and helium, respectively, undoped sample [161]
BaUO$_3$	N.A.	11.0 [1]	[1] Dilatometry and HT-XRD under reducing conditions and helium, respectively, undoped sample [161]
LaREO$_3$ series			
LaScO$_3$	4 in H$_2$ [1]	8 [2]	[1] $t_{OH} \approx 1$, doped with 10 mol% Sr [169] [2] Estimated from dilatometry plot, gas composition not specified, sample doped with 10 mol% Sr [170]
LaYO$_3$	0.8 in air [1]	11.1(1) [2] 11.4(3) [3]	[1] $t_{OH} \approx 1$, doped with 10 mol% Sr [171] [2] Dilatometry, dry (p_{H_2O} = 0.001 atm) air, sample doped with 10 mol% Sr [172] [3] Dilatometry, moist (p_{H_2O} = 0.025 atm) air, sample doped with 10 mol% Sr [172]
LaYbO$_3$	0.8 in air [1]	8.95 [2]	[1] $t_{OH} \approx 1$, doped with 10 mol% Sr [171] [2] Dilatometry, gas composition not specified, undoped sample [173]
LaLuO$_3$	0.7 in H$_2$ [1]	4.5 [2]	[1] $t_{OH} \approx 1$, doped with 10 mol% Sr [169] [2] Estimation of $\alpha_V/3$, undoped LaLuO$_3$ single crystal measured by HT XRD camera in temperature range from RT–1000 °C [174]
LaInO$_3$	3 in air [1]	6–9 [2]	[1] $t_{OH} \approx 0.9$, for 10 mol% Sr-doped specimen [171] [2] The range of true coefficient calculated from dilatometry in the temperature range 250–873 K, gas composition not specified, undoped sample [175]

As shown in Table 3, the A^{II}B^{IV}O$_3$ perovskites appear to exhibit similar thermal expansion coefficients to that of undoped and acceptor doped BaZrO$_3$ and BaCeO$_3$ (Table 2), with values ranging from 7 to 11 $\times$ 10^{-6} K^{-1}. For the LaREO$_3$ series on the other hand, the values scatter quite significantly, being especially low for LaLuO$_3$, and we can only speculate whether these large differences in TECs may stem from influences of chemical expansion due to hydration.

From the data presented in Tables 2 and 3, we see that there appears to be a general tendency for the cubic perovskites, for example, $BaZrO_3$, $BaSnO_3$ and $SrMoO_3$, to display lower thermal expansion coefficients compared to their orthorhombic counterparts. In an attempt to quantitatively describe this trend, we use the Goldschmidt tolerance factor, which effectively accounts for deviations from a cubic structure based on the ionic radii of the constituent ions [176]. Figure 6 presents experimentally determined linear thermal expansion coefficients for more than 20 Ba-based perovskite compositions as a function of the tolerance factor in (a), while (b) shows the same coefficients plotted versus the deviation from the perfect cubic structure, defined as the absolute value of (tolerance factor—1). Note that non Ba-based perovskites, such as $CaZrO_3$ and $LaScO_3$, have been disregarded to limit the number of discrepancies, although they appear to show some of the same tendencies. For $BaZr_{1-x}Y_xO_{3-\delta}$ (BZY) and solid solutions of $BaZr_{1-x}Y_xO_{3-\delta}$ and $BaCe_{1-x}Y_xO_{3-\delta}$ (BZY-BCY) from Table 2, where numerous expansion coefficients are reported, only values by Han et al. [78] and Lyagaeva et al. [151], respectively, are used, as their work cover a wide range of compositions. As demonstrated in the figure, the thermal expansion coefficient shows an apparent trend, increasing with decreasing tolerance factor, ranging from $\sim 7 \times 10^{-6}$ K^{-1} for a cubic structure (tolerance factor close to 1) to 11×10^{-6} K^{-1} for a tolerance factor of ~ 0.9. This trend can be rationalized on the basis of work by Zhao et al. [118,177], who previously described the thermal expansion of perovskites to be a combination of two effects: an elongation of the B-O bonds upon heating and a change in the octahedral tilting angle with temperature. The volumetric thermal expansion coefficient of an ABO_3 perovskite can then be expressed by

$$\alpha_V = \frac{3}{(B-O)}\frac{\partial(B-O)}{\partial T} + \frac{2}{\cos\phi}\frac{\partial(\cos\phi)}{\partial T} \tag{40}$$

where $(B-O)$ and ϕ represent the B-O bond length and octahedral tilting angle, respectively. For cubic perovskites, where the octahedral tilting angle is consistently zero, the equation reduces to only the first term and the expansion upon heating is then only related to changes in the B-O bond length. Perovskites of a lower symmetry, such as orthorhombic $SrZrO_3$, will therefore have a higher thermal expansion coefficient due to the inclusion of the second term of (40), accounting for the change in octahedral tilting with temperature. In fact, Zhao et al. [118] estimated that the change in the tilting angle accounts for 40–60% of the total thermal expansion of the tetragonal and orthorhombic polymorphs of $SrZrO_3$. Overall, the results shown in Figure 6 suggest that the tolerance factor may be used to estimate the thermal expansion coefficient of a perovskite, at least to a first approximation.

As a final part of this subsection on the ABO_3 perovskites, we will now briefly review some other structurally related proton conducting compounds, namely the complex perovskites and the brownmillerites. Both of these structural classes are related to one another, in which the complex perovskites consist of two or more ordered layers of the perovskite structure, for example, the double perovskite $A_2B'B''O_6$ consisting of alternating layers of $AB'O_3$ and $AB''O_3$, while the brownmillerites represent a heavily oxygen deficient perovskite with the general formula $ABO_{2.5}$. The brownmillerites can also be considered to be a subset of the complex perovskites, composed of alternating layers of ABO_3 and ABO_2, where the B-site cation is both tetrahedrally and octahedrally coordinated to the oxide ions within the same structure.

For the proton conducting complex perovskites, most of the work has been devoted to the $Ba_3Ca_{1+x}Nb_{2-x}O_{9-\delta}$ (BCN) series, where larger proportions of Ca introduce a higher oxygen nonstoichiometry and thus more protons upon hydration. The proton conductivity of these compounds peaks at ~ 3 mS cm^{-1} at around 400 °C [178], falling a little short of the performance of acceptor doped $BaZrO_3$ and $BaCeO_3$. The poorer proton conductivity is mainly attributed to the less favorable hydration thermodynamics of the BCN series compared to that of Y-doped $BaZrO_3$ and $BaCeO_3$ [73,179,180], destabilizing the protons at higher temperatures. Attempts at increasing the proton conductivity of the BCN series by Ce and K substitutions have therefore been tried, although this has only led to minor improvements [181,182]. The thermal expansion coefficients of BCN increase slightly with increasing x, ranging from 10×10^{-6} to $12 \times 10^{-6} \ K^{-1}$ [183]. These values are very similar to most regular ABO_3 perovskites, albeit in the higher end of the scale (see for example, Figure 6). Due to the fairly high proton concentrations of these materials, being up to 0.18 mol $OH_O^{\bullet}$ per mol $Ba_3Ca_{(1+x)/3}Nb_{(2-x)/3}O_{3-\delta}$, chemical expansion upon hydration also poses a significant challenge in terms of device fabrication. We estimate the chemical expansion coefficient upon hydration, β_{hydr}, to be consistently 0.14 per mol H_2O for the entire series based on lattice parameters determined by Schober and his coworkers [180,184,185], being comparable to that of BZY (cf. Figure 5).

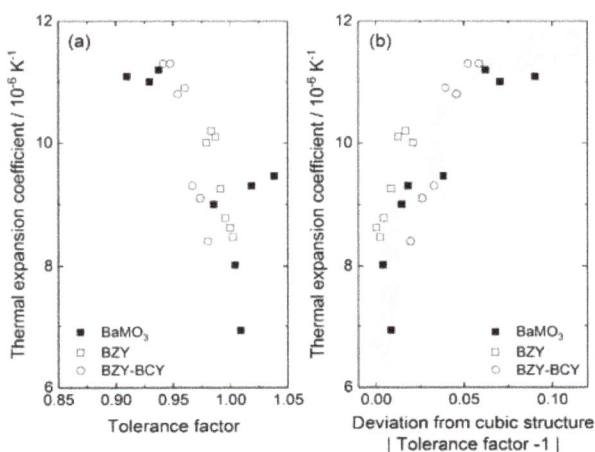

Figure 6. Linear thermal expansion coefficients for Ba-based perovskites from Tables 2 and 3 as a function of the Goldschmidt tolerance factor in (**a**), while (**b**) shows the same coefficients plotted versus the deviation from the perfect cubic structure, defined as the absolute value of (tolerance factor—1).

This similarity may be attributed to the fact that both BCN and BZY are both cubic perovskites, generally showing higher chemical expansion coefficients upon hydration compared to orthorhombic perovskites such as $BaCeO_3$ and $SrZrO_3$ [84].

Unlike the ABO_3 perovskites and the complex perovskites, which may at maximum incorporate ~20–30 mol% protons, the brownmillerites are capable of absorbing up to 1 mol H_2O per mol oxide due to their large oxygen deficiencies. As proton conductors, they perform reasonably well, with their proton conductivity peaking at around 4–6 mS cm^{-1} at 400–600 °C for Ti-substituted $Ba_2In_2O_5$ and $Ba_2Sc_2O_5$ [186–192]. However, due to the large concentrations of protons, these materials display significant chemical expansion, with corresponding volume expansions of more than +3% for pure $Ba_2In_2O_5$ upon hydration. Based on the work by Quarez et al. on $Ba_2In_{1.6}Ti_{0.4}O_{6-\delta}$ [189], β_{hydr} is estimated to be 0.067 per mol H_2O, that is, approximately half of the value determined for BZY (Figure 5). However, due to significantly higher proton concentrations in this system, the volume will chemically expand more than twice the amount compared to that of 20 mol% Y-doped $BaZrO_3$ (Figure 5 bottom panel). Thus, for the successful realization of brownmillerite type electrolytes in an electrochemical device, additional care must be taken to minimize chemical expansion. In terms of the thermal expansion of the brownmillerites, the coefficient has been shown to increase upon hydration, being $12.5 \times 10^{-6} \ K^{-1}$ for a completely dry specimen, increasing to $14.1 \times 10^{-6} \ K^{-1}$ when fully hydrated. Note that both of these values are larger than all other proton conducting perovskites that have been covered in this review (Tables 2 and 3) and the brownmillerites may as such represent an alternative electrolyte to Y-doped $BaZrO_3$, being more compatible with electrodes exhibiting higher thermal expansion coefficients.

5.1.2. ABO$_4$ Oxides

Several representatives of the ABO$_4$ oxides have been studied as potential candidates for proton conducting electrolytes. These materials are typically composed of a trivalent rare earth element on the A-site [65,193–199], while the B-site element is a pentavalent cation from group 5 [65,193–195] or 15 [196–199] of the periodic table. Unlike the perovskites (Section 5.1.1), these oxides show a much lower solubility for acceptor dopants, typically being limited to ~1–2 mol% on both the A- and B-site [65,194,200–205]. For this reason, chemical expansion upon hydration can be completely neglected for the ABO$_4$ oxides and we will hereafter only consider thermal expansion for this material system.

The ABO$_4$ oxides can adapt a number of different structures, where zircon ($I4_1/amd$), wolframite ($P2_1/c$), monazite ($P2_1/n$), scheelite ($I4_1/a$) and fergusonite ($I2/a$) are among the most commonly found in literature [206,207]. The thermal expansion coefficient of each of these structures has been shown to be distinctly different, with coefficients in the range of 4–6 × 10^{-6} K^{-1} for oxides exhibiting the wolframite structure, whereas the corresponding values for the scheelite- and zircon-type structured oxides will generally be above 10 × 10^{-6} K^{-1} [208]. Li et al. [208] demonstrated that these differences can be rationalized in terms of changes in the coordination number of the A-site cation and A-O bond length, while the BO$_4$ (scheelite and zircon) and BO$_6$ (wolframite) polyhedra remain rigid with respect to temperature and do therefore not contribute to thermal expansion. The increase in α upon going from the wolframite to the scheelite/zircon structure is therefore argued to stem from an increase in the coordination number of the A-site cations, varying from 6 to 8. Overall, their work suggests that the thermal expansion coefficient can be tailored by changing or partially substituting the A-site cation with a different element. Such a substitution will alter the A-O bond length and may also change its coordination number, both of which contribute to the thermal expansion of these oxides.

Another important aspect of the ABO$_4$ oxides is that they often undergo a structural phase transition, causing an abrupt change to the thermal expansion coefficient [209,210]. A notable example of this is demonstrated for LaNbO$_4$, which goes from a monoclinic fergusonite structure to a tetragonal scheelite-type structure at ~500 °C. The corresponding thermal expansion coefficient halves upon this transition, from 15–17 × 10^{-6} to 7–8 × 10^{-6} K^{-1}, respectively [63,103,200]. Such a large change in α poses a serious challenge in finding compatible materials for device fabrication, requiring additional care to minimize thermal mismatch.

Most of the work in the literature on the ABO$_4$ oxides has primarily been focused on acceptor doped LaNbO$_4$, following the promising proton conductivity of ~10^{-3} S cm^{-1} measured by Haugsrud and Norby in 2006 [65,194]. At the time, it was considered somewhat of a breakthrough, as LaNbO$_4$ represented a new oxide system posing as an alternative electrolyte material to the traditional Y-doped BaZrO$_3$ or BaCeO$_3$ perovskites. Following that paper, significant attempts went into optimizing the performance of the oxide by investigating alternative doping strategies [63,64,200,201,205,211–214], as well as different synthesis routes and conditions [202,215–219]. However, most of this work ultimately resulted in minimal improvements compared to the original work by Haugsrud and Norby. Although much less interest is currently being devoted to acceptor doped LaNbO$_4$, recent developments have shown great promise, reaching total conductivities of 2 × 10^{-3} S cm^{-1} at 720 °C [202], indicating that LaNbO$_4$ should not be written off as a potential proton conducting electrolyte yet. Table 4 lists thermal expansion coefficients for undoped and cation-substituted LaNbO$_4$, along with their total electrical conductivity at 800 °C in moist atmospheres and their structure type in specific temperature intervals for reference.

Table 4. Thermal expansion coefficients of undoped and cation-substituted LaNbO$_4$ compounds, along with their total conductivity at 800 °C under wet conditions and their structure type for reference. Note that acceptor dopants are generally on the La-site unless specified otherwise.

Compound	Conductivity at 800 °C (10^{-3} S cm^{-1})	Temp. Range (°C)	Structure Type	$<\alpha>$ (10^{-6} K^{-1})	Comments and References
Acceptor doped LaNbO$_4$ (maximum 2–3 mol% dopant)					
undoped	0.02 in H$_2$ [1]	RT to ~500	fergusonite	14 [2]/15.3 [3] 17.1 [4]/17.3 [5]	[1] t$_{OH}$ not specified [211] [2] Dilatometry, measured upon cooling, gas conditions not specified [200] [3] Dilatometry, measured upon cooling in moist Ar [103] [4] Expansion of the unit cell measured by ND [103] [5] Expansion of the unit cell measured or HT-XRD [63]
		~500–1000	scheelite	8.4 [2]/8.8 [3] 4.2 [4]/7.1 [5]	
Ca-doped	0.8 in O$_2$ [1] 0.7 in H$_2$ [1]	RT to ~500	fergusonite	12 [2]	[1] t$_{OH}$ ≈ 1; doped with 1 mol% Ca [65] [2] Dilatometry, measured upon cooling, gas conditions not specified, sample doped with 2 mol% Ca [200]
		~500–1000	scheelite	8.3 [2]	
Ba- or Sr-doped	0.4 in H$_2$ [1]	RT to ~500	fergusonite	14 [2]	[1] t$_{OH}$ ≈ 1; for 2 mol% Sr [211] [2] Dilatometry, averaged for all compositions, measurement upon cooling, gas conditions not specified [200]
		~500–1000	scheelite	8.6(5) [2]	
Mg-doped	2 in H$_2$ [1]	RT to 490	fergusonite	12 [2]/17.7 [3]	[1] at 720 °C, t$_{OH}$ not specified, for 2 mol% Mg [202] [2] Dilatometry, gas conditions not specified [102] [3] Expansion of the unit cell measured by HT-XRD [102]
		490–1000	scheelite	8 [2]/10.8 [3]	
Ti-doped (Nb-site)	0.1 in N$_2$ and O$_2$ [1]	RT to ~500	fergusonite	13 [2]	[1] t$_{OH}$ not specified, 2–3 mol% Ti [203] [2] Dilatometry, gas conditions not specified [203]
		~500–1000	scheelite	7 [2]	
Acceptor doped LaNb$_{1-x}$M$_x$O$_4$ (M = V, Ta or Sb)—M and Nb are isovalent					
LaNb$_{1-x}$V$_x$O$_4$					[1] t$_{OH}$ estimated to 0.7–0.9, sample acceptor doped with 2 mol% Sr [220] [2] Dilatometry, gas composition not specified [220]
x = 0.3	0.1 in O$_2$ [1]	RT–1000	scheelite	8 [2]	
LaNb$_{1-x}$Ta$_x$O$_4$					[1] t$_{OH}$ not specified, acceptor doped with 1 mol% Ca [221] [2] Expansion of the unit cell measured by HT-XRD, sample without acceptor dopants [63]
x = 0.2	0.3 in Ar [1]	RT–670	fergusonite	15.7(3) [2]	
		670–1000	scheelite	9.1(1) [2]	
x = 0.4	0.08 in Ar [1]	RT–800	fergusonite	14.0(2) [2]	
		800–1000	scheelite	11.5(1) [2]	
LaNb$_{1-x}$Sb$_x$O$_4$					[1] t$_{OH}$ ≈ 1, sample acceptor doped with 2 mol% Ca [214] [2] Dilatometry, dry Ar, measured on sample without acceptors [64]
x = 0.1	0.3 in air [1]	RT–351 351–1000	fergusonite scheelite	16.7(4) [2] 9.1(7) [2]	
x = 0.3	0.08 in air [1]	RT–1000	scheelite	8.1(3) [2]	
LaNb$_{1-x}$As$_x$O$_4$					[1] Dilatometry, dry Ar, measured on sample without acceptors [222]
x = 0.1	N/A	RT–340 340–1000	fergusonite scheelite	13.9(1) [1] 8.6(2) [1]	
x = 0.3	N/A	RT–1000	scheelite	8.3(1) [1]	

As already mentioned, undoped and acceptor doped LaNbO$_4$ undergoes a phase transition from fergusonite to scheelite at ~500 °C [103,210,223–225], whereupon the thermal expansion coefficient decreases by a factor of 2. As these two phases display significantly different thermal expansion coefficients, it is important to understand how various cation substitutions may influence this phase transition temperature. From the data given in Table 4, we see that acceptor doping (<3 mol%) appears to have no effect on the phase transition, while isovalent substitutions of ~30 mol% V, As and Sb on the Nb-site lower the transition temperature to below RT, such that these compositions consistently adopt the scheelite structure. Ta has the opposite effect and acts as fergusonite-stabilizer, shifting the phase transition temperature up to 800 °C for 40 mol% Ta. In terms of the thermal expansion coefficients, values for the specific polymorphs are generally quite similar, regardless of composition, being around 12–14 × 10^{-6} K^{-1} for the low temperature monoclinic fergusonite structure, while the corresponding coefficients for the high temperature tetragonal scheelite polymorph are in the range of 8–9 × 10^{-6} K^{-1}. Lastly, we want to underline that although isovalent substitutions on the Nb-site appear to be promising in that they can effectively change the phase transition temperature by several hundred degrees, the same compositions are often poorer proton conductors. For instance, LaNbO$_4$

substituted by either 40 mol% Ta or 30 mol% Sb both display total conductivities of 8×10^{-5} S cm^{-1}, which is 1 order of magnitude lower than corresponding compositions without Ta or Sb [65,194,202].

Although LaNbO$_4$ is the most well studied proton conducting oxide within the ABO$_4$ group, several other proton conductors with similar proton conductivities have also been identified, such as the rare-earth orthophosphates, REPO$_4$, LaAsO$_4$ and NdNbO$_4$ to mention a few. Table 5 summarizes the thermal expansion coefficients of these oxides, along with their total conductivity at 800 °C under wet conditions and their structure for reference. Note that in contrast to LaNbO$_4$, most of these oxides have the benefit of retaining their structure up to relatively high temperatures (>800 °C), that is, they do not undergo a phase transition. The table therefore only includes one structure for each of the oxides presented.

Table 5. Other ABO$_4$ compounds and their crystal structure, conductivity at 800 °C in moist conditions and thermal expansion coefficient.

Compounds	Conductivity at 800 °C (10^{-3} S cm^{-1})	Structure Type	$\langle\alpha\rangle$ (10^{-6} K^{-1})	Comments and References
Rare earth orthophosphates				
YPO$_4$	0.09 in O$_2$ [1]	xenotime	6.2 [2]/5.4 [3]	[1] $t_{OH} \approx 1$, doped with 3 mol% Ca [199] [2] Dilatometry, gas conditions not specified, undoped sample [226] [3] Expansion of the unit cell measured by HT-XRD [227]
LaPO$_4$	0.03 in air [1]	monazite	10.0 [2]	[1] $t_{OH} \approx 1$, doped with 4 mol% Sr [196] [2] Dilatometry, gas conditions not specified, undoped sample [228]
Lanthanide orthoniobates (La is excluded)				
NdNbO$_4$	0.5 in H$_2$ [1]	fergusonite	10.3 [2]	[1] $t_{OH} \approx 1$, doped with 1 mol% Ca [194] [2] Dilatometry, gas composition not specified, undoped sample [229]
ErNbO$_4$	0.04 in H$_2$ [1]	fergusonite	12.0 [2]	[1] $t_{OH} \approx 1$, doped with 1 mol% Ca [194] [2] Expansion of the unit cell measured by HT-XRD, undoped sample [230]
Lanthanide orthotantalates				
LaTaO$_4$	0.2 in H$_2$ [1]	layered perovskite	5.33 [2]	[1] $t_{OH} \approx 1$, doped with 1 mol% Ca [193] [2] Dilatometry, gas composition not specified, undoped sample [231]
NdTaO$_4$	0.07 in H$_2$ [1]	fergusonite	9.87 [2]	[1] $t_{OH} \approx 1$, doped with 1 mol% Ca [193] [2] Dilatometry, gas composition not specified, undoped sample [231]
GdTaO$_4$	0.03 in H$_2$ [1]	fergusonite	a 6.17 [2] b 12.12 [2] c 13.40 [2] average 10.6 [3]	[1] $t_{OH} \approx 0.7$, doped with 1 mol% Ca [193] [2] Dilatometry, gas conditions not specified, measured for an undoped single crystal along *a*, *b* and *c* axis [232] [3] Calculated for comparative reasons
Others				
LaAsO$_4$	0.03 in O$_2$ [1]	monazite	7.7 [2]	[1] $t_{OH} \approx 1$, doped with 1 mol% Sr [197] [2] Calculated from the chemical bond theory of dielectric description, undoped material [233]
LaVO$_4$	0.3 in O$_2$ [1]	monazite	6.1 [2]	[1] $t_{OH} \approx 0.15$, doped with 1 mol% Ca, at this temperature predominantly oxygen ion conductor, protonic conductivity dominates below 500 °C [234] [2] Calculated from the chemical bond theory of dielectric description, undoped material [235]

The table demonstrates distinct differences in the thermal expansion coefficients between the different structure types, generally increasing in the order xenotime → monazite → fergusonite. Further, for each structure type, the changes in the thermal expansion coefficients appear to scale inversely with the size of the A-site cation. For instance, for the fergusonite-structured oxides, α increases in the order NdTaO$_4$ → GdTaO$_4$ and in the order NdNbO$_4$ → ErNbO$_4$, i.e. with decreasing size of the lanthanide cation.

5.1.3. Pyrochlore- and Fluorite-Related Structures

Another family of oxides that has been investigated as possible alternatives to Y-doped $BaZrO_3$ and $BaCeO_3$ are the pyrochlores with the general formula $A_2B_2O_7$. These oxides typically consist of a trivalent rare-earth cation on the A-site, while the B-site cation is tetravalent. Similar to the ABO_4 oxides, the pyrochlores tend to display low proton concentrations, peaking at around 0.08 mol $OH_O^\bullet$ per mol oxide for Ca-doped $La_2Sn_2O_7$ [236], although this will be lower than 0.05 mol for most compositions [236–240]. Thus, for the majority of the systems presented here, chemical expansion upon hydration will be minimal and we will therefore only focus on thermal expansion for the $A_2B_2O_7$ oxides due to the limited data available. In terms of proton conduction, the best performing pyrochlore to date is acceptor doped $La_2Zr_2O_7$ with a maximum proton conductivity of ~10^{-3} S cm^{-1} [241,242]. Most studies on pyrochlore structured oxides are therefore devoted to $La_2Zr_2O_7$ or some derivative of this material. Table 6 summarizes linear thermal expansion coefficients (α) for various $A_2B_2O_7$ compounds in the 400 to 1000 °C range. It should be noted that $La_2Ce_2O_7$, unlike the other oxides in Table 6, is not a pyrochlore and instead adopts a disordered fluorite type structure, where the cations are randomly distributed in the cation sublattice [243–246].

Overall, the changes in the thermal expansion coefficients for the $A_2B_2O_7$ oxides given in Table 6 are small, with the majority of them having a coefficient of ~9×10^{-6} K^{-1}. Further, we find that there appears to be a linear correlation between the thermal expansion coefficient and the cation radius ratio, $r_{A^{3+}}/r_{B^{4+}}$, as shown in Figure 7. The figure indicates that the thermal expansion coefficient increases with decreasing $r_{A^{3+}}/r_{B^{4+}}$, ranging from 7.9×10^{-6} to 11.4×10^{-6} K^{-1} as the cation radius ratio decreases from 1.7 to 1.3, respectively. On the basis of this, we suggest that adjustments to the thermal expansion coefficients of the $A_2B_2O_7$ oxides can be achieved by partially substituting smaller or larger cations to alter the cation radius ratio, $r_{A^{3+}}/r_{B^{4+}}$.

Table 6. Linear thermal expansion coefficients (α) for various $A_2B_2O_7$ oxides in the temperature range 400 to 1000 °C.

Composition	$\langle\alpha\rangle$ 10^{-6} K^{-1}	Reference
$La_2Ce_2O_7$	10–12	[243]
$La_2Zr_2O_7$	8.7–9.6	[247]
$La_{1.8}Nd_{0.2}Zr_2O_7$	8.75–9.25	[247]
$La_{1.6}Nd_{0.4}Zr_2O_7$	8.25–9.25	[247]
$La_{1.4}Nd_{0.6}Zr_2O_7$	9–9.25	[247]
$La_{1.2}Nd_{0.8}Zr_2O_7$	9.2–9.75	[247]
$La_2Zr_2O_7$	8.5–9	[248]
$Nd_2Zr_2O_7$	9.2–9.5	[248]
$Eu_2Zr_2O_7$	10.5	[248]
$Gd_2Zr_2O_7$	9–10.5	[248]
$La_{1.4}Nd_{0.6}Zr_2O_7$	8.5–8.7	[248]
$La_{1.4}Eu_{0.6}Zr_2O_7$	9–9.2	[248]
$La_{1.4}Gd_{0.6}Zr_2O_7$	9.2–9.3	[248]
$La_{1.7}Dy_{0.3}Zr_2O_7$	8.5–9	[248]

Other related oxides that have recently emerged as promising proton conducting ceramics are the fluorite-related lanthanum tungstates (LWO), with the general formula $La_{28-x}W_{4+x}O_{54+\delta}$. This class of oxides has been shown to only be single phase for approximate La/W ratios between 5.3 and 5.7, correspondingly making them inherently oxygen deficient. Undoped samples display proton conductivities in the order of 3–5 $\times$ 10^{-3} S cm^{-1} [249], whereas acceptor doping on the B-site increases the ionic conductivity to ~0.01 S cm^{-1} at 800 °C [250] but it is not clear whether this increase stems from oxide ions or protons [251].

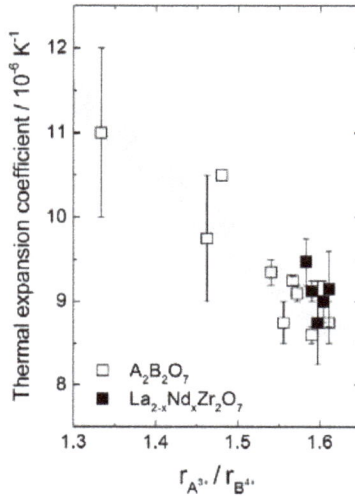

Figure 7. Linear thermal expansion coefficients of the $A_2B_2O_7$ oxides as a function of their cation radii ratio. The coefficients have been taken from Table 6 and the uncertainties represent the respective ranges in values given.

In terms of expansion, no quantitative study has determined the chemical expansion coefficient upon hydration for the lanthanum tungstates. However, we can estimate the coefficient using high temperature neutron diffraction (HT-ND) data on LWO56 (LWO where La/W = 5.6) by Magrasó et al. [252], along with corresponding proton concentrations from TG data by Hancke et al. [253]. The lattice parameter for LWO56 has been shown to increase by 0.14% upon hydration [252], while the proton concentration reaches an upper limit of 0.96 mol per mol oxide, that is, a hydration level where 66% of all oxygen vacancies (0.48 out of 0.73) in the structure hydrate [253]. Combining these data results in a chemical expansion coefficient due to hydration, β_{hydr}, of 0.0088 per mol oxygen vacancy, which is much lower than that of the acceptor doped perovskites (see Section 5.1.1). Note that this value is only a lower limit for β_{hydr}, as the HT-ND data by Magrasó et al. [252] may not be entirely at equilibrium at lower temperatures ($\leq$400 °C). This is based on the shorter equilibration times used for their measurements (1–2 h), whereas Hancke et al. [253] equilibrated their samples up to ~3.5 h at each temperature for their thermogravimetric analysis. Nonetheless, the data clearly indicates a much lower chemical expansion due to hydration for the lanthanum tungstates compared to Y-doped BaZrO₃ and BaCeO₃.

Thermal expansion coefficients of the lanthanum tungstates have been shown to be in the order of 11×10^{-6} K^{-1}, being slightly higher (~11.5×10^{-6} K^{-1}) under dry conditions [252,254–256]. These small differences could be related to small contributions from chemical expansion upon hydration, which reduce the expansion coefficients (see for example, Figure 1). Insignificant changes to α are observed upon doping on both the A- and B-site. Moreover, the coefficients are also independent of p_{O_2}, reflecting that LWO does not undergo a redox reaction, being primarily an ionic conductor [252].

5.2. Mixed Ionic Electronic Conductors

Moving away from the oxides that are predominantly proton conductors, this section focuses on Mixed Ionic Electronic Conductors (MIECs), which exhibit both electronic and ionic conductivity and are thus useful as electrodes or gas separation membranes. First, a list of MIEC electrodes (single-phase and composite) studied on proton conducting ceramics (PCCs) is provided, before summarizing the thermal and chemical expansion coefficients for the most widely studied MIECs.

5.2.1. Electrode Materials Used on PCCs

The topic of mixed protonic-electronic conductors and triple conducting oxides in high temperature electrochemical cells is much younger than the field of proton conducting oxides itself. First attempts to develop suitable electrodes for electrochemical cells operating with high temperature proton conductors as an electrolyte focused on, either mimicking, or even using the exact same materials as those used for SOFCs. Although this worked fairly well in the beginning, these electrodes generally do not conduct protons, restricting the electrode reactions to the triple phase boundaries—confined spatial sites where the electrode, electrolyte and gas are in contact. Thus, most of the current efforts in the PCFC/PCEC community are devoted to developing electrode materials that also conduct protons, ensuring that the whole electrode surface is electrochemically active [7,26,135,257]. Merkle et al. [258] estimated that a proton conductivity in the order of 10^{-5} S cm^{-1} should be enough to make large parts of the electrode active. Although this may seem simple to achieve, it is experimentally difficult to verify whether these material systems are actually mixed protonic electronic conductors, or just mixed oxide ion and electron conductors and this remains a source for discussion and controversy in the literature, being for instance the topic of a recent paper by Téllez et al. [259]. Much of this controversy is related to the fact that the proton concentrations appear to be very low, peaking in the region of 1–3 mol% for $BaGd_{0.8}La_{0.2}Co_2O_{6-\delta}$ and $BaCo_{0.4}Fe_{0.4}Zr_{0.2}O_{3-\delta}$ [260,261]. Our intention is not to contribute in this discussion but to rather, highlight some of the ongoing discussion regarding proton conducting electrode materials.

Table 7 summarizes various electrodes (non-exhaustive list) that have been studied on proton-conducting ceramic electrolytes, separated by their specific application, for example, as anodes for Proton Ceramic Electrolyser Cells (PCECs). As the table demonstrates, a significant number of materials have been investigated over the years, underlining the immense workload involved. As it is virtually impossible to go through each of these studies individually, we will only describe some of the general trends for these materials, treating a few of the more successful electrode systems as examples.

Table 7. List of materials used as electrodes in electrochemical cells based on proton conducting ceramics.

Type	Composition	Reference	Application
	Materials with perovskite structure		
	$BaCoO_{3-\delta}$		
	$Ba_{0.5}Sr_{0.5}CoO_{3-\delta}$	[150]	
	$Ba_{0.5}La_{0.5}CoO_{3-\delta}$	[150,262]	
	$Ba_{0.5}Sr_{0.5}Co_{0.8}Fe_{0.2}O_{3-\delta}$	[263–265]	
	$BaCo_{0.4}Fe_{0.4}Zr_{0.2}O_{3-\delta}$	[266]	
	$BaCo_{0.4}Fe_{0.4}Zr_{0.1}Y_{0.1}O_{3-\delta}$	[7]	
	$BaCo_{0.4}Fe_{0.4}Zr_{0.2}O_{3-\delta}$	[261]	
	$BaCe_xFe_{1-x}O_{3-\delta}$	[267]	
	$Ba_{0.5}Sr_{0.5}Fe_{0.8}Zn_{0.2}O_{3-\delta}$	[268,269]	
Single phase	$BaPr_{0.8}Gd_{0.2}O_{3-\delta}$	[270]	PCFC cathode
	$BaCe_{0.4}Pr_{0.4}Y_{0.2}O_{3-\delta}$	[271]	
	$BaCe_{0.5}Bi_{0.5}O_{3-\delta}$	[272]	
	$SrCoO_{3-\delta}$	[150]	
	$Sr_{0.5}Sm_{0.5}CoO_{3-\delta}$	[273]	
	$Sr_{0.5}La_{0.5}CoO_{3-\delta}$	[150]	
	$LaCoO_{3-\delta}$		
	$La_{0.6}Sr_{0.4}Co_{0.2}Fe_{0.8}O_{3-\delta}$	[264,265]	
	$La_{0.58}Sr_{0.4}Co_{0.2}Fe_{0.8}O_{3-\delta}$	[96]	
	$La_{0.8}Sr_{0.2}MnO_{3-\delta}$	[265]	
	$La_{0.6}Sr_{0.4}Fe_{0.8}Ni_{0.2}O_{3-\delta}$	[265,274]	
	$SrFe_{0.75}Mo_{0.25}O_{3-\delta}$	[16]	PCFC cathode/PCFC anode
	$Sr_{0.5}Sm_{0.5}CoO_{3-\delta}$	[17]	PCEC anode
	$(La_{0.75}Sr_{0.25})_{0.97}Cr_{0.5}Mn_{0.5}O_{3-\delta}$	[275]	

Table 7. *Cont.*

Type	Composition	Reference	Application
	Layered perovskites and perovskite-related structures		
	$PrBaCo_2O_{6-\delta}$	[264]	
	$PrBa_{0.5}Sr_{0.5}Co_{1.5}Fe_{0.5}O_{6-\delta}$	[26]	
	$SmBaCo_2O_{6-\delta}$	[276]	
	$NdBa_{0.5}Sr_{0.5}Co_{1.5}Fe_{0.5}O_{6-\delta}$	[277]	
	$GdBaCo_2O_{6-\delta}$	[278,279]	
	$LaBaCo_2O_{6-\delta}$	[262]	
	$PrBaCo_2O_{6-\delta}$	[280]	PCFC cathode
	$SmBa_{0.5}Sr_{0.5}Co_2O_{6-\delta}$	[281]	
	$SmBaCuCoO_{6-\delta}$	[282]	
	$PrBaCuFeO_{6-\delta}$	[283]	
	$SmBaCuFeO_{6-\delta}$	[282]	
	$PrBaCuFeO_{6-\delta}$	[283]	
Single phase	$BaPrCo_2O_{6-\delta}$		
	$BaGd_{0.8}La_{0.2}Co_2O_{6-\delta}$	[260]	PCFC cathode/PCEC anode
	$BaGdCo_{1.8}Fe_{0.2}O_{6-\delta}$		
	$BaPrCo_{1.4}Fe_{0.6}O_{6-\delta}$		
	Ruddlesden-Popper structure $A_2NiO_{4+\delta}$		
	$Pr_2NiO_{4+\delta}$	[264,265,284–286]	
	$La_2NiO_{4+\delta}$		
	$Nd_2NiO_{4+\delta}$	[265,287]	PCFC cathode
	$LaSrNiO_{4\pm\delta}$	[265]	
	$LaSrCo_{0.5}Fe_{0.5}O_{4-\delta}$	[288]	
	Other structures		
	$BaGa_2O_4$		
	$Ba_3Co_2O_6(CO_3)_{0.6}$	[289]	
	$YBaCo_{3.5}Zn_{0.5}O_{7+\delta}$	[290]	
	$Ba_3In_{1.4}Y_{0.3}M_{0.3}ZrO_8$ (M = Gd^{3+}/Ga^{3+})	[291]	PCFC cathode
	$Ca_{3-x}La_xCo_4O_{9+\delta}$ (x = 0, 0.3)	[292]	
	$La_4BaCu_5O_{13+\delta}$		
	$La_{6.4}Sr_{1.6}Cu_8O_{20\pm\delta}$	[293]	
	Cercer composites		
	$Ba_{0.5}Sr_{0.5}Co_{0.5}Fe_{0.5}O_{3-\delta}$/BZCY (3:2)	[294]	
	$Sm_{0.5}Sr_{0.5}CoO_{3-\delta}$/BZCY (7:3)	[295]	
	$Sm_{0.5}Sr_{0.5}CoO_{3-\delta}$/$Ce_{0.8}Sm_{0.2}O_{2-\delta}$ (6:4)	[296]	
	$La_{0.7}Sr_{0.3}FeO_{3-\delta}$/$Ce_{0.8}Sm_{0.2}O_{2-\delta}$	[297]	
	$SrCe_{0.9}Yb_{0.1}O_{3-\delta}$ or $BaCe_{0.9}Yb_{0.1}O_{3-\delta}$/$La_{0.6}Sr_{0.4}Co_{0.2}Fe_{0.8}O_{3-\delta}$	[298]	
	50% $PrBaCo_2O_{6-\delta}$/50% BZCY and 75–25 graded	[299]	PCFC cathode
	$Ba(Zr_{0.1}Ce_{0.7}Y_{0.2})O_{3-\delta}$ and $La_{0.6}Sr_{0.4}Co_{0.2}Fe_{0.8}O_{3-\delta}$	[271]	
	$La_{0.6}Sr_{0.4}Co_{0.2}Fe_{0.8}O_{3-\delta}$/$BaZr_{0.5}Pr_{0.3}Y_{0.2}O_{3-\delta}$	[300]	
	$Pr_{0.58}Sr_{0.4}Fe_{0.8}Co_{0.2}O_{3-\delta}$/$BaCe_{0.9}Yb_{0.1}O_{3-\delta}$	[301]	
	$Sm_{0.5}Sr_{0.5}CoO_{3-\delta}$/$BaCe_{0.8}Sm_{0.2}O_{3-\delta}$	[257]	
	$BaZr_{0.1}Ce_{0.7}Y_{0.1}Yb_{0.1}O_{3-\delta}$/$Nd_{1.95}NiO_{4+\delta}$	[302]	
	$BaCe_{0.8}Zr_{0.1}Y_{0.1}O_{3-\delta}$/$Sm_{0.5}Sr_{0.5}CoO_{3-\delta}$	[303]	
	$Gd_{0.1}Ce_{0.9}O_{2-\delta}$ infiltrated $PrBaCo_2O_{6-\delta}$ and $BaZr_{0.1}Ce_{0.7}Y_{0.2}O_{3-\delta}$	[304]	
Composite by mixing	$La_{0.6}Sr_{0.4}Co_{0.2}Fe_{0.8}O_3$/BZCY72		
	$La_2NiO_{4+\delta}$/BZCY72	[305]	
	$La_{0.87}Sr_{0.13}CrO_4$/BZCY72		PCEC anode
	$La_{0.8}Sr_{0.2}MnO_{3-\delta}$/BZCY72		
	$La_{0.8}Sr_{0.2}MnO_{3-\delta}$/$BaCe_{0.5}Zr_{0.3}Y_{0.16}Zn_{0.04}O_{3-\delta}$ with Fe_2O_3 nanoparticles	[306]	
	$Ba_2FeMoO_{6-\delta}$/$BaZr_{0.1}Ce_{0.7}Y_{0.1}Yb_{0.1}O_{3-\delta}$ for methane dehydro-aromatization	[307]	Electrodes for membrane reactors
	Cermet composites		
	65 wt.% NiO–35 wt.% BZCY72	[308]	
	65 wt.% NiO–35 wt.% BZCY72	[309]	
	$BaCe_{0.8}Y_{0.2}O_{3-\delta}$ and $BaCe_{0.6}Zr_{0.2}Y_{0.2}O_{3-\delta}$ with NiO 50:50	[310]	
	BZCY72-Ni	[8]	
	NiO-BZO and NiO-BZY15 (40 vol% Ni)	[311]	PCFC anode/PCEC cathode
	65 wt.% NiO–35 wt.% BZCY72	[28]	
	NiO-BZY cermet anodes (40 vol% Ni)	[312]	
	60 wt% NiO + 40 wt% BCGCu with functional layer 45 wt% NiO + 55 wt% BCGCu and BCGCu electrolyte $BaCe_{0.89}Gd_{0.1}Cu_{0.01}O_{3-\delta}$	[313]	
	$Ba(Zr_{0.85}Yb_{0.15})O_{3-\delta}$ and NiO the volume ratio of Ni to the total bulk volume in the reduced composite was designed to be around 40 vol%	[314]	
	NiO-$BaZr_{0.1}Ce_{0.7}Y_{0.1}Yb_{0.1}O_{3-\delta}$ (60:40 wt.%)	[315]	

Table 7. *Cont.*

Type	Composition	Reference	Application
Composite by infiltration	*Cercer composites* $(Pr_{0.9}La_{0.1})_2(Ni_{0.74}Cu_{0.21}Nb_{0.05})O_{4+\delta}$ infiltrated in $BaZr_{0.1}Ce_{0.7}Y_{0.2}O_{3-\delta}$	[316]	PCFC cathode
	$La_{0.58}Sr_{0.4}Co_{0.2}Fe_{0.8}O_{3-\delta}$ infiltrated in BZCY72	[96]	
	$LaCoO_3$ infiltrated in BZCY72	[317]	
	$Ba_{0.5}Sr_{0.5}Co_{0.8}Fe_{0.2}O_{3-\delta}$ infiltrated in BZCY72	[305]	
	$La_2NiO_{4+\delta}$ infiltrated in BZCY72	[318]	PCEC anode
	$Ba_{0.5}Gd_{0.8}La_{0.7}Co_2O_{6-\delta}$ infiltrated in BZCY72	[319]	
	$(La_{0.7}Sr_{0.3})V_{0.90}O_{3-\delta}$ infiltrated in $Ba(Ce_{0.51}Zr_{0.30}Y_{0.15}Zn_{0.04})O_{3-\delta}$	[320]	PCFC anode/PCEC cathode
	Cermet composites Fe infiltrated in $BaCe_{0.2}Zr_{0.6}Y_{0.2}O_{2.9}$	[321]	PCFC anode/PCEC cathode
	Ni infiltrated in $BaCe_{0.9}Y_{0.1}O_{3-\delta}$ and $BaZr_{1-x}Y_xO_{3-\delta}$ (x = 0.10, 0.20)	[322]	
Exsolution	*Cercer composites* $La_{0.5}Ba_{0.5}Co_{1/3}Mn_{1/3}Fe_{1/3}O_{3-\delta}/BaZr_{1-z}Y_zO_{3-\delta}$	[323]	PCFC cathode

5.2.2. Expansion of Air Electrodes

Air-side electrodes are typically composed of at least one transition metal (cf. Table 7), which means that they may chemically expand due to reduction at higher temperatures (cf. Sections 2.2.2 and 4.1.2 for details). Further, a few of the materials may also hydrate, resulting in a chemical expansion at lower temperatures (cf. Sections 2.2.1 and 4.1.1). However, due to low proton concentrations, the volume expansion due to hydration will generally be negligible. In total, the volume (or lattice parameter) of air electrodes can therefore be considered to only change due to thermal expansion and/or chemical expansion upon reduction.

One of the challenges when reviewing thermal and chemical expansion coefficients in the literature, lies in the ambiguity in the use of the term "thermal expansion". Some authors simply use it as a collective term to describe the change in volume (or lattice parameter) upon a change in temperature. However, as already detailed in Section 4.1, such an approach completely neglects effects of chemical expansion, which in the case of electrode materials, results in a non-linearity beyond a certain temperature, T^*. At higher temperatures, where $T > T^*$, the expansion of the material stems from a sum of both chemical and thermal contributions, while only thermal expansion contributes when $T < T^*$ (cf. Figure 2). For that reason, it has become common practice for many authors to divide expansion data in different temperature intervals, reporting average expansion coefficients for each region [119]. In what follows, we use the general term 'expansion coefficient' when data are not sufficient to extract whether it is thermal or chemical, or when it is the sum of thermal and chemical.

5.2.2.1. General Trends

Although the list of studied electrode materials on PCCs is extensive (Table 7), some general trends in terms of their expansion can be identified. For instance, Co is notorious for resulting in high thermal expansion coefficients, although it often improves the catalytic activity needed for the surface exchange reactions [273,324–327]. This is clearly demonstrated for $La_{0.8}Sr_{0.2}Co_{1-y}Fe_yO_{3-\delta}$, where the thermal expansion coefficient generally decreases with increasing Fe-content (y), ranging from 20.7 to 12.6×10^{-6} K^{-1} throughout the entire range $0 \leq y \leq 1$ [328]. Ni and Mn as substituents for Co have also been shown to decrease α, even though this is usually also at the expense of the electrochemical performance [329–332]. One of the reasons why Co-containing oxides have higher thermal expansion coefficients than other transition metals is that the Co ions may undergo a spin transition as the temperature is raised. Such a spin transition, for example from low to high spin, will increase the ionic radius of the Co-ions, resulting in an effectively higher thermal expansion coefficient [333,334]. In fact, Radaelli et al. [333] demonstrated that the spin transition from high to low spin in $LaCoO_3$ accounts for approximately half of the total lattice expansion at elevated temperatures. Thus, most strategies to lower the thermal expansion coefficient have attempted to substitute Co for another element, although this tends to have an adverse effect on the conductivity and catalytic activity of the material. In fact, such a trade-off is very typically found for electrode-type materials and was well demonstrated in a review by Ullmann et al. [335], who showed an inverse empirical relationship between the thermal expansion coefficient and the oxide ion conductivity, expressed by the following expression

$$\log \sigma_{O^{2-}} = -7.08 + 2.9 \times 10^5 \alpha_L \tag{41}$$

where α_L is the linear thermal expansion coefficient in 10^{-6} K^{-1}, while $\sigma_{O^{2-}}$ is the oxide ion conductivity in S cm^{-1}. Note that this correlation specifically applies to the SOFC/SOEC community but it is tempting to consider whether similar correlations may exist for the proton conductivity of PCFCs/PCECs electrodes. Nevertheless, the correlation does suggest a potential problem, in that the better performing electrodes generally tend to be less thermally compatible with the best electrolyte materials. An alternative and popular strategy to mitigate this problem is therefore to use composite materials, composed of a good-performing electrode and the electrolyte [16,317,319,320,336–339].

Other efforts have also indicated that the thermal expansion coefficient can be linked to ionicity, where more ionic bonds tend to result in higher α-values. For instance, in the system BaRECo$_2$O$_{6-\delta}$, α decreases with decreasing size of the rare earth cation (La > Nd > Sm > Gd > Y), reflecting a reduction in the ionic character of the RE-O bond [340]. A similar trend has also been demonstrated for Sr$_{0.4}$RE$_{0.6}$CoO$_{3-\delta}$, Sr$_{0.2}$RE$_{0.8}$Fe$_{0.8}$Co$_{0.2}$O$_{3-\delta}$, Sr$_{0.3}$RE$_{0.65}$MnO$_{3-\delta}$ and BaRECuFeO$_{6-\delta}$ [34,324,341–345]. Ionicity can also be used to rationalize the general tendency for Sr-containing perovskites to exhibit high thermal expansion coefficients. For instance, in the system La$_{1-x}$Sr$_x$MnO$_{3-\delta}$, a larger proportion of Sr results in higher thermal expansion coefficients [346]. This can be explained on the basis of electronegativity, where the less electronegative element, which in this case is Sr (0.99 versus 1.08 for La in the Allred-Rochow scale), will induce a higher ionic character of the A-O bond. However, it should be noted that increasing the amount of Sr will also result in a larger oxygen nonstoichiometry, subsequently leading to a higher chemical expansion upon reduction at higher temperatures.

5.2.2.2. Selected Air Side Electrode Materials

This section will cover some of the more well-studied compositions used as air side electrodes on PCFCs and PCECs, starting with Ba$_{1-x}$Sr$_x$Co$_{0.8}$Fe$_{0.2}$O$_{3-\delta}$ (BSCF).

Ba$_{1-x}$Sr$_x$Co$_{0.8}$Fe$_{0.2}$O$_{3-\delta}$ (BSCF) Based Materials

Ba$_{0.5}$Sr$_{0.5}$Co$_{0.8}$Fe$_{0.2}$O$_{3-\delta}$ was originally introduced as a dense ceramic membrane for oxygen separation [347,348] but was soon after shown to also display excellent properties as a SOFC cathode on Sm$_{0.15}$Ce$_{0.85}$O$_{2-\delta}$ [349]. This sparked an immediate interest within the research community to study related compositions, making BSCF one of the most studied electrode materials to this day. A few years later, in 2008, Lin et al. [350] also showed that BSCF could perform exceptionally well as a PCFC cathode, using it on a proton conducting Y-doped BaCeO$_3$ electrolyte. Due to its popularity, several studies have described its thermal and chemical expansion.

Figure 8 shows the thermal expansion coefficients determined for BSCF as a function of the Sr-content (x), along with the specific values and references given in Table 8. In studies reporting more than one coefficient, we only use values determined at lower temperatures to avoid influences of chemical expansion upon reduction. As the figure and table demonstrate, the literature values for α vary tremendously, spanning a range from 11.5 to 24.4 × 10^{-6} K^{-1} [351–356] with the majority lying in the range of ~20 × 10^{-6} K^{-1}. While the higher values (>20 × 10^{-6} K^{-1}) can be rationalized on the basis of an additional chemical expansion due to reduction at higher temperatures, the lower expansion coefficients, reported for instance by Patra et al. [351], appear to be anomalous. We suspect that their measurements could stem from an unusual thermal history, where the samples have already been reduced, causing them to re-oxidize upon heating at lower temperatures (<300 °C). This would effectively reduce the expansion coefficient, thus explaining why their values are particularly low. The thermal expansion coefficient of Ba$_{1-x}$Sr$_x$Co$_{0.8}$Fe$_{0.2}$O$_{3-\delta}$ appears to otherwise be fairly independent with respect to changes in the A-site composition, that is, α is similar for the entire $BaCo_{0.8}Fe_{0.2}O_{3-\delta}$—$SrCo_{0.8}Fe_{0.2}O_{3-\delta}$ system.

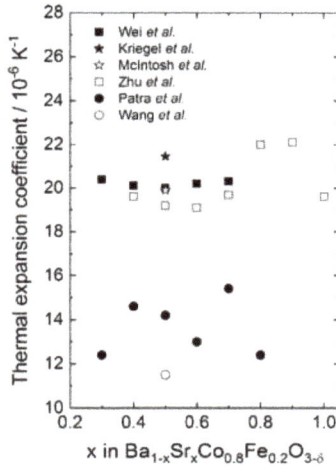

Figure 8. Linear thermal expansion coefficients of $Ba_{1-x}Sr_xCo_{0.8}Fe_{0.2}O_{3-\delta}$ (BSCF) as a function of Sr-content, x. All values are summarized in Table 8 along with the references used. We have used average values for the data by Kriegel et al. [354] and McIntosh et al. [353].

Table 8. Thermal expansion coefficients of $Ba_{1-x}Sr_xCo_{0.8}Fe_{0.2}O_{3-\delta}$ (BSCF).

x	Thermal Expansion Coefficient/10^{-6} K^{-1}	Temperature Range	References
0.3	12.4 [1]	RT–300 °C	[1] Patra et al. [351]
	20.4 [2]	50–1000 °C	[2] Wei et al. [352]
0.4	14.6 [1]	RT–300 °C	[1] Patra et al. [351]
	20.1 [2]	50–1000 °C	[2] Wei et al. [352]
	19.6 [3]	RT–850 °C	[3] Zhu et al. [355]
0.5	14.2 [1]	RT–300 °C	[1] Patra et al. [351]
	20.0 [2]	50–1000 °C	[2] Wei et al. [352]
	19.0–20.8 [3] *	50–1000 °C	[3] McIntosh et al. [353]
	18.5–24.4 [4] *	600–900 °C	[4] Kriegel et al. [354]
	19.2 [5]	RT–1000 °C	[5] Zhu et al. [355]
	11.5 [6]	RT–850 °C	[6] Wang et al. [356]
0.6	13.0 [1]	RT–300 °C	[1] Patra et al. [351]
	20.2 [2]	50–1000 °C	[2] Wei et al. [352]
	19.1 [3]	RT–850 °C	[3] Zhu et al. [355]
0.7	15.4 [1]	RT–300 °C	[1] Patra et al. [351]
	20.3 [2]	50–1000 °C	[2] Wei et al. [352]
	19.7 [3]	RT–850 °C	[3] Zhu et al. [355]
0.8	12.4 [1]	RT–300 °C	[1] Patra et al. [351]
	22.0 [2]	RT–850 °C	[2] Zhu et al. [355]
0.9	22.1 [1]	RT–850 °C	[1] Zhu et al. [355]
1.0	19.6 [1]	RT–850 °C	[1] Zhu et al. [355]

* These studies have determined thermal and chemical expansion coefficients, and are thus less prone to overestimating the thermal expansion coefficient, which can be an issue at higher temperatures due to the influence of a chemical expansion upon reduction.

In terms of the linear chemical expansion of BSCF, values for β_{red} are also scattered, varying in the region of 0.008–0.026 per mol oxygen vacancy [353,354]. Interestingly, we see that these values differ depending on the specific nonstoichiometry of the oxide and tend to average to around 0.012–0.016 for larger changes in δ. Thus, for most work dealing with modelling the thermochemical expansion of BCSF, values of 0.012–0.016 should be appropriate. Note that these β_{red}-values are very small, being just slightly more than a quarter of the corresponding chemical expansion coefficient upon hydration determined for BZY (linear $\beta_{hydr} = 0.045$, see Section 5.1.1.1 for details). Overall, this suggests that chemical expansion upon reduction should not pose a large issue in terms of thermally matching BCSF with other device components.

$La_{1-x}Sr_xCo_yFe_{1-y}O_{3-\delta}$ (LSCF) Based Materials

Another popular electrode material, which has received considerable interest in both the SOFC/SOEC and PCFC/PCEC communities, is $La_{1-x}Sr_xCo_yFe_{1-y}O_{3-\delta}$ (LSCF), where it is especially the composition $La_{0.6}Sr_{0.4}Co_{0.2}Fe_{0.8}O_{3-\delta}$ (LSCF-6428) that has been studied extensively [95,96,271,300,357]. As for most cobalt-containing electrode materials, increasing the Co/Fe ratio tends to increase the thermal expansion coefficient, being no different in case of LSCF. This is exemplified in a study by Tai et al. [328], where the thermal expansion coefficient increases from 12.6 to 19.7×10^{-6} K^{-1} with increasing Co-content for the $La_{0.8}Sr_{0.2}Co_yFe_{1-y}O_{3-\delta}$ system.

The effect of A-site compositional changes on the thermal expansion coefficient in LSCF is far less understood, although some authors suggest that the coefficient increases with increasing Sr-content [327,358]. However, most reviews and studies on the subject are far more careful in attempting to draw such links, as Sr also increases the oxygen nonstoichiometry of the oxide. This will subsequently lead to a larger influence of chemical expansion upon reduction, which will very easily distort the results. To shed some light on the subject, all available thermal expansion coefficients for varying La/Sr ratios for a fixed B-site composition of 20 mol% and 80 mol% Co and Fe, respectively, are summarized in Table 9 and plotted as a function of the Sr-content in Figure 9. This compositional range is particularly sensible to choose as it takes advantage of the large number of studies on LSCF-6428 [47,99,327,359–363]. Further, we only choose thermal expansion coefficients determined at lower temperatures for studies reporting more than one coefficient to limit the influence of chemical expansion. As seen in Figure 9, there appears to be no apparent trend in the thermal expansion coefficient with respect to Sr-content, averaging out to be around $15–16 \times 10^{-6}$ K^{-1} for all compositions. This suggests that variations in the La/Sr ratio will therefore only affect the amount of chemical expansion in LSCF. We otherwise note that the thermal expansion coefficient of LSCF is clearly lower than that of BSCF (cf. Figure 8 and Table 8) but still somewhat higher than typical electrolyte materials, where the coefficients are in the range of $8–12 \times 10^{-6}$ K^{-1} (cf. Figures 6 and 7 and Tables 2–6). A potential way of mitigating this small thermal mismatch between LSCF and potential electrolytes, such as BZY, could be to intentionally have a small A-site deficiency in LSCF (La and/or Sr), as this has been demonstrated to lower the thermal expansion coefficient [363]. However, an A-site deficiency will undoubtedly again lead to a higher oxygen nonstoichiometry, subsequently resulting in more chemical expansion upon reduction.

The chemical expansion coefficient upon reduction, β_{red}, for LSCF is high, lying in the region of 0.02–0.06 per mol $v_O^{\bullet\bullet}$ (versus ~0.012–0.016 for BSCF) [47,99,121–123,364,365]. Table 10 lists all linear chemical expansion coefficients of LSCF in the order of increasing Co-content. Although it is hard to navigate through the specific values presented, there appears to be a general tendency for LSCF to have a higher β_{red}-value for higher Fe-contents. Thus, the role of Fe in LSCF seems to be rather complicated, in that more Fe increases β_{red}, while a higher proportion of Fe also lowers the thermal expansion coefficient. Although this indicates a possible trade-off between the two expansion processes, we will not dwell on this further due to the limited data available. β_{red} seems to otherwise be insensitive to changes in the La/Sr ratio, which means that the amount of chemical expansion of LSCF can easily be decreased by increasing the La/Sr ration (less Sr). This is for instance demonstrated in Figure 2 showing the thermochemical expansion of LSCF upon heating in air ($p_{O_2} = 0.21$ atm). β_{red} is fixed to 0.031, while the amount of Sr is varied from 0.2 to 0.4 mole fractions, clearly demonstrating less chemical expansion for lower Sr-content due to a smaller oxygen nonstoichiometry.

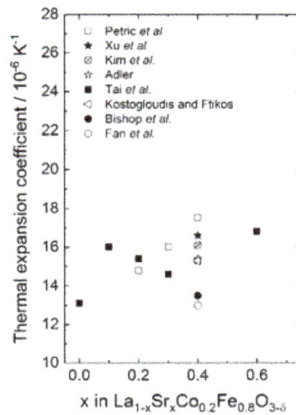

Figure 9. Linear thermal expansion coefficients of $La_{1-x}Sr_xCo_{0.2}Fe_{0.8}O_{3-\delta}$ (LSCF) as a function of Sr-content, x. All values are summarized in Table 9 along with the references used. Average values have been used for studies where only ranges in the thermal expansion coefficient are provided.

Table 9. Linear thermal expansion coefficients of $La_{1-x}Sr_xCo_{0.2}Fe_{0.8}O_{3-\delta}$ (LSCF).

x	Thermal Expansion Coefficient/10^{-6} K^{-1}	Temperature Range	References
0	13.1 [1]	100–500 °C	[1] Tai et al. [359]
0.1	16.0 [1]	300–900 °C	[1] Tai et al. [359]
0.2	15.4 [1]	100–800 °C	[1] Tai et al. [359]
	14.8 [2]	30–1000 °C	[2] Petric et al. [327]
0.3	16.0 [1]	30–1000 °C	[1] Petric et al. [327]
0.4	15.3 [1]	100–600 °C	[1] Tai et al. [359]
	17.5 [2]	30–1000 °C	[2] Petric et al. [327]
	16.6 [3]	100–650 °C	[3] Xu et al. [360]
	11–15 [4]	100–700 °C	[4] Fan et al. [361]
	13–14 [5] *	RT–950 °C	[5] Bishop et al. [99]
	15.4 [6] *	100–600 °C	[6] Adler [47]
	16.1 [7]	Not given	[7] Kim et al. [362]
	15.3 [8]	20–700 °C	[8] Kostgloudis and Ftikos [363]
0.6	16.8 [1]	100–400 °C	[1] Tai et al. [359]

* These studies have determined thermal and chemical expansion coefficients and are thus less prone to overestimating the thermal expansion coefficient, which can be an issue at higher temperatures due to the influence of a chemical expansion upon reduction.

Table 10. Linear chemical expansion coefficients upon reduction, β_{red} , for $La_{1-x}Sr_xCo_yFe_{1-y}O_{3-\delta}$ (LSCF) in the order of increasing Co-content (y).

y	Composition	Linear Chemical Expansion Coefficient	Temperature	References
0	$La_{0.5}Sr_{0.5}FeO_{3-\delta}$	0.059 [1]	800 °C	[1] Fossdal et al. [123]
	$La_{0.3}Sr_{0.7}FeO_{3-\delta}$	0.017–0.047 [2]	650–875 °C	[2] Kharton et al. [364]
0.2	$La_{0.6}Sr_{0.4}Co_{0.2}Fe_{0.8}O_{3-\delta}$	0.031 [1]	700–900 °C	[1] Bishop et al. [99]
		0.032 [2]	700–890 °C	[2] Adler [47]
0.5	$La_{0.5}Sr_{0.5}Co_{0.5}Fe_{0.5}O_{3-\delta}$	0.036–0.039 [1]	800–1000 °C	[1] Lein et al. [121]
0.8	$La_{0.6}Sr_{0.4}Co_{0.8}Fe_{0.2}O_{3-\delta}$	0.022 [1]	800 °C	[1] Wang et al. [365]
1	$La_{1-x}Sr_xCoO_{3-\delta}$ (x = 0.2–0.7)	0.023–0.024 [1]	600–900 °C	[1] Chen et al. [122]
	$La_{0.5}Sr_{0.5}CoO_{3-\delta}$	0.035 [2]	800 °C	[2] Lein et al. [121]

Other Single Perovskite Materials

Table 11 lists the thermal expansion coefficients for other single perovskite materials commonly used as air electrodes, based on lanthanum or barium on the A-site and transition metals on the

B-site (Ni, Co, Fe, Mn). Most of the thermal expansion coefficients summarized in the table are averaged over a wide temperature range and are thus overestimated as they also include a chemical expansion contribution. As observed for most other oxides, increasing the Co content leads to an increase in the expansion coefficient, being no different in the case of these perovskites. The lowest expansion coefficients are thus obtained for cobalt free compositions such as $La_{0.8}Sr_{0.2}MnO_{3-\delta}$ and $LaNi_{1-x}Fe_xO_{3-\delta}$. It is otherwise interesting to note that α decreases with increasing Fe content in the $LaNi_{1-x}Fe_xO_{3-\delta}$ system.

Table 11. Linear thermal expansion coefficients of single perovskite electrode materials, other than LSCF and BSCF.

Composition	Thermal Expansion Coefficient/10^{-6} K^{-1}	Temperature Range	References
$BaCo_{0.4}Fe_{0.4}Zr_{0.2}O_{3-\delta}$	21.9	300–700 °C	Duan et al. [366]
$BaCo_{0.4}Fe_{0.4}Zr_{0.1}Y_{0.1}O_{3-\delta}$	21.6	300–700 °C	Duan et al. [366]
$LaNi_{0.4}Co_{0.2}Fe_{0.4}O_{3-\delta}$	14.3		
$LaNi_{0.3}Co_{0.3}Fe_{0.4}O_{3-\delta}$	15.5	RT–900 °C	Taguchi et al. [325]
$LaNi_{0.2}Co_{0.4}Fe_{0.4}O_{3-\delta}$	17.2		
$LaCo_{0.6}Ni_{0.4}O_{3-\delta}$	17.2	RT–900 °C	Wang et al. [367]
$LaCo_{0.8}Fe_{0.1}Ni_{0.1}O_{3-\delta}$	20.8		
$LaCo_{0.7}Fe_{0.1}Ni_{0.2}O_{3-\delta}$	18.6	RT–827 °C	Kharton et al. [368]
$LaCo_{0.6}Fe_{0.2}Ni_{0.2}O_{3-\delta}$	19.2		
$LaCo_{0.5}Fe_{0.2}Ni_{0.3}O_{3-\delta}$	18		
$LaNiO_{3-\delta}$	12.3		
$LaNi_{0.8}Fe_{0.2}O_{3-\delta}$	11.5		
$LaNi_{0.6}Fe_{0.4}O_{3-\delta}$	11	RT–800 °C	Chiba et al. [369]
$LaNi_{0.4}Fe_{0.6}O_{3-\delta}$	9.9		
$LaNi_{0.2}Fe_{0.8}O_{3-\delta}$	9.4		
$LaFeO_{3-\delta}$	9.7		
$La_{0.8}Sr_{0.2}MnO_{3-\delta}$	11.8	30–1000 °C	Tietz et al. [326]
$La_{0.8}Sr_{0.2}MnO_{3+\delta}$	14.2 *	RT–400 °C	Grande et al. [370]
$La_{0.7}Sr_{0.3}MnO_{3+\delta}$	14.2 *	RT–500 °C	

* These studies have determined expansion coefficients in two temperature ranges: we only use values determined at lower temperatures to avoid influences of chemical expansion upon reduction.

Complex Perovskites

Although we briefly introduced the complex perovskites in Section 5.1.1.2 as electrolyte type materials, their usage is far more common in the field of electronics. Within this family of oxides, we find the so-called double perovskites, with the general formula $A'A''B'B''O_{6-\delta}$, where the A-sites are typically occupied by alkaline earth elements (often Ba or Sr), and/or by rare earth elements such as La and Y, while the B-site cations tend to be a transition metal, for example, Co, Fe, Mn, Mo, Ni or Cu [329,371,372]. A distinguishing feature of the double perovskites, compared to that of the simple perovskites, ABO_3, is that they exhibit cation ordering, either on the A- or B-site. Such ordering means that the cations are distributed in separate alternating crystallographic layers, where only the A' or A'' cations (or B' or B'' cations) are present in each layer. It should be noted that double perovskites cannot have cation ordering on both sites simultaneously as this results in a higher order complex perovskite. Similarly, no cation ordering reduces the double perovskite to a simple perovskite, ABO_3, where the cations are randomly distributed in each of the cation sublattices. Double perovskites are therefore more conveniently expressed by $A'A''B_2O_{6-\delta}$ or $A_2B'B''O_{6-\delta}$ for A- and B-site ordered systems, respectively.

Before continuing it is important to clarify that for the double perovskites two alternative notations of the composition are used in the literature. The first one is based on the fact that a high understoichiometry of the oxygen sublattice is expected in these compounds, such that the oxygen content is often closer to 5 than 6. Thus, many authors choose to express the composition as $A'A''B'B''O_{5+\delta}$. An alternative approach, which is exclusively used here, is based on the premise that all deviations in stoichiometry use the ideal defect-free structure as a frame of reference and the composition is then expressed as $A'A''B'B''O_{6-\delta}$. This distinction is important, as δ in both notations will have different numerical values, which can easily lead to some confusion.

Out of all the double perovskites, none have been studied more intensively than the Ba-based layered cobaltites, with the general formula $BARECo_{2-x}M_xO_{6-\delta}$ (RE: rare earth, M: other transition metal). These have been shown to offer high catalytic activity and excellent electrochemical performances [329,373]. However, as with all cobalt-containing electrodes, their thermal expansion coefficients are high, typically lying in the range of ~20 × 10^{-6} K^{-1}. Additionally, they also exhibit considerable chemical expansion upon reduction, which effectively increases the total lattice expansion at higher temperatures (see for example, Section 4.1.2 for more details). Several studies have therefore been devoted to understanding the role of various cation substitutions on the thermal expansion of $BARECo_{2-x}M_xO_{6-\delta}$ and we will therefore focus on these efforts here.

Thermal expansion coefficients of $BARECo_{2-x}M_xO_{6-\delta}$ and some related compositions from the literature are listed in Table 12. Note that due to the influence of chemical expansion upon reduction, values have been separated to reflect whether the coefficients are purely thermal, extracted from different temperature intervals, for example, low and high temperatures, or simply averages over a wide range. For the discussion of trends in thermal expansion coefficients for this class of oxides, we will primarily use values that are purely thermal or that are determined at lower temperatures.

Using the values presented in Table 12, we see that the thermal expansion coefficient of $BARECo_2O_{6-\delta}$ decreases with decreasing size of the rare earth elements (RE), in the order La → Nd → Sm → Gd → Y, varying from 24.3 to 15.8 × 10^{-6} K^{-1} in the entire series. This can be attributed to a reduction of the ionic character of the RE-O bond [34,324,329,341,344,345]. Other work investigating A-site compositional changes have involved partially substituting Sr for Ba. This has been largely ineffective, with larger proportions of Sr in the system $Ba_{1-x}Sr_xLnCoCuO_{6-\delta}$ ($0 \leq x \leq 0.75$) (Ln = Nd or Gd) resulting in insignificant changes to the thermal expansion coefficient [374,375]. However, the introduction of Sr seems to induce a small increase of the chemical expansion upon reduction, although this was only demonstrated for a few of the compositions studied.

Most of the work involved in altering the thermal expansion coefficient of $BARECo_{2-x}M_xO_{6-\delta}$ has been devoted to B-site substitutions, as the reduction of the Co-content generally reduces α. This has been demonstrated for a myriad of elements (M) including but not limited to Sc, Mn, Ni and Cu, where a larger proportion of M reduces the coefficient [376–382]. The attentive reader may here question why Fe has been excluded from the list of examples but this was done intentionally as the effect of Fe substitutions on thermal expansion has been contradictory [373,383–388]. We will therefore now focus on the system $BARECo_{2-x}Fe_xO_{6-\delta}$ in attempt to rationalize the contradicting studies. Although the majority of work studying the thermal expansion of Fe substituted $BARECo_2O_{6-\delta}$ demonstrates that α either decreases or remains unchanged with increasing Fe-content, there are some notable exceptions. For instance, Zhao et al. [385] report that in $BaPrCo_{2-x}Fe_xO_{6-\delta}$, the thermal expansion coefficient first increases as x is increased from 0 to 0.5, from 24.6 to 26 × 10^{-6} K^{-1}, whereupon it starts to readily decrease finally reaching a coefficient of 17.2 × 10^{-6} K^{-1} for x = 2. The problem with this erratic trend is that the values determined by Zhao et al. [385] are average coefficients, taken over a wide temperature range (RT–800 °C) and thus reflect a combination of thermal and chemical expansion. Estimating pure thermal expansion coefficients from a lower temperature regime (before an apparent chemical expansion onset of ~250 °C) using the dilatometry data presented, we find a clear systematic trend with respect to Fe content, decreasing from 19.3 to 13.4 × 10^{-6} K^{-1} upon going from x = 0 to 2. Similarly, in a study by Xue et al. [389], α was also reported to increase for higher Fe content in $BaYCo_{2-x}Fe_xO_{6-\delta}$ increasing from 16.3 to 18.0 × 10^{-6} K^{-1} when x increased from 0 to 0.4. However, again we find that these coefficients are just average values for the entire temperature range studied (30–900 °C) and the lower temperature value of α (<250 °C) appears to be constant (~17.9 × 10^{-6} K^{-1}), irrespective of the Fe concentration. Thus, most of the confusion regarding the effect of Fe on the thermal expansion of $BARECo_{2-x}Fe_xO_{6-\delta}$ seems to stem from the misusage of the term "thermal expansion coefficient".

Table 12. Linear thermal expansion coefficients of $BaRECo_{2-x}M_xO_{6-\delta}$ and some related double perovskites. The data are separated in different categories: no chemical expansion reported (thermal only, linear over the entire temperature range), separation in different temperature ranges (low temperature LT and high temperature HT) and average over a large temperature range.

Composition	Thermal Expansion Coefficient/10^{-6} K^{-1}	Temperature Range	References
	Thermal only		
$LaSrMnCoO_{6-\delta}$	15.8	RT–1000 °C	Zhou et al. [390]
$BaSmCo_{1.4}Fe_{0.6}O_{6-\delta}$	21.1	RT–1100 °C	Volkova et al. [388]
$BaSmCo_{1.2}Fe_{0.8}O_{6-\delta}$	21.0		
$BaSmCo_2O_{6-\delta}$	21.0	RT–1100 °C	Aksenova et al. [391]
$BaPr_{0.7}Y_{0.3}Co_2O_{6-\delta}$	17.6	50–800 °C	Zhao et al. [392]
$BaPr_{0.5}Y_{0.5}Co_2O_{6-\delta}$	17.2		
$BaPrCoCuO_{6-\delta}$	15.2	RT–1000 °C	Zhao et al. [393]
	LT/HT		
$BaLaCo_2O_{6-\delta}$	24.3	80–900 °C	
	29.8	500–900 °C	
$BaNdCo_2O_{6-\delta}$	19.1	80–900 °C	
	21.1	500–900 °C	
$BaSmCo_2O_{6-\delta}$	17.1	80–900 °C	Kim et al. [340]
	18.7	500–900 °C	
$BaGdCo_2O_{6-\delta}$	16.6	80–900 °C	
	16.8	500–900 °C	
$BaYCo_2O_{6-\delta}$	15.8	80–900 °C	
	14.9	500–900 °C	
$BaPrCo_{1.95}Sc_{0.05}O_{6-\delta}$	17.0	30–300 °C	
	27.0	300–1000 °C	
$BaPrCo_{1.9}Sc_{0.10}O_{6-\delta}$	16.0	30–300 °C	Li et al. [376]
	26.0	300–1000 °C	
$BaPrCo_{1.8}Sc_{0.20}O_{6-\delta}$	15.8	30–300 °C	
	25.8	300–1000 °C	
$BaSmCo_{1.6}Fe_{0.4}O_{6-\delta}$	19.0	30–400 °C	
	20.8	400–1100 °C	
$BaSmCo_{1.8}Fe_{0.2}O_{6-\delta}$	18.4	30–470 °C	Volkova et al. [388]
	20.4	470–1100 °C	
$BaSmCo_{1.9}Fe_{0.1}O_{6-\delta}$	18.7	30–500 °C	
	21.1	500–1100 °C	
$BaGdCo_2O_{6-\delta}$	18.0	100–350 °C	
	19.6	600–900 °C	
$BaGdCo_{2/3}Fe_{2/3}Ni_{2/3}O_{6-\delta}$	15.2	100–350 °C	
	16.8	600–900 °C	Jo et al. [381]
$BaGdCo_{2/3}Fe_{2/3}Cu_{2/3}O_{6-\delta}$	13.6	100–350 °C	
	15.7	600–900 °C	
$BaGdCoCuO_{6-\delta}$	14.0	100–350 °C	
	12.8	600–900 °C	
$BaGdCo_2O_{6-\delta}$	17.6	RT–450 °C	Mogni et al. [394]
	19.0	500–900 °C	
$Ba_{0.5}Sr_{0.5}GdCo_{1.5}Fe_{0.5}O_{6-\delta}$	15.4	RT–200 °C	Kuroda et al. [395]
	21.8	300–800 °C	
$BaNdCo_2O_{6-\delta}$	17.9	RT–300 °C	
	22.9	RT–900 °C	Kim et al. [375]
$Ba_{1.5}Sr_{0.5}NdCo_2O_{6-\delta}$	17.2	RT–300 °C	
	26.3	RT–900 °C	
$BaNdCo_2O_{6-\delta}$	18.3	RT–250 °C	
	23.8	350–1000 °C	
$BaNdCo_{1.8}Fe_{0.2}O_{6-\delta}$	18.8	RT–250 °C	
	21.9	350–1000 °C	
$BaNdCo_{1.6}Fe_{0.4}O_{6-\delta}$	18.9	RT–250 °C	Cherepanov et al. [373]
	21.9	350–1000 °C	
$BaNdCo_{1.4}Fe_{0.6}O_{6-\delta}$	18.3	RT–250 °C	
	22.1	350–1000 °C	
$BaNdCo_{1.2}Fe_{0.8}O_{6-\delta}$	18.4	RT–250 °C	
	21.9	350–1000 °C	

Table 12. *Cont.*

Composition	Thermal Expansion Coefficient/10^{-6} K^{-1}	Temperature Range	References
	Average over whole T range		
BaPrCo$_{2/3}$Fe$_{2/3}$Cu$_{2/3}$O$_{6-\delta}$	16.6	30–850 °C	Jin et al. [383]
BaPrCo$_{1.6}$Ni$_{0.4}$O$_{6-\delta}$	20.6		
BaNdCo$_{1.6}$Ni$_{0.4}$O$_{6-\delta}$	19.4	30–1000 °C	Che et al. [345]
BaSmCo$_{1.6}$Ni$_{0.4}$O$_{6-\delta}$	16.6		
BaPrCo$_2$O$_{6-\delta}$	21.5	50–800 °C	Zhao et al. [392]
BaPrCuCoO$_{6-\delta}$	24.1	30–1000 °C	Zhao et al. [393]
Ba$_{0.5}$Sr$_{0.5}$PrCo$_{1.5}$Fe$_{0.5}$O$_{6-\delta}$	21.3	RT–900 °C	Jiang et al. [384]
Ba$_{0.5}$Sr$_{0.5}$PrCo$_{0.5}$Fe$_{1.5}$O$_{6-\delta}$	19.2		
BaPrCo$_2$O$_{6-\delta}$	24.6		
BaPrCo$_{1.5}$Fe$_{0.5}$O$_{6-\delta}$	26.0		
BaPrCoFeO$_{6-\delta}$	25.0	20–800 °C	Zhao et al. [385]
BaPrCo$_{0.5}$Fe$_{1.5}$O$_{6-\delta}$	19.1		
BaPrFe$_2$O$_{6-\delta}$	17.2		
Ba$_{0.5}$Sr$_{0.5}$PrCo$_{1.9}$Ni$_{0.1}$O$_{6-\delta}$	21.9	30–900 °C	Liu et al. [380]
Ba$_{0.5}$Sr$_{0.5}$PrCo$_{1.7}$Ni$_{0.3}$O$_{6-\delta}$	19.7		
BaPrCoFeO$_{6-\delta}$	21.0	30–1000 °C	Jin et al. [344]
BaNdCoFeO$_{6-\delta}$	19.5		
BaNdCo$_2$O$_{6-\delta}$	23.0	RT–1100 °C	Aksenova et al. [391]
Ba$_{0.5}$Sr$_{0.5}$NdCo$_2$O$_{6-\delta}$	20.3		
Ba$_{0.5}$Sr$_{0.5}$NdCo$_{1.75}$Mn$_{0.25}$O$_{6-\delta}$	16.4	RT–900 °C	Kim et al. [377]
Ba$_{0.5}$Sr$_{0.5}$NdCo$_{1.5}$Mn$_{0.5}$O$_{6-\delta}$	14.3		
BaNdCo$_2$O$_{6-\delta}$	21.5		
BaNdFe$_2$O$_{6-\delta}$	18.3	80–900 °C	Kim et al. [387]
BaGdCo$_2$O$_{6-\delta}$	19.9		
BaGdFe$_2$O$_{6-\delta}$	18.8		
BaGdCoCuO$_{6-\delta}$	16.3		
Ba$_{0.25}$Sr$_{0.75}$GdCoCuO$_{6-\delta}$	16.0	RT–900 °C	West et al. [374]
Ba$_{0.25}$Sr$_{0.75}$NdCoCuO$_{6-\delta}$	17.0		
BaGdCo$_{1.9}$Ni$_{0.1}$O$_{6-\delta}$	18.9	30–900 °C	Wei et al. [331]
BaGdCo$_{1.7}$Ni$_{0.3}$O$_{6-\delta}$	15.5		
BaYCo$_2$O$_{6-\delta}$	17.8		
BaYCo$_{1.8}$Cu$_{0.2}$O$_{6-\delta}$	16.7		
BaYCo$_{1.6}$Cu$_{0.4}$O$_{6-\delta}$	15.7	30–850 °C	Zhang et al. [382]
BaYCo$_{1.4}$Cu$_{0.6}$O$_{6-\delta}$	14.7		
BaYCo$_{1.2}$Cu$_{0.8}$O$_{6-\delta}$	13.4		
BaYCo$_2$O$_{6-\delta}$	16.3	30–900 °C	Xue et al. [389]
BaYCo$_{1.6}$Fe$_{0.4}$O$_{6-\delta}$	18.0		
SrSmCo$_2$O$_{6-\delta}$	22.6		
SrSmCo$_{1.8}$Mn$_{0.2}$O$_{6-\delta}$	21.0		
SrSmCo$_{1.6}$Mn$_{0.4}$O$_{6-\delta}$	20.8	RT–850 °C	Wang et al. [378]
SrSmCo$_{1.4}$Mn$_{0.6}$O$_{6-\delta}$	18.0		
SrSmCo$_{1.2}$Mn$_{0.8}$O$_{6-\delta}$	17.5		
SrSmCoMnO$_{6-\delta}$	13.7		

Although there are several studies that have investigated the lattice expansion of BaRECo$_{2-x}$M$_x$O$_{6-\delta}$, there is still a surprising lack of systematic work where the extent of chemical expansion has been determined. Most studies will just report differences in expansion coefficients (sum of thermal and chemical) upon varying the composition, but such changes can stem from a change in the chemical expansion coefficient, β_{red}, and/or a change in the oxygen nonstoichiometry, δ. We can estimate β_{red} for the system BaNdCo$_{2-x}$Fe$_x$O$_{6-\delta}$ from the data by Cherepanov et al. [373] and find that the linear coefficient lies in the region of 0.005–0.008 per mol $v_O^{\bullet\bullet}$. If we assume that these compositions are representative for the layered double perovskite cobaltites, it means that this class of oxides displays a significantly lower chemical expansion coefficient upon reduction compared to the well-established electrodes BSCF ($\beta_{red} = \sim$0.012–0.016) and LSCF ($\beta_{red} = $ 0.02–0.06).

Ruddlesden-Popper Structures

Ruddlesden-Popper is the name of a whole class of structures with general formula A$_{n+1}$B$_n$O$_{3n+1}$, where A is either an alkaline earth or a rare earth element, while B is generally a transition metal.

Ruddlesden-Popper structures are composed of alternating layers of n rock-salt layers (AO) wedged in between a single perovskite slab (ABO_3). Thus, the number n expresses the ratio of rock salt to perovskite layers (n = AO/ABO_3).

From the transition metal point of view the most popular electrode materials with Ruddlesden-Popper structure would be ferrites, cobaltites, nickelates and cuprites [396]. The coefficients of thermal expansion for compounds of the Ruddlesden-Popper family are summarized in Table 13. The highest expansion coefficients are observed for cobaltites ($>20 \times 10^{-6}$ K^{-1}) [397], slightly lower for ferrites (around 20×10^{-6} K^{-1}) [398], whereas in case of nickelates and cuprates the coefficients are typically in the range of $11-14 \times 10^{-6}$ K^{-1} [399–403]. Among all of the compounds with Ruddlesden-Popper structure nickelates are the most widely studied as the cathode materials due to their high diffusion coefficient of the oxygen ions and good electronic conductivity [400].

Similar to other compounds, the thermal expansion coefficient can be decreased by changing the ionicity of the A-O bond. This can be done by partial or full replacement of the alkali metal by a rare earth element [397,398]. The effect of decreasing the ionic radius of the rare earth element is not as clear as it is, for example, in the case of double perovskites. In the Ln_2NiO_4 series the thermal expansion coefficient decreases in the order Pr < La < Nd [399]. In another study, however [404], substitution of La by Pr in $La_{2-x}Pr_xNiO_{4+\delta}$ ($0 \leq x \leq 2$) had almost no effect on the thermal expansion coefficient which was in the range of $13.0-13.5 \times 10^{-6}$ K^{-1} for all studied compositions. The reason for this discrepancy can be a result of two effects. First of all, Pr can either display a +3 or +4 oxidation state, each of which have different ionic radii. The second reason lies in changes in the crystal symmetry observed for this material system. As indicated by Flura et al. [399], the $Ln_2NiO_{4+\delta}$ structure can exist in at least three different space group symmetries: two orthorhombic types—Bmab (no. 64) or Fmmm (no. 69)—and a tetragonal one—I4/mmm (no. 139). Starting with La, the symmetry of the $La_2NiO_{4+\delta}$ cell changes from orthorhombic Fmmm to tetragonal I4/mmm but the coefficient stays constant and equal to 13×10^{-6} K^{-1}. $Pr_2NiO_{4+\delta}$ on the other hand undergoes a structural change from Bmab to I4/mmm at around 400 °C, which results in a decrease in the expansion coefficient from 14.2 to 13.4×10^{-6} K^{-1}. A transition similar to the case of the La-based compound is observed for $Nd_2NiO_{4+\delta}$ at around 400 °C, where the structure changes from an orthorhombic Fmmm to tetragonal I4/mmm. However, in this case the transition is associated with a change in the thermal expansion coefficient from 11.1 to 12.4×10^{-6} K^{-1}.

All in all, the general rule for the nickelates is that thermal expansion coefficient is lower than in the case of other electrode materials. This seems to be true not only for Ruddlesden-Popper structures with n = 1 but also for higher order structures. Amow et al. [401] performed a comparative study of $La_{n+1}Ni_nO_{3n+1}$ for n = 1, 2 and 3 and found that the thermal expansion coefficient is 13.8×10^{-6} K^{-1} for n = 1, while it hardly changes for n = 2 or 3, displaying values of 13.2×10^{-6} K^{-1}.

Interestingly, the chemical expansion upon reduction in Ruddlesden-Popper electrode materials is very low [399,405], differentiating them from other electrode materials. It is important to note that there is a fundamental difference between oxygen nonstoichiometry in Ruddlesden-Popper structure types and, for instance, perovskite electrodes. The former has an excess of oxygen ions, in the form of oxygen interstitials, whereas in the latter oxygen vacancies are formed and the material exhibits oxygen deficiency. The interstitial defects are formed to compensate high structural stress arising from competing A-O and B-O bonds, which are present between perovskite and rock salt layers. The affinity of oxygen incorporation is so high that the compounds can incorporate excess oxygen even at room temperature [403]. As a result fast oxygen diffusion paths for oxygen transport are formed in the material and the unit cell changes anisotropically as δ increases. Kharton et al. [405] showed that the lattice parameter a increases while the corresponding lattice parameter c decreases with increasing oxygen nonstoichiometry. As a result of this, the total chemical expansion of Ruddlesden-Popper compounds is nearly zero—the chemical contribution to the apparent thermochemical expansion coefficient does not exceed 5%.

Table 13. Linear thermal expansion coefficients of Ruddlesden-Popper structured oxides.

Composition		Thermal Expansion Coefficient/10^{-6} K^{-1}	Temperature Range	References
$LaSrCoO_{4-\delta}$		25.3		
$La_{0.9}Sr_{1.1}CoO_{4-\delta}$		24.3	300–700 °C	Hu et al. [397]
$La_{1.1}Sr_{0.9}CoO_{4-\delta}$		25.4		
$Sr_3Fe_2O_{6+\delta}$		20	RT–900 °C	Prado et al. [398]
$Pr_2NiO_{4+\delta}$		13.4		
$La_2NiO_{4+\delta}$		13	400–1000 °C	Flura et al. [399]
$Nd_2NiO_{4+\delta}$		12.4		
$La_2NiO_{4+\delta}$		13		
$Pr_2NiO_{4+\delta}$		13.6	RT–950 °C	Boehm et al. [406]
$Nd_2NiO_{4+\delta}$		12.7		
$La_2NiO_{4+\delta}$		13.7	RT–900 °C	Skinner et al. [402]
$La_2Ni_{0.5}Cu_{0.5}O_{4+\delta}$		12.8	RT–950 °C	
$La_2NiCuO_{4+\delta}$		8.6	RT–250 °C	Boehm et al. [403]
		13.6	250–900 °C	
$La_{n+1}Ni_nO_{3n+1}$	n = 1	13.8		
	n = 2	13.2	RT–900 °C	Amow et al. [400,401]
	n = 3	13.2		
$La_{2-x}Pr_xNiO_{4+\delta}$ $0\leq x \leq 0.2$		13–13.5	RT–1000 °C	Vibhu et al. [404]

5.2.3. Expansion of Fuel Side Electrodes

Unlike the air-side electrodes, which have been studied extensively, much less has been done on the fuel side electrodes. A major reason for this is from the benefit of having reducing atmospheres, enabling the use of composites composed of the proton or oxide ion conducting electrolyte and an electronically conducting metal. Under reducing atmospheres, the metal will no longer run the risk of oxidizing, which would very easily lower its performance and possibly lead to device failure. The majority of reducing atmosphere electrodes is therefore based on ceramic-metal composites, cermets, and these will be the focus of the remaining part of this section.

A key benefit of using cermets as electrodes is that they tend to display minimal thermal mismatch with the electrolyte, as the ceramic part is generally based on the same composition. The metallic component is typically a transition metal, with Ni being by far the most popular choice, mostly due to its excellent catalytic activity towards H_2 oxidation. Unfortunately, limited work has investigated the thermal and chemical expansion of these cermets, relying mostly on the use of simple expressions, for example, (36) in Section 4.2. As many success stories from the SOFC/SOEC community tend to make their way into the PCFC/PCEC community, we choose to take advantage of the work that has been done on the widely successful YSZ/Ni cermet and will draw parallels to the proton conducting analogue, BZCY/Ni cermet.

Before continuing, it is important to describe how these fuel side electrode cermets are fabricated. The most common practice is to co-sinter a thin electrolyte (YSZ or BZCY) onto a composite support composed of NiO and the electrolyte material [32,407–410], YSZ on YSZ/NiO for SOFCs, or BZCY on BZCY/NiO in the case of PCFCs. The co-sintering takes place in oxidizing atmospheres at temperatures above 1300 °C, followed by a reduction step to reduce NiO to Ni. Consequently, the expansion of the whole device needs to be understood with both NiO and Ni.

Figure 10a shows the thermal expansion coefficients as a function temperature for the case of an SOFC/SOEC system and all its components, including Ni, NiO, YSZ, as well as the composites Ni/YSZ and NiO/YSZ, while (b) provides the PCFC/PCEC equivalent system replacing YSZ with BZCY. For the YSZ-system in (a), the data has been taken from Mori et al. [131], in which the composites had 52 vol% and 40 vol% NiO and Ni, respectively. For the BZCY equivalent, we use the data from Figure 4b for BZCY and (36) to calculate the expansion of NiO/BZCY composite. We used the same proportions between NiO and BZCY phases as in the case of NiO/YSZ for the sake of simplicity. Unfortunately, such a simple model cannot be used to calculate the expansion of Ni/BCZY and this will be explained in detail in the next paragraphs upon dealing with the thermal expansion of the Ni/YSZ cermet as an example.

As expected, a linear dependence between the amount of NiO in the anode and the thermal expansion of NiO/YSZ composites was observed in the experimental data presented [131,411]. This goes in line with the model described by Equation (36). This assures the validity of calculation of thermal expansion coefficients of NiO/BZCY composite. Therefore, in both cases, the effective expansion coefficient will be proportional to the thermal expansion of both NiO and the electrolyte material, and the relative volume proportion between the two. It can be seen that the antiferromagnetic-to-paramagnetic transition of NiO strongly affects also the properties of both NiO/YSZ and NiO/BZCY.

Cermets obtained after reduction of the composites will exhibit different properties. The expansion of Ni/YSZ is dominated by the electrolyte backbone. The anomalous peak due to the ferromagnetic-to-paramagnetic transition of Ni is not observed in the thermal expansion coefficient of the cermet and as the temperature increases the influence of Ni on the total expansion of cermet diminishes completely, such that the thermal expansion coefficient of the cermet approaches that of the electrolyte. This is caused by the ductility of metallic Ni. The strain caused by the thermal mismatch between the ceramic and metallic part of the composite causes the metal part to plastically deform. The situation is very different than in the case where two ceramic materials form a composite, as in the case of the NiO-based composites. For that reason, Equation (36) could not be used to calculate thermal expansion of Ni/BZCY, because the model on which it is based does not account for such effects. However, it can be assumed that Ni/BZCY will behave similar to Ni/YSZ. Namely, at lower temperatures some effects of Ni presence will be observed, whereas at higher temperatures the thermal expansion of the cermet will equal that of BZCY. All in all, in the whole temperature range the thermal expansion coefficient of Ni/BZCY should be expected to be rather close to the values of BZCY.

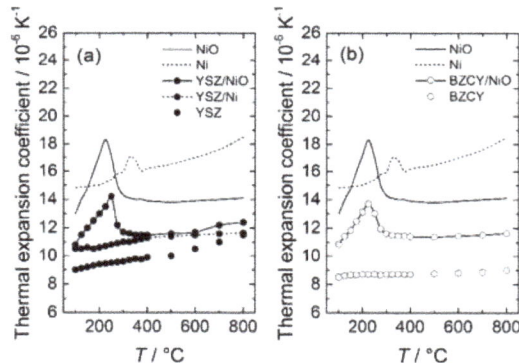

Figure 10. Thermal expansion coefficients (α) of (**a**) NiO, YSZ, YSZ/NiO with 52 vol% NiO, Ni and YSZ/Ni with 40 vol% Ni [131] and (**b**) NiO, BZCY72 (from Figure 4b), BZCY72/NiO with 52 vol% NiO and Ni. Note that the thermal expansion coefficient for the cermet BZCY72/Ni is not included as described in the text.

One of the main challenges of these electrodes lies in the redox cycles caused by an interruption of fuel [412]. Based on their molar volumes, the oxidation of Ni to NiO is accompanied by an increase of solid volume by 69.9%. Strategies to alleviate this problem are to increase the porosity of the anode or to use an uneven distribution of the electrolyte grain size. Nasani et al. [413] studied the redox behavior of the anode Ni/BZY cermet and showed a similar expansion of NiO upon reoxidation as the one observed for YSZ/NiO. Their microstructural analysis proved that the microcracks and delamination observed were due to the nickel expansion phase upon reoxidation.

A recent study focused on the preparation on Ni/BZY20 anodes, varying the amount of NiO [414]. The co-sintering of the BZY20 onto BZY20/NiO supports failed for 80 wt.% of NiO, due to the difference in thermal expansion coefficients between NiO and BZY20. Indeed, the NiO-BZY20 supports shrink more than the thin BZY20 electrolyte upon cooling. The authors recommend supports BZY20/NiO with <70 wt.% NiO. While the ratio NiO to BZCY is of importance, the composition of the PCC material also appears to be an important parameter. Stability of Ni-BaCe$_{0.8}$Y$_{0.2}$O$_{3-d}$ (Ni-BCY) and Ni-BaCe$_{0.6}$Zr$_{0.2}$Y$_{0.2}$O$_{3-d}$ (Ni-BCZY) after reduction and oxidation (redox) cycles at 800 °C showed that NiO particles had a stronger bonding with BCZY particles than with BCY particles. SEM images after redox cycles showed larger and more cracks for BCY than BCZY [310].

One difference to consider is that contrary to YSZ, proton conducting oxides used in the cermet electrodes of PCFCs exhibit chemical expansion upon hydration (see Sections 2.2.1, 4.1.1 and 5.1.1), i.e., the lattice expands upon the incorporation of protonic defects into the lattice. The hydration of BZY/NiO supports with BZY thin membranes is very critical. Onishi et al. reported that the hydration of BZY20 should be carried out at higher temperatures to alleviate compressive stress [414]. This observation is based on the observation of cells (60 wt.%NiO/BZY20 with thin BZY20 electrolyte) heated to 100 or 200 °C in dry Ar and then up to 600 °C in 3% H$_2$O in Ar. Cracks are observed only on the cells hydrated at 100 °C.

Finally, the hydration of the proton conducting ceramic also influences the strength of the material. Sazinas et al. [415] reported enhanced fracture toughness for hydrated samples compared to dehydrated samples.

6. Thermal Compatibility and Mismatch: Symmetries, Issues, Methods of Mitigation

The previous sections compiled expansion coefficients for different classes of electrolytes and electrode materials (both air side and reducing atmosphere electrodes). In this part, guidelines to minimize expansion coefficient mismatch between the different layers of a PCFC, PCEC or membrane reactor are proposed. First, different parameters that can lead to cell failure are listed, before moving to mitigating strategies summarized in Figure 11.

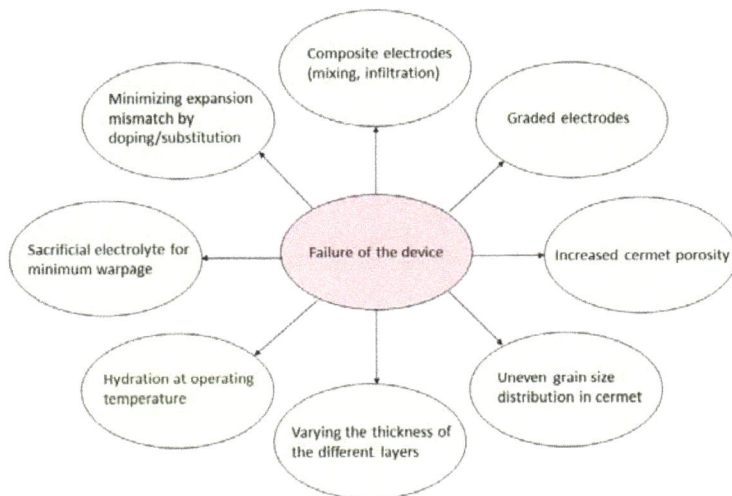

Figure 11. Summary of the different strategies when cells are failing due to mismatch between the expansion coefficients of the different layers.

6.1. Issues That Can Lead to Cell Failure

Results on both SOFC and PCFC studies are used for the following list, as work on the PCFCs is scarcer. It is, however, important to note that the operating temperature of protonic ceramic based devices is significantly lower than that of SOFCs (500–600 °C versus 800–1000 °C). As a result, the chemical expansion upon reduction observed in typical SOFC cathodes (BSCF and LSCF) is much lower at 500–600 °C than at 800–1000 °C (see Figure 2) and should thus pose much less of a problem.

Typically, the most critical issue lays in the large mismatch between the coefficient of expansion of the different layers. As indicated in previous sections thermal expansion coefficients of different layers of the electrochemical cell can span from the range 8–12×10^{-6} K^{-1} for the electrolytes (Section 5.1) up to 25×10^{-6} K^{-1} for the electrodes (Section 5.2). This will result in cracks or delamination between the different layers.

Another issue stems from chemical expansion caused either by hydration or reduction of a material. The former will appear in PCCs during either the treatment in conditions where water vapor partial pressures are changing, or as a result of exposing material with thermochemical history to conditions of different humidity (e.g., material pre-dried or pre-saturated with protons exposed respectively to humid or dry gas). In the case of varying water vapor pressure PCC will expand/contract due to hydration/drying. As a result, stresses will appear on the interfaces between the PCC and the other layers of the electrochemical device. In another case, device failure can be observed when a PCC with a thermal history is exposed to conditions different to the one it was previously treated in. An example could be a dry proton conductor, that is, completely dehydrated, being co-sintered with an electrode material in humid conditions. This will cause the PCC material to expand due to hydration. Chemical expansion upon reduction will result in similar issues as those presented for hydration, although this will be typically restricted to electrode materials and will be present at higher temperatures.

Although cell failures tend to stem from mismatches in expansion coefficients between the different components, microstructural changes can also pose a significant problem for the integrity of the electrochemical cell. This is typically encountered when electrodes composed of composites are used. A degradation of the cell can thus be caused by significant changes in the composite's microstructure during fabrication, as the volume fractions of the individual components will change, which subsequently alters the overall thermal expansion coefficient. Further, one may experience chemical mixing due to ion diffusion at elevated temperatures, resulting in a poorer performance of the composite.

6.2. Mitigating Strategies

The best strategy to mitigate the mismatch depends on the reason of the interfacial stresses. If the reason lies only in the difference of thermal expansion coefficients, then the simplest strategy would be choosing materials with a smaller difference. Expansion coefficients for the different electrolyte materials and electrodes materials can be extracted from Figures 5–9 and Tables 2–13. The coefficients can be tailored by choosing an adequate dopant, dopant content, substituent and tolerance factor. As already mentioned in Section 4.1.1, the history of the materials used for cell fabrication will also influence their thermal and chemical expansions due to kinetic effects. It is hard to predict the exact thermal expansion coefficients mismatch that a cell can withstand, because this is related not only to the difference of coefficient values but also to other mechanical properties of ceramics, such as Young's modulus, tensile strength or fracture toughness and these can vary significantly from one case to another. However, Zhang and Xia [416] presented a model for calculating the durability of a fuel cell on the basis of the performance of the electrode. They conclude that with a high restriction of durability (performance loss expressed as an increase of polarization resistance that does not exceed 1% over 100 thermal cycles from RT to 800 °C), the upper limit for the thermal expansion coefficient of an electrode material should be 15.1×10^{-6} K^{-1} for the case of an electrolyte with α equal to

$10.8 \times 10^{-6} \text{ K}^{-1}$. On this basis, we can estimate that the safety limit in which the performance of an electrochemical device should not be obstructed is for a thermal expansion mismatch of less than 50%.

Another strategy of mitigating the mismatch problem is to fabricate electrolyte-electrode composites. Several methods have previously been used to synthesize such composites, including direct mixing of the two phases as well as infiltration/impregnation based techniques [16,317,319,320,336–339]. These composites can be used as interlayers with intermediate thermal expansion coefficient, which will reduce the mismatch that would appear if the electrode was in direct contact with the electrolyte.

A further development of that idea would be to produce a composite electrode with a graded composition. Graded electrodes have been studied for both SOFC and PCFC anodes [417–420] and have been reported to decrease the probability of failure [421]. Continuously graded BaCe$_{0.5}$Zr$_{0.35}$Y$_{0.15}$O$_{3-d}$ (BCZY)/NiO anode functional layers were prepared by electrostatic slurry spray deposition with a rotation stage [420]. In addition to extending the active three-phase boundary (TPB) length, this functionally and continuously gradient layer also helps to reduce the undesirable discontinuities of thermal expansion coefficients, which would contribute to enhance chemical and mechanical stabilities.

An approach to limit crack formation and warping during the half-cell fabrication consists in depositing an electrolyte layer on both sides of the electrode material. After co-sintering, the sacrificial electrolyte layer is removed [32,318,422].

The stability of the cell can be further tailored by controlling microstructure of its layers. For instance, as already mentioned in Section 5.2.3, high porosity reducing atmosphere electrodes show improved stability upon redox cycles. Moreover, an uneven distribution of the electrolyte grain size in the composite also improves the stability of the electrochemical cell upon redox cycling [412].

The stability can be also controlled by geometry features. Majumdar et al. reported that varying the thickness and thickness ratio of the cell components can be used to tolerate thermal expansion mismatch [411]. For instance, it was observed that 600-mm-thick samples Ni/YSZ were approximately 60% stronger after 8 h of reduction than 900-mm-thick samples [423].

The problem of chemical expansion can be tackled by a careful control of the fabrication conditions in terms of temperature, oxygen and water vapor partial pressures. This requires detailed insight to the thermal and chemical expansion of each of the cell components. This knowledge can be used to mitigate the problem of degradation of the cell. For instance, to prevent cracks on BZCY/Ni with a thin BZCY electrolyte, it has been advised to heat the sample in dry atmosphere and then switch to moist conditions when the operating temperature has been reached [119,414,424]. The stability of Ni-based composites can also be improved by controlling the reduction process: Ni et al. [386] studied the influence of the anode reduction temperature (8 h 700 °C, 6 h 800 °C, 4 h 900 °C and 2 h at 1000 °C) on the mechanical properties of Ni-YSZ and found improved strength (~12%) and enhanced elastic moduli at elevated reduction temperatures. This phenomenon was explained by promoted mass diffusion among Ni particles at higher temperature and healing of the defects/flaws within the material.

Lastly using supporting layers should be considered as a potential improvement of the cell stability. Metal supported cells have been developed for SOFCs and a similar approach is also being developed for PCFC/PCECs [425]. Tucker [339] discussed the coefficient of thermal expansion mismatch between the electrolyte and metal support as a function of temperature. At the operating temperature, initial thermal stresses in the cell due to coefficient of thermal expansion mismatch are expected to dissipate via metal creep and a "zero-stress" state is achieved after a sufficient time period. As the thermal expansion coefficient of the metal support is generally higher than that of the electrolyte (also valid for PCFC), the electrolyte is held in compression during cooling, a desirable situation for the mechanical integrity of the cell.

7. Summary

This work provides a set of theory and data for thermochemical expansion of Proton Conducting Ceramics (PCCs). The theory of thermal expansion of a solid from microscopic and macroscopic points of view is provided. The influence of chemical expansion upon hydration and oxidation/reduction of PCCs is analyzed and presented so that the chemical and thermal expansion can be decoupled. The importance of precisely controlling the atmosphere conditions, while studying the expansion of PCCs is explained in the context of chemical expansion.

Experiment conducted within this study show that, as long as atmospheric conditions are controlled and the rate is not too fast (<20 K min^{-1}), the heating/cooling rate has little to no effect in determining thermal expansion coefficients experimentally.

The literature data is given for PCCs, which in this study are divided into two groups: oxides in which protons are dominating charged species—the electrolyte,s and mixed ionic and electronic conductors (MIECs). Data is predominantly gathered from the studies presenting dilatometry results, since this measurement gives the bulk expansion coefficient, which, as discussed in the theory section, is more useful from an application point of view.

The aspect of thermal compatibility in the electrochemical cells based on PCCs is discussed and measures to avoid the thermal mismatch problem are given.

The interest in the research field of PCCs has been continuously increasing and efforts are now directed towards upscaling, where expansion and thermal mismatch become more critical. It is clear from this review that there is only limited available work on the expansion behavior of electrodes compatible with PCCs (both reducing and air sides). This area thus represents an opportunity for high impact research.

Funding: The Polish contribution was supported by two projects: 2015/17/N/ST5/02813 funded by the National Science Centre, Poland and 2016/22/Z/ST5/00691 funded by the European Commission and National Science Centre, Poland via M-era.Net network. AL wishes to express gratitude to the Centre for Earth Evolution and Dynamics, through the CoE grant from the Research Council of Norway, for funding and support. The US contribution was supported by the Colorado School of Mines Foundation via the Angel Research Fund.

Acknowledgments: The authors would like to thank Tadeusz Miruszewski and Bartosz Kamecki from Gdańsk University of Technology for their support during this work.

Conflicts of Interest: The authors declare no conflicts of interest.

Abbreviations and Symbols List

Abbr.	Meaning
a, b, c	lattice parameters
Acc	acceptor dopant
BCN	$Ba_3Ca_{1+x}Nb_{2-x}O_{9-\delta}$
BCY	$BaCe_{1-x}Y_xO_{3-\delta}$
BZCY	$BaZr_{1-x-y}Ce_xY_yO_{3-\delta}$
BZCY72	$BaZr_{0.7}Ce_{0.2}Y_{0.1}O_{3-\delta}$
BSCF	$Ba_{1-x}Sr_xCo_{0.8}Fe_{0.2}O_{3-\delta}$
BZO	barium zirconate
BZY	$BaZr_{1-x}Y_xO_{3-\delta}$
cercer	ceramic ceramic composite
cermet	ceramic metal composite
CTE	coefficient of thermal expansion
DFT	density functional theory
DIL	Dilatometry
EoS	equation of state
G	Gibbs free energy
HA	harmonic approximation
HT	high temperature
HTPC	high temperature proton conductor
HT-XRD	high temperature X-ray diffraction
hydr	hydration/hydrated
K	bulk modulus

Abbr.	Meaning
L	Length
LSCF	$La_{1-x}Sr_xCo_yFe_{1-y}O_{3-\delta}$
Ln	lanthanide
LT	low temperature
LWO	lanthanum tungstate
MD	molecular dynamics
MIEC	mixed ionic electronic conductor
ND	neutron diffraction
PCC	proton conducting ceramic
PCE	proton ceramic electrolyte
PCEC	proton ceramic electrolyser cell
PCFC	proton ceramic fuel cell
QHA	quasi-harmonic approximation
RE	rare earth
RT	room temperature
SOEC	solid oxide electrolyser cell
SOFC	solid oxide fuel cell
$\Delta_{hydr}H$	standard hydration enthalpy
$\Delta_{hydr}S^\circ$	standard hydration entropy
T	temperature
t	transport number
TCO	triple conducting oxide
TEC	thermal expansion coefficient
TG	thermogravimetric analysis
TPB	triple-phase boundary
V	volume
XRD	X-ray diffraction
YSZ	yttria stabilized zirconia
α	coefficient of thermal expansion
β	coefficient of chemical expansion
δ	oxygen non-stoichiometry
ε	uniaxial strain
μ	shear modulus
v	volume fraction

References

1. Desaguliers, J.T. *A Course of Experimental Philosophy*; W. Innys: London, UK, 1745.
2. Touloukian, Y.S.; Kirby, R.K.; Taylor, R.E.; Desai, P.D. *Thermal Expansion*; Springer US: Boston, MA, USA, 1975; ISBN 978-1-4757-1624-5.
3. Marrony, M.; Berger, P.; Mauvy, F.; Grenier, J.C.; Sata, N.; Magrasó, A.; Haugsrud, R.; Slater, P.R.; Taillades, G.; Roziere, J.; et al. *Proton-Conducting Ceramics. From Fundamentals to Applied Research*; Marrony, M., Ed.; Pan Stanford Publishing: Singapore, 2016; ISBN 978-981-4613-84-2.
4. Colomban, P. *Proton Conductors: Solids, Membranes and Gels-Materials and Devices*; Cambridge University Press: Cambridge, UK, 1992.
5. Iwahara, H.; Yajima, T.; Hibino, T.; Ozaki, K.; Suzuki, H. Protonic conduction in calcium, strontium and barium zirconates. *Solid State Ion.* **1993**, *61*, 65–69. [CrossRef]
6. Iwahara, H.; Uchida, H.; Ono, K.; Ogaki, K. Proton Conduction in Sintered Oxides Based on $BaCeO_3$. *J. Electrochem. Soc.* **1988**, *135*, 529–533. [CrossRef]
7. Duan, C.; Tong, J.; Shang, M.; Nikodemski, S.; Sanders, M.; Ricote, S.; Almansoori, A.; OHayre, R.; O'Hayre, R. Readily processed protonic ceramic fuel cells with high performance at low temperatures. *Science* **2015**, *349*, 1321–1326. [CrossRef] [PubMed]

8. Morejudo, S.H.; Zanón, R.; Escolástico, S.; Yuste-Tirados, I.; Malerød-Fjeld, H.; Vestre, P.K.; Coors, W.G.; Martínez, A.; Norby, T.; Serra, J.M.; et al. Direct conversion of methane to aromatics in a catalytic co-ionic membrane reactor. *Science* **2016**, *353*, 563–566. [CrossRef] [PubMed]

9. Coors, W.G. Protonic ceramic fuel cells for high-efficiency operation with methane. *J. Power Sources* **2003**, *118*, 150–156. [CrossRef]

10. Malerød-Fjeld, H.; Clark, D.; Yuste-Tirados, I.; Zanón, R.; Catalán-Martinez, D.; Beeaff, D.; Morejudo, S.H.; Vestre, P.K.; Norby, T.; Haugsrud, R.; et al. Thermo-electrochemical production of compressed hydrogen from methane with near-zero energy loss. *Nat. Energy* **2017**, *2*, 923–931. [CrossRef]

11. Iwahara, H. High temperature proton conducting oxides and their applications to solid electrolyte fuel cells and steam electrolyzer for hydrogen production. *Solid State Ion.* **1988**, *28–30*, 573–578. [CrossRef]

12. Molenda, J.; Kupecki, J.; Baron, R.; Blesznowski, M.; Brus, G.; Brylewski, T.; Bucko, M.; Chmielowiec, J.; Cwieka, K.; Gazda, M.; et al. Status report on high temperature fuel cells in Poland—Recent advances and achievements. *Int. J. Hydrog. Energy* **2017**, *42*, 4366–4403. [CrossRef]

13. Iwahara, H. Proton conducting ceramics and their applications. *Solid State Ion.* **1996**, *86–88*, 9–15. [CrossRef]

14. Norby, T. Solid-state protonic conductors: Principles, properties, progress and prospects. *Solid State Ion.* **1999**, *125*, 1–11. [CrossRef]

15. Zagórski, K.; Wachowski, S.; Szymczewska, D.; Mielewczyk-Gryń, A.; Jasiński, P.; Gazda, M. Performance of a single layer fuel cell based on a mixed proton-electron conducting composite. *J. Power Sources* **2017**, *353*, 230–236. [CrossRef]

16. Wachowski, S.; Li, Z.; Polfus, J.M.; Norby, T. Performance and stability in H_2S of $SrFe_{0.75}Mo_{0.25}O_{3-\delta}$ as electrode in proton ceramic fuel cells. *J. Eur. Ceram. Soc.* **2018**, *38*, 163–171. [CrossRef]

17. Sakai, T.; Matsushita, S.; Matsumoto, H.; Okada, S.; Hashimoto, S.; Ishihara, T. Intermediate temperature steam electrolysis using strontium zirconate-based protonic conductors. *Int. J. Hydrog. Energy* **2009**, *34*, 56–63. [CrossRef]

18. Katahira, K.; Matsumoto, H.; Iwahara, H.; Koide, K.; Iwamoto, T. Solid electrolyte hydrogen sensor with an electrochemically-supplied hydrogen standard. *Sens. Actuators B Chem.* **2001**, *73*, 130–134. [CrossRef]

19. Yajima, T.; Koide, K.; Takai, H.; Fukatsu, N.; Iwahara, H. Application of hydrogen sensor using proton conductive ceramics as a solid electrolyte to aluminum casting industries. *Solid State Ion.* **1995**, *79*, 333–337. [CrossRef]

20. Serret, P.; Colominas, S.; Reyes, G.; Abellà, J. Characterization of ceramic materials for electrochemical hydrogen sensors. *Fusion Eng. Des.* **2011**, *86*, 2446–2449. [CrossRef]

21. Volkov, A.; Gorbova, E.; Vylkov, A.; Medvedev, D.; Demin, A.; Tsiakaras, P. Design and applications of potentiometric sensors based on proton-conducting ceramic materials. A brief review. *Sens. Actuators B Chem.* **2017**, *244*, 1004–1015. [CrossRef]

22. Phair, J.W.; Badwal, S.P.S. Review of proton conductors for hydrogen separation. *Ionics* **2006**, *12*, 103–115. [CrossRef]

23. Lundin, S.T.B.; Patki, N.S.; Fuerst, T.F.; Ricote, S.; Wolden, C.A.; Way, J.D. Dense Inorganic Membranes for Hydrogen Separation. In *Membranes for Gas Separations*; World Scientific: Singapore, 2017; pp. 271–363.

24. Tao, Z.; Yan, L.; Qiao, J.; Wang, B.; Zhang, L.; Zhang, J. A review of advanced proton-conducting materials for hydrogen separation. *Prog. Mater. Sci.* **2015**, *74*, 1–50. [CrossRef]

25. Fontaine, M.L.; Norby, T.; Larring, Y.; Grande, T.; Bredesen, R. Oxygen and Hydrogen Separation Membranes Based on Dense Ceramic Conductors. *Membr. Sci. Technol.* **2008**, *13*, 401–458. [CrossRef]

26. Choi, S.; Kucharczyk, C.J.; Liang, Y.; Zhang, X.; Takeuchi, I.; Ji, H.I.; Haile, S.M. Exceptional power density and stability at intermediate temperatures in protonic ceramic fuel cells. *Nat. Energy* **2018**, *3*, 202–210. [CrossRef]

27. Iwahara, H.; Esaka, T.; Uchida, H.; Maeda, N. Proton conduction in sintered oxides and its application to steam electrolysis for hydrogen production. *Solid State Ion.* **1981**, *3–4*, 359–363. [CrossRef]

28. Vasileiou, E.; Kyriakou, V.; Garagounis, I.; Vourros, A.; Stoukides, M. Ammonia synthesis at atmospheric pressure in a $BaCe_{0.2}Zr_{0.7}Y_{0.1}O_{2.9}$ solid electrolyte cell. *Solid State Ion.* **2015**, *275*, 110–116. [CrossRef]

29. Marnellos, G.; Stoukides, M. Ammonia Synthesis at Atmospheric Pressure. *Science* **1998**, *282*, 98–100. [CrossRef] [PubMed]

30. Gocha, A. *CeramicTechToday from The American Ceramic Society*; The American Ceramic Society: Westerville, OH, USA, 2017.

31. Dubois, A.; Ricote, S.; Braun, R.J. Benchmarking the expected stack manufacturing cost of next generation, intermediate-temperature protonic ceramic fuel cells with solid oxide fuel cell technology. *J. Power Sources* **2017**, *369*, 65–77. [CrossRef]

32. Duan, C.; Kee, R.J.; Zhu, H.; Karakaya, C.; Chen, Y.; Ricote, S.; Jarry, A.; Crumlin, E.J.; Hook, D.; Braun, R.; et al. Highly durable, coking and sulfur tolerant, fuel-flexible protonic ceramic fuel cells. *Nature* **2018**, *557*, 217–222. [CrossRef] [PubMed]

33. Nakajo, A.; Stiller, C.; Härkegård, G.; Bolland, O. Modeling of thermal stresses and probability of survival of tubular SOFC. *J. Power Sources* **2006**, *158*, 287–294. [CrossRef]

34. Tietz, F. Thermal expansion of SOFC materials. *Ionics* **1999**, *5*, 129–139. [CrossRef]

35. Selimovic, A.; Kemm, M.; Torisson, T.; Assadi, M. Steady state and transient thermal stress analysis in planar solid oxide fuel cells. *J. Power Sources* **2005**, *145*, 463–469. [CrossRef]

36. Lin, C.K.; Chen, T.T.; Chyou, Y.P.; Chiang, L.K. Thermal stress analysis of a planar SOFC stack. *J. Power Sources* **2007**, *164*, 238–251. [CrossRef]

37. Laurencin, J.; Delette, G.; Lefebvre-Joud, F.; Dupeux, M. A numerical tool to estimate SOFC mechanical degradation: Case of the planar cell configuration. *J. Eur. Ceram. Soc.* **2008**, *28*, 1857–1869. [CrossRef]

38. Carter, B.; Norton, G. *Ceramic Materials*; Springer: New York, NY, USA, 2007; ISBN 978-0-387-46270-7.

39. Kingery, W.D.; Bowen, H.K.; Uhlmann, D.R. *Introduction to Ceramics*, 2nd ed.; Wiley: Hoboken, NJ, USA, 1976; ISBN 978-0-471-47860-7.

40. Kittel, C.; McEuen, P. *Introduction to Solid State Physics*; Willey: Hoboken, NJ, USA, 1998; Volume 8, ISBN 047141526X.

41. Levy, R.A. *Principles of Solid State Physics*; Academic Press: Cambridge, MA, USA, 1968; ISBN 9780124457508.

42. Brown, F.C. *The Physics of Solids*; W.A. Benjamin: New York, NY, USA, 1967.

43. Krishnan, R.S.; Srinivasan, R.; Devanarayan, S. *Thermal Expansion of Crystals*; Pergamon Press: Oxford, UK, 1979; ISBN 0-08-021405-3.

44. Belousov, R.I.; Filatov, S.K. Algorithm for calculating the thermal expansion tensor and constructing the thermal expansion diagram for crystals. *Glas. Phys. Chem.* **2007**, *33*, 271–275. [CrossRef]

45. Paufler, P.; Weber, T. On the determination of linear expansion coefficients of triclinic crystals using X-ray diffraction. *Eur. J. Mineral.* **1999**, *11*, 721–730. [CrossRef]

46. Branson, D.L. Thermal Expansion Coefficients of Zirconate Ceramics. *J. Am. Ceram. Soc.* **1965**, *48*, 441–442. [CrossRef]

47. Adler, S.B. Chemical Expansivity of Electrochemical Ceramics. *J. Am. Ceram. Soc.* **2004**, *84*, 2117–2119. [CrossRef]

48. Garai, J. Correlation between thermal expansion and heat capacity. *Calphad* **2006**, *30*, 354–356. [CrossRef]

49. Mohazzabi, P.; Behroozi, F. Thermal expansion of solids: A simple classical model. *Eur. J. Phys.* **1997**, *18*, 237–240. [CrossRef]

50. Suzuki, I. Thermal expansion of periclase and olivine and their anharmonic properties. In *Elastic Properties and Equations of State*; American Geophysical Union: Washington, DC, USA, 1988; Volume 23, pp. 361–375, ISBN 0875902405.

51. Samara, G.A.; Morosin, B. Anharmonic Effects in $KTaO_3$: Ferroelectric Mode, Thermal Expansion, and Compressibility. *Phys. Rev. B* **1973**, *8*, 1256–1264. [CrossRef]

52. Li, C.W.; Tang, X.; Muñoz, J.A.; Keith, J.B.; Tracy, S.J.; Abernathy, D.L.; Fultz, B. Structural Relationship between Negative Thermal Expansion and Quartic Anharmonicity of Cubic ScF_3. *Phys. Rev. Lett.* **2011**, *107*, 195504. [CrossRef] [PubMed]

53. Janio de Castro Lima, J.; Paraguassu, A.B. Linear thermal expansion of granitic rocks: Influence of apparent porosity, grain size and quartz content. *Bull. Eng. Geol. Environ.* **2004**, *63*, 215–220. [CrossRef]

54. Parker, F.J.; Rice, R.W. Correlation between Grain Size and Thermal Expansion for Aluminum Titanate Materials. *J. Am. Ceram. Soc.* **1989**, *72*, 2364–2366. [CrossRef]

55. Antal, D.; Húlan, T.; Štubňa, I.; Záleská, M.; Trník, A. The influence of texture on elastic and thermophysical properties of kaolin- and illite-based ceramic bodies. *Ceram. Int.* **2017**, *43*, 2730–2736. [CrossRef]

56. Paulik, S.W.; Faber, K.T.; Fuller, E.R. Development of Textured Microstructures in Ceramics with Large Thermal Expansion Anisotropy. *J. Am. Ceram. Soc.* **1994**, *77*, 454–458. [CrossRef]

57. Mogensen, M.; Sammes, N.M.; Tompsett, G.A. Physical, chemical and electrochemical properties of pure and doped ceria. *Solid State Ion.* **2000**, *129*, 63–94. [CrossRef]

58. Marrocchelli, D.; Perry, N.H.; Bishop, S.R. Understanding chemical expansion in perovskite-structured oxides. *Phys. Chem. Chem. Phys.* **2015**, *17*, 10028–10039. [CrossRef] [PubMed]

59. Marrocchelli, D.; Bishop, S.R.; Tuller, H.L.; Yildiz, B. Understanding Chemical Expansion in Non-Stoichiometric Oxides: Ceria and Zirconia Case Studies. *Adv. Funct. Mater.* **2012**, *22*, 1958–1965. [CrossRef]

60. Marrocchelli, D.; Bishop, S.R.; Tuller, H.L.; Watson, G.W.; Yildiz, B. Charge localization increases chemical expansion in cerium-based oxides. *Phys. Chem. Chem. Phys.* **2012**, *14*, 12070–12074. [CrossRef] [PubMed]

61. Haugsrud, R. On the high-temperature oxidation of nickel. *Corros. Sci.* **2003**, *45*, 211–235. [CrossRef]

62. Richardson, J.T.; Scates, R.; Twigg, M.V. X-ray diffraction study of nickel oxide reduction by hydrogen. *Appl. Catal. A Gen.* **2003**, *246*, 137–150. [CrossRef]

63. Vullum, F.; Nitsche, F.; Selbach, S.M.; Grande, T. Solid solubility and phase transitions in the system LaNb$_{1-x}$Ta$_x$O$_4$. *J. Solid State Chem.* **2008**, *181*, 2580–2585. [CrossRef]

64. Wachowski, S.; Mielewczyk-Gryn, A.; Gazda, M. Effect of isovalent substitution on microstructure and phase transition of LaNb$_{1-x}$M$_x$O$_4$ (M = Sb, V or Ta; x = 0.05–0.3). *J. Solid State Chem.* **2014**, *219*, 201–209. [CrossRef]

65. Haugsrud, R.; Norby, T. High-temperature proton conductivity in acceptor-doped LaNbO$_4$. *Solid State Ion.* **2006**, *177*, 1129–1135. [CrossRef]

66. Vegard, L. Die Konstitution der Mischkristalle und die Raumfüllung der Atome. *Z. Phys.* **1921**, *5*, 17–26. [CrossRef]

67. Yamazaki, Y.; Yang, C.K.; Haile, S.M. Unraveling the defect chemistry and proton uptake of yttrium-doped barium zirconate. *Scr. Mater.* **2011**, *65*, 102–107. [CrossRef]

68. Shannon, R.D. Revised effective ionic radii and systematic studies of interatomic distances in halides and chalcogenides. *Acta Crystallogr. Sect. A* **1976**, *32*, 751–767. [CrossRef]

69. Omata, T.; Noguchi, Y.; Otsuka-Yao-Matsuo, S. Infrared Study of High-Temperature Proton-Conducting Aliovalently Doped SrZrO$_3$ and BaZrO$_3$. *J. Electrochem. Soc.* **2005**, *152*, E200–E205. [CrossRef]

70. Imashuku, S.; Uda, T.; Nose, Y.; Awakura, Y. To journal of phase equilibria and diffusion phase relationship of the BaO-ZrO$_2$-YO$_{1.5}$ system at 1500 and 1600 °C. *J. Phase Equilibria Diffus.* **2010**, *31*, 348–356. [CrossRef]

71. Giannici, F.; Shirpour, M.; Longo, A.; Martorana, A.; Merkle, R.; Maier, J. Long-range and short-range structure of proton-conducting Y:BaZrO$_3$. *Chem. Mater.* **2011**, *23*, 2994–3002. [CrossRef]

72. Shirpour, M.; Rahmati, B.; Sigle, W.; Van Aken, P.A.; Merkle, R.; Maier, J. Dopant segregation and space charge effects in proton-conducting BaZrO$_3$ perovskites. *J. Phys. Chem. C* **2012**, *116*, 2453–2461. [CrossRef]

73. Kreuer, K.D.; Adams, S.; Münch, W.; Fuchs, A.; Klock, U.; Maier, J. Proton conducting alkaline earth zirconates and titanates for high drain electrochemical applications. *Solid State Ion.* **2001**, *145*, 295–306. [CrossRef]

74. Oikawa, I.; Takamura, H. Correlation among Oxygen Vacancies, Protonic Defects, and the Acceptor Dopant in Sc-Doped BaZrO$_3$ Studied by 45Sc Nuclear Magnetic Resonance. *Chem. Mater.* **2015**, *27*, 6660–6667. [CrossRef]

75. Han, D.; Shinoda, K.; Uda, T. Dopant Site Occupancy and Chemical Expansion in Rare Earth-Doped Barium Zirconate. *J. Am. Ceram. Soc.* **2014**, *97*, 643–650. [CrossRef]

76. Hong, S.J.; Virkar, A.V. Lattice Parameters and Densities of Rare-Earth Oxide Doped Ceria Electrolytes. *J. Am. Ceram. Soc.* **1995**, *78*, 433–439. [CrossRef]

77. Andersson, A.K.E.; Selbach, S.M.; Knee, C.S.; Grande, T. Chemical Expansion Due to Hydration of Proton-Conducting Perovskite Oxide Ceramics. *J. Am. Ceram. Soc.* **2014**, *97*, 2654–2661. [CrossRef]

78. Han, D.; Hatada, N.; Uda, T. Chemical Expansion of Yttrium-Doped Barium Zirconate and Correlation with Proton Concentration and Conductivity. *J. Am. Ceram. Soc.* **2016**, *99*, 3745–3753. [CrossRef]

79. Yamaguchi, S.; Yamada, N. Thermal lattice expansion behavior of Yb-doped BaCeO$_3$. *Solid State Ion.* **2003**, *162–163*, 23–29. [CrossRef]

80. Kreuer, K.D. Proton-conducting oxides. *Annu. Rev. Mater. Res.* **2003**, *33*, 333–359. [CrossRef]

81. Kinyanjui, F.G.; Norberg, S.T.; Ahmed, I.; Eriksson, S.G.; Hull, S. In-situ conductivity and hydration studies of proton conductors using neutron powder diffraction. *Solid State Ion.* **2012**, *225*, 312–316. [CrossRef]

82. Jedvik, E.; Lindman, A.; Benediktsson, M.P.; Wahnström, G. Size and shape of oxygen vacancies and protons in acceptor-doped barium zirconate. *Solid State Ion.* **2015**, *275*, 2–8. [CrossRef]

83. Bjørheim, T.S.; Kotomin, E.A.; Maier, J. Hydration entropy of BaZrO$_3$ from first principles phonon calculations. *J. Mater. Chem. A* **2015**, *3*, 7639–7648. [CrossRef]

84. Bjørheim, T.S.; Løken, A.; Haugsrud, R. On the relationship between chemical expansion and hydration thermodynamics of proton conducting perovskites. *J. Mater. Chem. A* **2016**, *4*, 5917–5924. [CrossRef]

85. Løken, A.; Saeed, S.W.; Getz, M.N.; Liu, X.; Bjørheim, T.S. Alkali metals as efficient A-site acceptor dopants in proton conducting BaZrO$_3$. *J. Mater. Chem. A* **2016**, *4*, 9229–9235. [CrossRef]

86. Løken, A.; Haugsrud, R.; Bjørheim, T.S. Unravelling the fundamentals of thermal and chemical expansion of BaCeO$_3$ from first principles phonon calculations. *Phys. Chem. Chem. Phys.* **2016**, *18*, 31296–31303. [CrossRef] [PubMed]

87. Løken, A.; Bjørheim, T.S.; Haugsrud, R. The pivotal role of the dopant choice on the thermodynamics of hydration and associations in proton conducting BaCe$_{0.9}$X$_{0.1}$O$_{3-\delta}$ (X = Sc, Ga, Y, In, Gd and Er). *J. Mater. Chem. A* **2015**, *3*, 23289–23298. [CrossRef]

88. Kim, H.S.; Jang, A.; Choi, S.Y.; Jung, W.; Chung, S.Y. Vacancy-Induced Electronic Structure Variation of Acceptors and Correlation with Proton Conduction in Perovskite Oxides. *Angew. Chem. Int. Ed.* **2016**, *55*, 13499–13503. [CrossRef] [PubMed]

89. Løken, A. Hydration Thermodynamics of Oxides. Effects of Defect Associations. Ph.D. Thesis, University of Oslo, Oslo, Norway, 2017.

90. Bishop, S.R.; Duncan, K.L.; Wachsman, E.D. Defect equilibria and chemical expansion in non-stoichiometric undoped and gadolinium-doped cerium oxide. *Electrochim. Acta* **2009**, *54*, 1436–1443. [CrossRef]

91. Marrocchelli, D.; Chatzichristodoulou, C.; Bishop, S.R. Defining chemical expansion: The choice of units for the stoichiometric expansion coefficient. *Phys. Chem. Chem. Phys.* **2014**, *16*, 9229–9232. [CrossRef] [PubMed]

92. Atkinson, A.; Ramos, T.M.G.M. Chemically-induced stresses in ceramic oxygen ion-conducting membranes. *Solid State Ion.* **2000**, *129*, 259–269. [CrossRef]

93. Jiang, S.P. A comparison of O$_2$ reduction reactions on porous (La,Sr)MnO$_3$ and (La,Sr)(Co,Fe)O$_3$ electrodes. *Solid State Ion.* **2002**, *146*, 1–22. [CrossRef]

94. Esquirol, A.; Brandon, N.P.; Kilner, J.A.; Mogensen, M. Electrochemical Characterization of La$_{0.6}$Sr$_{0.4}$Co$_{0.2}$Fe$_{0.8}$O$_3$ Cathodes for Intermediate-Temperature SOFCs. *J. Electrochem. Soc.* **2004**, *151*, A1847–A1855. [CrossRef]

95. Tietz, F.; Haanappel, V.A.C.; Mai, A.; Mertens, J.; Stöver, D. Performance of LSCF cathodes in cell tests. *J. Power Sources* **2006**, *156*, 20–22. [CrossRef]

96. Ricote, S.; Bonanos, N.; Rørvik, P.M.; Haavik, C. Microstructure and performance of La$_{0.58}$Sr$_{0.4}$Co$_{0.2}$Fe$_{0.8}$O$_{3-\delta}$ cathodes deposited on BaCe$_{0.2}$Zr$_{0.7}$Y$_{0.1}$O$_{3-\delta}$ by infiltration and spray pyrolysis. *J. Power Sources* **2012**, *209*, 172–179. [CrossRef]

97. Sun, S.; Cheng, Z. Electrochemical Behaviors for Ag, LSCF and BSCF as Oxygen Electrodes for Proton Conducting IT-SOFC. *J. Electrochem. Soc.* **2017**, *164*, F3104–F3113. [CrossRef]

98. Bishop, S.R.; Duncan, K.L.; Wachsman, E.D. Surface and Bulk Defect Equilibria in Strontium-Doped Lanthanum Cobalt Iron Oxide. *J. Electrochem. Soc.* **2009**, *156*, B1242–B1248. [CrossRef]

99. Bishop, S.R.; Duncan, K.L.; Wachsman, E.D. Thermo-Chemical Expansion in Strontium-Doped Lanthanum Cobalt Iron Oxide. *J. Am. Ceram. Soc.* **2010**, *93*, 4115–4121. [CrossRef]

100. Kuhn, M.; Hashimoto, S.; Sato, K.; Yashiro, K.; Mizusaki, J. Thermo-chemical lattice expansion in La$_{0.6}$Sr$_{0.4}$Co$_{1-y}$Fe$_y$O$_{3-\delta}$. *Solid State Ion.* **2013**, *241*, 12–16. [CrossRef]

101. James, J.D.; Spittle, J.A.; Brown, S.G.R.; Evans, R.W. A review of measurement techniques for the thermal expansion coefficient of metals and alloys at elevated temperatures. *Meas. Sci. Technol.* **2001**, *12*, R1–R15. [CrossRef]

102. Mielewczyk-Gryn, A.; Gdula-Kasica, K.; Kusz, B.; Gazda, M. High temperature monoclinic-to-tetragonal phase transition in magnesium doped lanthanum ortho-niobate. *Ceram. Int.* **2013**, *39*, 4239–4244. [CrossRef]

103. Huse, M.; Skilbred, A.W.B.; Karlsson, M.; Eriksson, S.G.; Norby, T.; Haugsrud, R.; Knee, C.S. Neutron diffraction study of the monoclinic to tetragonal structural transition in LaNbO$_4$ and its relation to proton mobility. *J. Solid State Chem.* **2012**, *187*, 27–34. [CrossRef]

104. Rietveld, H.M. A profile refinement method for nuclear and magnetic structures. *J. Appl. Crystallogr.* **1969**, *2*, 65–71. [CrossRef]

105. Le Bail, A.; Duroy, H.; Fourquet, J.L. Ab-initio structure determination of LiSbWO$_6$ by X-ray powder diffraction. *Mater. Res. Bull.* **1988**, *23*, 447–452. [CrossRef]

106. Tsvetkov, D.S.; Sereda, V.V.; Zuev, A.Y. Oxygen nonstoichiometry and defect structure of the double perovskite GdBaCo$_2$O$_{6-\delta}$. *Solid State Ion.* **2010**, *180*, 1620–1625. [CrossRef]

107. Zuev, A.Y.; Tsvetkov, D.S. Conventional Methods for Measurements of Chemo-Mechanical Coupling. In *Electro-Chemo-Mechanics of Solids*; Bishop, S.R., Perry, N.H., Marrocchelli, D., Sheldon, B., Eds.; Springer: New York, NY, USA, 2017; pp. 5–33, ISBN 9783319514055.

108. Nedeltcheva, T.; Simeonova, P.; Lovchinov, V. Improved iodometric method for simultaneous determination of non-stoichiometric oxygen and total copper content in YBCO superconductors. *Anal. Chim. Acta* **1995**, *312*, 227–229. [CrossRef]

109. Rørmark, L.; Wiik, K.; Stølen, S.; Grande, T. Oxygen stoichiometry and structural properties of $La_{1-x}A_xMnO_{3\pm\delta}$ (A = Ca or Sr and $0 \leq x \leq 1$). *J. Mater. Chem.* **2002**, *12*, 1058–1067. [CrossRef]

110. Baroni, S.; de Gironcoli, S.; Dal Corso, A.; Giannozzi, P. Phonons and related crystal properties from density-functional perturbation theory. *Rev. Mod. Phys.* **2001**, *73*, 515–562. [CrossRef]

111. Togo, A.; Tanaka, I. First principles phonon calculations in materials science. *Scr. Mater.* **2015**, *108*, 1–5. [CrossRef]

112. Vinet, P.; Smith, J.R.; Ferrante, J.; Rose, J.H. Temperature effects on the universal equation of state of solids. *Phys. Rev. B* **1987**, *35*, 1945–1953. [CrossRef]

113. Birch, F. Finite Elastic Strain of Cubic Crystals. *Phys. Rev.* **1947**, *71*, 809–824. [CrossRef]

114. Nosé, S. A unified formulation of the constant temperature molecular dynamics methods. *J. Chem. Phys.* **1984**, *81*, 511–519. [CrossRef]

115. Nosé, S. A molecular dynamics method for simulations in the canonical ensemble. *Mol. Phys.* **1984**, *52*, 255–268. [CrossRef]

116. Hoover, W.G. Canonical dynamics: Equilibrium phase-space distributions. *Phys. Rev. A* **1985**, *31*, 1695–1697. [CrossRef]

117. Berendsen, H.J.C.; Postma, J.P.M.; van Gunsteren, W.F.; DiNola, A.; Haak, J.R. Molecular dynamics with coupling to an external bath. *J. Chem. Phys.* **1984**, *81*, 3684–3690. [CrossRef]

118. Zhao, Y.; Weidner, D.J. Thermal expansion of $SrZrO_3$ and $BaZrO_3$ perovskites. *Phys. Chem. Miner.* **1991**, *18*, 294–301. [CrossRef]

119. Hudish, G.; Manerbino, A.; Coors, W.G.; Ricote, S. Chemical expansion in $BaZr_{0.9-x}Ce_xY_{0.1}O_{3-\delta}$ (x = 0 and 0.2) upon hydration determined by high-temperature X-ray diffraction. *J. Am. Ceram. Soc.* **2018**, *101*, 1298–1309. [CrossRef]

120. Hiraiwa, C.; Han, D.; Kuramitsu, A.; Kuwabara, A.; Takeuchi, H.; Majima, M.; Uda, T. Chemical Expansion and Change in Lattice Constant of Y-Doped $BaZrO_3$ by Hydration/Dehydration Reaction and Final Heat-Treating Temperature. *J. Am. Ceram. Soc.* **2013**, *96*, 879–884. [CrossRef]

121. Lein, H.L.; Wiik, K.; Grande, T. Thermal and chemical expansion of mixed conducting $La_{0.5}Sr_{0.5}Fe_{1-x}Co_xO_{3-\delta}$ materials. *Solid State Ion.* **2006**, *177*, 1795–1798. [CrossRef]

122. Chen, X.; Yu, J.; Adler, S.B. Thermal and Chemical Expansion of Sr-Doped Lanthanum Cobalt Oxide $(La_{1-x}Sr_xCoO_{3-\delta})$. *Chem. Mater.* **2005**, *17*, 4537–4546. [CrossRef]

123. Fossdal, A.; Menon, M.; Waernhus, I.; Wiik, K.; Einarsrud, M.A.; Grande, T. Crystal Structure and Thermal Expansion of $La_{1-x}Sr_xFeO_{3-\delta}$ Materials. *J. Am. Ceram. Soc.* **2005**, *87*, 1952–1958. [CrossRef]

124. Hashimoto, S.; Fukuda, Y.; Kuhn, M.; Sato, K.; Yashiro, K.; Mizusaki, J. Thermal and chemical lattice expansibility of $La_{0.6}Sr_{0.4}Co_{1-y}Fe_yO_{3-\delta}$ (y = 0.2, 0.4, 0.6 and 0.8). *Solid State Ion.* **2011**, *186*, 37–43. [CrossRef]

125. Kuhn, M.; Hashimoto, S.; Sato, K.; Yashiro, K.; Mizusaki, J. Oxygen nonstoichiometry, thermo-chemical stability and lattice expansion of $La_{0.6}Sr_{0.4}FeO_{3-\delta}$. *Solid State Ion.* **2011**, *195*, 7–15. [CrossRef]

126. Mather, G.C.; Heras-Juaristi, G.; Ritter, C.; Fuentes, R.O.; Chinelatto, A.L.; Pérez-Coll, D.; Amador, U. Phase Transitions, Chemical Expansion, and Deuteron Sites in the $BaZr0.7Ce0.2Y0.1O3-\delta$ Proton Conductor. *Chem. Mater.* **2016**, *28*, 4292–4299. [CrossRef]

127. Kerner, E.H. The Elastic and Thermo-elastic Properties of Composite Media. *Proc. Phys. Soc. Sect. B* **1956**, *69*, 808–813. [CrossRef]

128. Pratihar, S.K.; Dassharma, A.; Maiti, H.S. Properties of Ni/YSZ porous cermets prepared by electroless coating technique for SOFC anode application. *J. Mater. Sci.* **2007**, *42*, 7220–7226. [CrossRef]

129. Coble, R.L.; Kingery, W.D. Effect of Porosity on Physical Properties of Sintered Alumina. *J. Am. Ceram. Soc.* **1956**, *39*, 377–385. [CrossRef]

130. Shyam, A.; Bruno, G.; Watkins, T.R.; Pandey, A.; Lara-curzio, E.; Parish, C.M.; Stafford, R.J. Journal of the European Ceramic Society The effect of porosity and microcracking on the thermomechanical properties of cordierite. *J. Eur. Ceram. Soc.* **2015**, *35*, 4557–4566. [CrossRef]

131. Mori, M.; Yamamoto, T.; Itoh, H.; Inaba, H.; Tagawa, H. Thermal Expansion of Nickel-Zirconia Anodes in Solid Oxide Fuel Cells during Fabrication and Operation. *J. Electrochem. Soc.* **1998**, *145*, 1374–1381. [CrossRef]

132. Elomari, S.; Skibo, M.D.; Sundarrajan, A.; Richards, H. Thermal expansion behavior of particulate metal-matrix composites. *Compos. Sci. Technol.* **1998**, *58*, 369–376. [CrossRef]

133. Sevostianov, I. On the thermal expansion of composite materials and cross-property connection between thermal expansion and thermal conductivity. *Mech. Mater.* **2012**, *45*, 20–33. [CrossRef]

134. Hayashi, H.; Saitou, T.; Maruyama, N.; Inaba, H.; Kawamura, K.; Mori, M. Thermal expansion coefficient of yttria stabilized zirconia for various yttria contents. *Solid State Ion.* **2005**, *176*, 613–619. [CrossRef]

135. Fabbri, E.; Pergolesi, D.; Traversa, E. Materials challenges toward proton-conducting oxide fuel cells: A critical review. *Chem. Soc. Rev.* **2010**, *39*, 4355–4369. [CrossRef] [PubMed]

136. Malavasi, L.; Fisher, C.A.J.; Islam, M.S. Oxide-ion and proton conducting electrolyte materials for clean energy applications: Structural and mechanistic features. *Chem. Soc. Rev.* **2010**, *39*, 4370–4387. [CrossRef] [PubMed]

137. Norby, T. Proton Conductivity in Perovskite Oxides. In *Perovskite Oxide for Solid Oxide Fuel Cells*; Ishihara, T., Ed.; Springer: Boston, MA, USA, 2009; pp. 217–241, ISBN 978-0-387-77708-5.

138. Hossain, S.; Abdalla, A.M.; Jamain, S.N.B.; Zaini, J.H.; Azad, A.K. A review on proton conducting electrolytes for clean energy and intermediate temperature-solid oxide fuel cells. *Renew. Sustain. Energy Rev.* **2017**, *79*, 750–764. [CrossRef]

139. Wang, S.; Zhao, F.; Zhang, L.; Chen, F. Synthesis of $BaCe_{0.7}Zr_{0.1}Y_{0.1}Yb_{0.1}O_{3-\delta}$ proton conducting ceramic by a modified Pechini method. *Solid State Ion.* **2012**, *213*, 29–35. [CrossRef]

140. Lagaeva, J.; Medvedev, D.; Demin, A.; Tsiakaras, P. Insights on thermal and transport features of $BaCe_{0.8-x}Zr_xY_{0.2}O_{3-\delta}$ proton-conducting materials. *J. Power Sources* **2015**, *278*, 436–444. [CrossRef]

141. Yamazaki, Y.; Hernandez-Sanchez, R.; Haile, S.M. High Total Proton Conductivity in Large-Grained Yttrium-Doped Barium Zirconate. *Chem. Mater.* **2009**, *21*, 2755–2762. [CrossRef]

142. Yamazaki, Y.; Blanc, F.; Okuyama, Y.; Buannic, L.; Lucio-Vega, J.C.; Grey, C.P.; Haile, S.M. Proton trapping in yttrium-doped barium zirconate. *Nat. Mater.* **2013**, *12*, 647–651. [CrossRef] [PubMed]

143. Ryu, K.H.; Haile, S.M. Chemical stability and proton conductivity of doped $BaCeO_3$–$BaZrO_3$ solid solutions. *Solid State Ion.* **1999**, *125*, 355–367. [CrossRef]

144. Haugsrud, R. High Temperature Proton Conductors—Fundamentals and Functionalities. *Diffus. Found.* **2016**, *8*, 31–79. [CrossRef]

145. Akbarzadeh, A.R.; Kornev, I.; Malibert, C.; Bellaiche, L.; Kiat, J.M. Combined theoretical and experimental study of the low-temperature properties of $BaZrO_3$. *Phys. Rev. B* **2005**, *72*, 205104. [CrossRef]

146. Yamanaka, S.; Fujikane, M.; Hamaguchi, T.; Muta, H.; Oyama, T.; Matsuda, T.; Kobayashi, S.; Kurosaki, K. Thermophysical properties of $BaZrO_3$ and $BaCeO_3$. *J. Alloys Compd.* **2003**, *359*, 109–113. [CrossRef]

147. Mathews, M.D.; Mirza, E.B.; Momin, A.C. High-temperature X-ray diffractometric studies of $CaZrO_3$, $SrZrO_3$ and $BaZrO_3$. *J. Mater. Sci. Lett.* **1991**, *10*, 305–306. [CrossRef]

148. Taglieri, G.; Tersigni, M.; Villa, P.L.; Mondelli, C. Synthesis by the citrate route and characterisation of $BaZrO_3$, a high tech ceramic oxide: Preliminary results. *Int. J. Ind. Chem.* **1999**, *1*, 103–110. [CrossRef]

149. Braun, A.; Ovalle, A.; Pomjakushin, V.; Cervellino, A.; Erat, S.; Stolte, W.C.; Graule, T. Yttrium and hydrogen superstructure and correlation of lattice expansion and proton conductivity in the $BaZr_{0.9}Y_{0.1}O_{2.95}$ proton conductor. *Appl. Phys. Lett.* **2009**, *95*, 224103. [CrossRef]

150. Goupil, G.; Delahaye, T.; Gauthier, G.; Sala, B.; Joud, F.L. Stability study of possible air electrode materials for proton conducting electrochemical cells. *Solid State Ion.* **2012**, *209–210*, 36–42. [CrossRef]

151. Lyagaeva, Y.G.; Medvedev, D.A.; Demin, A.K.; Tsiakaras, P.; Reznitskikh, O.G. Thermal expansion of materials in the barium cerate-zirconate system. *Phys. Solid State* **2015**, *57*, 285–289. [CrossRef]

152. Han, D.; Majima, M.; Uda, T. Structure analysis of $BaCe_{0.8}Y_{0.2}O_{3-\delta}$ in dry and wet atmospheres by high-temperature X-ray diffraction measurement. *J. Solid State Chem.* **2013**, *205*, 122–128. [CrossRef]

153. Malavasi, L.; Ritter, C.; Chiodelli, G. Correlation between Thermal Properties, Electrical Conductivity, and Crystal Structure in the $BaCe_{0.80}Y_{0.20}O_{2.9}$ Proton Conductor. *Chem. Mater.* **2008**, *20*, 2343–2351. [CrossRef]

154. Zhu, Z.; Tao, Z.; Bi, L.; Liu, W. Investigation of $SmBaCuCoO_{5+\delta}$ double-perovskite as cathode for proton-conducting solid oxide fuel cells. *Mater. Res. Bull.* **2010**, *45*, 1771–1774. [CrossRef]

155. Zhou, X.; Liu, L.; Zhen, J.; Zhu, S.; Li, B.; Sun, K.; Wang, P. Ionic conductivity, sintering and thermal expansion behaviors of mixed ion conductor $BaZr_{0.1}Ce_{0.7}Y_{0.1}Yb_{0.1}O_{3-\delta}$ prepared by ethylene diamine tetraacetic acid assisted glycine nitrate process. *J. Power Sources* **2011**, *196*, 5000–5006. [CrossRef]

156. Gorelov, V.P.; Balakireva, V.B.; Kuz'min, A.V.; Plaksin, S.V. Electrical conductivity of $CaZr_{1-x}Sc_xO_{3-delta}$ (x = 0.01–0.20) in dry and humid air. *Inorg. Mater.* **2014**, *50*, 495–502. [CrossRef]

157. Yajima, T.; Suzuki, H.; Yogo, T.; Iwahara, H. Protonic conduction in $SrZrO_3$. based oxides. *Solid State Ion.* **1992**, *51*, 101–107. [CrossRef]

158. Hibino, T.; Mizutani, K.; Yajima, T.; Iwahara, H. Evaluation of proton conductivity in $SrCeO_3$, $BaCeO_3$, $CaZrO_3$ and $SrZrO_3$ by temperature programmed desorption method. *Solid State Ion.* **1992**, *57*, 303–306. [CrossRef]

159. Matsuda, T.; Yamanaka, S.; Kurosaki, K.; Kobayashi, S. High temperature phase transitions of $SrZrO_3$. *J. Alloys Compd.* **2003**, *351*, 43–46. [CrossRef]

160. Iwahara, H.; Esaka, T.; Uchida, H.; Yamauchi, T.; Ogaki, K. High temperature type protonic conductor based on $SrCeO_3$ and its application to the extraction of hydrogen gas. *Solid State Ion.* **1986**, *18–19*, 1003–1007. [CrossRef]

161. Yamanaka, S.; Kurosaki, K.; Maekawa, T.; Matsuda, T.; Kobayashi, S.; Uno, M. Thermochemical and thermophysical properties of alkaline-earth perovskites. *J. Nucl. Mater.* **2005**, *344*, 61–66. [CrossRef]

162. Li, L.; Nino, J.C. Proton-conducting barium stannates: Doping strategies and transport properties. *Int. J. Hydrog. Energy* **2013**, *38*, 1598–1606. [CrossRef]

163. Maekawa, T.; Kurosaki, K.; Yamanaka, S. Thermal and mechanical properties of polycrystalline $BaSnO_3$. *J. Alloys Compd.* **2006**, *416*, 214–217. [CrossRef]

164. Snijkers, F.M.M.; Buekenhoudt, A.; Luyten, J.J.; Cooymans, J.; Mertens, M. Proton conductivity in perovskite type yttrium doped barium hafnate. *Scr. Mater.* **2004**, *51*, 1129–1134. [CrossRef]

165. Maekawa, T.; Kurosaki, K.; Yamanaka, S. Thermal and mechanical properties of perovskite-type barium hafnate. *J. Alloys Compd.* **2006**, *407*, 44–48. [CrossRef]

166. Furøy, K.A.; Haugsrud, R.; Hänsel, M.; Magrasó, A.; Norby, T. Role of protons in the electrical conductivity of acceptor-doped $BaPrO_3$, $BaTbO_3$, and $BaThO_3$. *Solid State Ion.* **2007**, *178*, 461–467. [CrossRef]

167. Purohit, R.D.; Tyagi, A.K.; Mathews, M.D.; Saha, S. Combustion synthesis and bulk thermal expansion studies of Ba and Sr thorates. *J. Nucl. Mater.* **2000**, *280*, 51–55. [CrossRef]

168. Fu, W.T.; Visser, D.; Knight, K.S.; IJdo, D.J.W. Temperature-induced phase transitions in $BaTbO_3$. *J. Solid State Chem.* **2004**, *177*, 1667–1671. [CrossRef]

169. Nomura, K.; Takeuchi, T.; Tanase, S.; Kageyama, H.; Tanimoto, K.; Miyazaki, Y. Proton conduction in $(La_{0.9}Sr_{0.1})MIIIO_{3-d}$ (MIII = Sc, In, and Lu) perovskites. *Solid State Ion.* **2002**, *155*, 647–652. [CrossRef]

170. Gorelov, V.P.; Stroeva, A.Y. Solid proton conducting electrolytes based on $LaScO_3$. *Russ. J. Electrochem.* **2012**, *48*, 949–960. [CrossRef]

171. Okuyama, Y.; Kozai, T.; Ikeda, S.; Matsuka, M.; Sakai, T.; Matsumoto, H. Incorporation and conduction of proton in Sr-doped $LaMO_3$ (M = Al, Sc, In, Yb, Y). *Electrochim. Acta* **2014**, *125*, 443–449. [CrossRef]

172. Danilov, N.; Vdovin, G.; Reznitskikh, O.; Medvedev, D.; Demin, A.; Tsiakaras, P. Physico-chemical characterization and transport features of proton-conducting Sr-doped $LaYO_3$ electrolyte ceramics. *J. Eur. Ceram. Soc.* **2016**, *36*, 2795–2800. [CrossRef]

173. Dietrich, M.; Vassen, R.; Stover, D. $LaYbO_3$, A Candidate for Thermal Barrier Coating Materials. In *27th Annual Cocoa Beach Conference on Advanced Ceramics and Composites: A: Ceramic Engineering and Science Proceedings, Volume 24, Issue 3*; John Wiley & Sons, Inc.: Hoboken, NJ, USA, 2003; pp. 637–643.

174. Ovanesyan, K.; Petrosyan, A.; Shirinyan, G.; Pedrini, C.; Zhang, L. Czochralski single crystal growth of Ce- and Pr-doped $LaLuO_3$ double oxide. *J. Cryst. Growth* **1999**, *198–199*, 497–500. [CrossRef]

175. Inaba, H.; Hayashi, H.; Suzuki, M. Structural phase transition of perovskite oxides $LaMO_3$ and $La_{0.9}Sr_{0.1}MO_3$ with different size of B-site ions. *Solid State Ion.* **2001**, *144*, 99–108. [CrossRef]

176. Goldschmidt, V.M. Die Gesetze der Krystallochemie. *Naturwissenschaften* **1926**, *14*, 477–485. [CrossRef]

177. Zhao, Y.; Weidner, D.J.; Parise, J.B.; Cox, D.E. Thermal expansion and structural distortion of perovskite—Data for $NaMgF_3$ perovskite. Part I. *Phys. Earth Planet. Inter.* **1993**, *76*, 1–16. [CrossRef]

178. Bohn, H.; Schober, T.; Mono, T.; Schilling, W. The high temperature proton conductor $Ba_3Ca_{1.18}Nb_{1.82}O_{9-\delta}$. I. Electrical conductivity. *Solid State Ion.* **1999**, *117*, 219–228. [CrossRef]

179. Krug, F.; Schober, T. The high-temperature proton conductor $Ba_3(Ca_{1.18}Nb_{1.82})O_{9-gd}$: Thermogravimetry of the water uptake. *Solid State Ion.* **1996**, *92*, 297–302. [CrossRef]
180. Schober, T.; Friedrich, J. The mixed perovskites $BaCa_{(1+x)/3}Nb_{(2-x)/3}O_{3-x/2}$ (x = 0 ... 0.18): Proton uptake. *Solid State Ion.* **2000**, *136–137*, 161–165. [CrossRef]
181. Bhella, S.S.; Thangadurai, V. Investigations on the thermo-chemical stability and electrical conductivity of K-doped $Ba_{3-x}K_xCaNb_2O_{9-\delta}$ (x = 0.5, 0.75, 1, 1.25). *Solid State Ion.* **2011**, *192*, 229–234. [CrossRef]
182. Wang, S.; Zhao, F.; Zhang, L.; Brinkman, K.; Chen, F. Doping effects on complex perovskite $Ba_3Ca_{1.18}Nb_{1.82}O_{9-\delta}$ intermediate temperature proton conductor. *J. Power Sources* **2011**, *196*, 7917–7923. [CrossRef]
183. Hassan, D.; Janes, S.; Clasen, R. Proton-conducting ceramics as electrode/electrolyte materials for SOFC's-part I: Preparation, mechanical and thermal properties of sintered bodies. *J. Eur. Ceram. Soc.* **2003**, *23*, 221–228. [CrossRef]
184. Mono, T.; Schober, T. Lattice parameter change in water vapor exposed $Ba_3Ca_{1.18}Nb_{1.82}O_{9-\delta}$. *Solid State Ion.* **1996**, *91*, 155–159. [CrossRef]
185. Schober, T.; Friedrich, J.; Triefenbach, D.; Tietz, F. Dilatometry of the high-temperature proton conductor $Ba_3Ca_{1.18}Nb_{1.82}O_{9-\delta}$. *Solid State Ion.* **1997**, *100*, 173–181. [CrossRef]
186. Jayaraman, V.; Magrez, A.; Caldes, M.; Joubert, O.; Ganne, M.; Piffard, Y.; Brohan, L. Characterization of perovskite systems derived from $Ba_2In_2O_5\square$: Part I: The oxygen-deficient $Ba_2In_{2(1-x)}Ti_{2x}O_{5+x}\square_{1-x}$ ($0 \leq x \leq 1$) compounds. *Solid State Ion.* **2004**, *170*, 17–24. [CrossRef]
187. Bjørheim, T.S.; Rahman, S.M.H.; Eriksson, S.G.; Knee, C.S.; Haugsrud, R. Hydration Thermodynamics of the Proton Conducting Oxygen-Deficient Perovskite Series $BaTi_{1-x}M_xO_{3-x/2}$ with M = In or Sc. *Inorg. Chem.* **2015**, *54*, 2858–2865. [CrossRef] [PubMed]
188. Rahman, S.M.H.; Knee, C.S.; Ahmed, I.; Eriksson, S.G.; Haugsrud, R. 50 mol% indium substituted $BaTiO_3$: Characterization of structure and conductivity. *Int. J. Hydrog. Energy* **2012**, *37*, 7975–7982. [CrossRef]
189. Quarez, E.; Noirault, S.; Caldes, M.T.; Joubert, O. Water incorporation and proton conductivity in titanium substituted barium indate. *J. Power Sources* **2010**, *195*, 1136–1141. [CrossRef]
190. Rahman, S.M.H.; Ahmed, I.; Haugsrud, R.; Eriksson, S.G.; Knee, C.S. Characterisation of structure and conductivity of $BaTi_{0.5}Sc_{0.5}O_{3-\delta}$. *Solid State Ion.* **2014**, *255*, 140–146. [CrossRef]
191. Rahman, S.M.H.; Norberg, S.T.; Knee, C.S.; Biendicho, J.J.; Hull, S.; Eriksson, S.G. Proton conductivity of hexagonal and cubic $BaTi_{1-x}Sc_xO_{3-\delta}$ ($0.1 \leq x \leq 0.8$). *Dalton Trans.* **2014**, *43*, 15055–15064. [CrossRef] [PubMed]
192. Noirault, S.; Quarez, E.; Piffard, Y.; Joubert, O. Water incorporation into the $Ba_2(In_{1-x}M_x)_2O_5$ (M = Sc^{3+} $0 \leq x < 0.5$ and M = Y^{3+} $0 \leq x < 0.35$) system and protonic conduction. *Solid State Ion.* **2009**, *180*, 1157–1163. [CrossRef]
193. Haugsrud, R.; Norby, T. High-Temperature Proton Conductivity in Acceptor-Substituted Rare-Earth Ortho-Tantalates, $LnTaO_4$. *J. Am. Ceram. Soc.* **2007**, *90*, 1116–1121. [CrossRef]
194. Haugsrud, R.; Norby, T. Proton conduction in rare-earth ortho-niobates and ortho-tantalates. *Nat. Mater.* **2006**, *5*, 193–196. [CrossRef]
195. Bi, Z.; Bridges, C.A.; Kim, J.H.; Huq, A.; Paranthaman, M.P. Phase stability and electrical conductivity of Ca-doped $LaNb_{1-x}Ta_xO_{4-\delta}$ high temperature proton conductors. *J. Power Sources* **2011**, *196*, 7395–7403. [CrossRef]
196. Norby, T.; Christiansen, N. Proton conduction in Ca- and Sr-substituted $LaPO_4$. *Solid State Ion.* **1995**, *77*, 240–243. [CrossRef]
197. Bjørheim, T.S.; Norby, T.; Haugsrud, R. Hydration and proton conductivity in $LaAsO_4$. *J. Mater. Chem.* **2012**, *22*, 1652–1661. [CrossRef]
198. Toyoura, K.; Matsunaga, K. Hydrogen Bond Dynamics in Proton-Conducting Lanthanum Arsenate. *J. Phys. Chem. C* **2013**, *117*, 18006–18012. [CrossRef]
199. Amezawa, K.; Tomii, Y.; Yamamoto, N. High temperature protonic conduction in Ca-doped YPO_4. *Solid State Ion.* **2003**, *162–163*, 175–180. [CrossRef]
200. Mokkelbost, T.; Lein, H.L.; Vullum, P.E.; Holmestad, R.; Grande, T.; Einarsrud, M.A. Thermal and mechanical properties of $LaNbO_4$-based ceramics. *Ceram. Int.* **2009**, *35*, 2877–2883. [CrossRef]
201. Fjeld, H.; Kepaptsoglou, D.M.; Haugsrud, R.; Norby, T. Charge carriers in grain boundaries of 0.5% Sr-doped $LaNbO_4$. *Solid State Ion.* **2010**, *181*, 104–109. [CrossRef]

202. Mielewczyk-Gryn, A.; Wachowski, S.; Zagórski, K.; Jasiński, P.; Gazda, M. Characterization of magnesium doped lanthanum orthoniobate synthesized by molten salt route. *Ceram. Int.* **2015**, *41*, 7847–7852. [CrossRef]

203. Brandão, A.D.; Gracio, J.; Mather, G.C.; Kharton, V.V.; Fagg, D.P. B-site substitutions in $LaNb_{1-x}M_xO_{4-\delta}$ materials in the search for potential proton conductors (M = Ga, Ge, Si, B, Ti, Zr, P, Al). *J. Solid State Chem.* **2011**, *184*, 863–870. [CrossRef]

204. Syvertsen, G.E.; Magrasó, A.; Haugsrud, R.; Einarsrud, M.A.; Grande, T. The effect of cation non-stoichiometry in $LaNbO_4$ materials. *Int. J. Hydrog. Energy* **2012**, *37*, 8017–8026. [CrossRef]

205. Huse, M.; Norby, T.; Haugsrud, R. Effects of A and B site acceptor doping on hydration and proton mobility of $LaNbO_4$. *Int. J. Hydrog. Energy* **2012**, *37*, 8004–8016. [CrossRef]

206. Depero, L.E.; Sangaletti, L. Cation Sublattice and Coordination Polyhedra in ABO_4 type of Structures. *J. Solid State Chem.* **1997**, *129*, 82–91. [CrossRef]

207. Errandonea, D.; Manjon, F. Pressure effects on the structural and electronic properties of ABX_4 scintillating crystals. *Prog. Mater. Sci.* **2008**, *53*, 711–773. [CrossRef]

208. Li, H.; Zhou, S.; Zhang, S. The relationship between the thermal expansions and structures of ABO_4 oxides. *J. Solid State Chem.* **2007**, *180*, 589–595. [CrossRef]

209. Ishibashi, Y.; Hara, K.; Sawada, A. The ferroelastic transition in some scheelite-type crystals. *Phys. B+C* **1988**, *150*, 258–264. [CrossRef]

210. David, W.I.F. High Resolution Neutron Powder Diffraction Studies of the Ferroelastic Phase Transition in LaNbO4. *MRS Proc.* **1989**, *166*, 203. [CrossRef]

211. Mokkelbost, T.; Kaus, I.; Haugsrud, R.; Norby, T.; Grande, T.; Einarsrud, M.A. High-Temperature Proton-Conducting Lanthanum Ortho-Niobate-Based Materials. Part II: Sintering Properties and Solubility of Alkaline Earth Oxides. *J. Am. Ceram. Soc.* **2008**, *91*, 879–886. [CrossRef]

212. Ivanova, M.; Ricote, S.; Meulenberg, W.A.; Haugsrud, R.; Ziegner, M. Effects of A- and B-site (co-)acceptor doping on the structure and proton conductivity of $LaNbO_4$. *Solid State Ion.* **2012**, *213*, 45–52. [CrossRef]

213. Mielewczyk-Gryn, A.; Wachowski, S.; Strychalska, J.; Zagórski, K.; Klimczuk, T.; Navrotsky, A.; Gazda, M. Heat capacities and thermodynamic properties of antimony substituted lanthanum orthoniobates. *Ceram. Int.* **2016**, *42*, 7054–7059. [CrossRef]

214. Wachowski, S.; Mielewczyk-Gryń, A.; Zagórski, K.; Li, C.; Jasiński, P.; Skinner, S.J.; Haugsrud, R.; Gazda, M. Influence of Sb-substitution on ionic transport in lanthanum orthoniobates. *J. Mater. Chem. A* **2016**, *4*, 11696–11707. [CrossRef]

215. Syvertsen, G.E.; Estournès, C.; Fjeld, H.; Haugsrud, R.; Einarsrud, M.; Grande, T.; Menon, M. Spark Plasma Sintering and Hot Pressing of Hetero-Doped LaNbO4. *J. Am. Ceram. Soc.* **2012**, *95*, 1563–1571. [CrossRef]

216. Mokkelbost, T.; Andersen, Ø.; Strøm, R.A.; Wiik, K.; Grande, T.; Einarsrud, M. High-Temperature Proton-Conducting LaNbO4-Based Materials: Powder Synthesis by Spray Pyrolysis. *J. Am. Ceram. Soc.* **2007**, *90*, 3395–3400. [CrossRef]

217. Magrasó, A.; Xuriguera, H.; Varela, M.; Sunding, M.F.; Strandbakke, R.; Haugsrud, R.; Norby, T. Novel Fabrication of Ca-Doped LaNbO4 Thin-Film Proton-Conducting Fuel Cells by Pulsed Laser Deposition. *J. Am. Ceram. Soc.* **2010**, *93*, 1874–1878. [CrossRef]

218. Amsif, M.; Marrero-López, D.; Ruiz-Morales, J.C.; Savvin, S.; Núñez, P. Low temperature sintering of LaNbO4 proton conductors from freeze-dried precursors. *J. Eur. Ceram. Soc.* **2012**, *32*, 1235–1244. [CrossRef]

219. Mielewczyk-Gryń, A.; Gdula, K.; Molin, S.; Jasinski, P.; Kusz, B.; Gazda, M. Structure and electrical properties of ceramic proton conductors obtained with molten-salt and solid-state synthesis methods. *J. Non. Cryst. Solids* **2010**, *356*, 1976–1979. [CrossRef]

220. Brandão, A.D.; Antunes, I.; Frade, J.R.; Torre, J.; Kharton, V.V.; Fagg, D.P. Enhanced Low-Temperature Proton Conduction in $Sr_{0.02}La_{0.98}NbO_{4-\delta}$ by Scheelite Phase Retention. *Chem. Mater.* **2010**, *22*, 6673–6683. [CrossRef]

221. Santibáñez-Mendieta, A.B.; Fabbri, E.; Licoccia, S.; Traversa, E. Tailoring phase stability and electrical conductivity of $Sr_{0.02}La_{0.98}Nb_{1-x}Ta_xO_4$ for intermediate temperature fuel cell proton conducting electrolytes. *Solid State Ion.* **2012**, *216*, 6–10. [CrossRef]

222. Wachowski, S.L.; Kamecki, B.; Winiarz, P.; Dzierzgowski, K.; Mielewczyk-Gryń, A.; Gazda, M.; Wachowski, S.L.; Jasiński, P.; Witkowska, A.; Gazda, M.; et al. Tailoring structural properties of lanthanum orthoniobates through an isovalent substitution on the Nb-site. *Inorg. Chem. Front.* **2018**, *24*, 1–16. [CrossRef]

223. Jian, L.; Wayman, C.M. Compressive behavior and domain-related shape memory effect in LaNbO$_4$ ceramics. *Mater. Lett.* **1996**, *26*, 1–7. [CrossRef]

224. Parlinski, K.; Hashi, Y.; Tsunekawa, S.; Kawazoe, Y. Computer simulation of ferroelastic phase transition in LaNbO$_4$. *J. Mater. Res.* **1997**, *12*, 2428–2437. [CrossRef]

225. Sarin, P.; Hughes, R.W.; Lowry, D.R.; Apostolov, Z.D.; Kriven, W.M. High-Temperature Properties and Ferroelastic Phase Transitions in Rare-Earth Niobates (LnNbO$_4$). *J. Am. Ceram. Soc.* **2014**, *97*, 3307–3319. [CrossRef]

226. Hikichi, Y.; Ota, T.; Daimon, K.; Hattori, T. Thermal, Mechanical, and Chemical Properties of Sintered Xenotime-Type RPO$_4$ (R = Y, Er, Yb, or Lu). *J. Am. Ceram. Soc.* **1998**, *81*, 2216–2218. [CrossRef]

227. Bayer, G. Thermal expansion of ABO$_4$-compounds with zircon-and scheelite structures. *J. Less Common Met.* **1972**, *26*, 255–262. [CrossRef]

228. Hikichi, Y.; Ota, T.; Hattori, T. Thermal, mechanical and chemical properties of sintered monazite-(La, Ce, Nd or Sm). *Mineral. J.* **1997**, *19*, 123–130. [CrossRef]

229. Omori, M.; Kobayashi, Y.; Hirai, T. Dilatometric behavior of martensitic transformation of NdNbO$_4$ polycrystals. *J. Mater. Sci.* **2000**, *35*, 719–721. [CrossRef]

230. Filatov, S.K. General concept of increasing crystal symmetry with an increase in temperature. *Crystallogr. Rep.* **2011**, *56*, 953–961. [CrossRef]

231. Akiyama, K.; Nagano, I.; Shida, M.; Ota, S. Thermal Barrier Coating Material. U.S. Patent No. 7,622,411, 24 November 2009.

232. Yang, H.; Peng, F.; Zhang, Q.; Guo, C.; Shi, C.; Liu, W.; Sun, G.; Zhao, Y.; Zhang, D.; Sun, D.; et al. A promising high-density scintillator of GdTaO$_4$ single crystal. *CrystEngComm* **2014**, *16*, 2480–2485. [CrossRef]

233. Li, H.; Zhang, S.; Zhou, S.; Cao, X. Bonding characteristics, thermal expansibility, and compressibility of RXO$_4$ (R = Rare Earths, X = P, As) within monazite and zircon structures. *Inorg. Chem.* **2009**, *48*, 4542–4548. [CrossRef] [PubMed]

234. Huse, M.; Norby, T.; Haugsrud, R. Proton Conductivity in Acceptor-Doped LaVO$_4$. *J. Electrochem. Soc.* **2011**, *158*, B857–B865. [CrossRef]

235. Zhang, S.; Zhou, S.; Li, H.; Li, N. Investigation of thermal expansion and compressibility of rare-earth orthovanadates using a dielectric chemical bond method. *Inorg. Chem.* **2008**, *47*, 7863–7867. [CrossRef] [PubMed]

236. Bjørheim, T.S.; Besikiotis, V.; Haugsrud, R. Hydration thermodynamics of pyrochlore structured oxides from TG and first principles calculations. *Dalton Trans.* **2012**, *41*, 13343–13351. [CrossRef] [PubMed]

237. Omata, T.; Ikeda, K.; Tokashiki, R.; Otsuka-Yao-Matsuo, S. Proton solubility for La$_2$Zr$_2$O$_7$ with a pyrochlore structure doped with a series of alkaline-earth ions. *Solid State Ion.* **2004**, *167*, 389–397. [CrossRef]

238. Eurenius, K.E.J.; Ahlberg, E.; Knee, C.S. Proton conductivity in Ln$_{1.96}$Ca$_{0.04}$Sn$_2$O$_{7-\delta}$ (Ln = La, Sm, Yb) pyrochlores as a function of the lanthanide size. *Solid State Ion.* **2010**, *181*, 1258–1263. [CrossRef]

239. Eurenius, K.E.J.; Ahlberg, E.; Ahmed, I.; Eriksson, S.G.; Knee, C.S. Investigation of proton conductivity in Sm$_{1.92}$Ca$_{0.08}$Ti$_2$O$_{7-\delta}$ and Sm$_2$Ti$_{1.92}$Y$_{0.08}$O$_{7-\delta}$ pyrochlores. *Solid State Ion.* **2010**, *181*, 148–153. [CrossRef]

240. Eurenius, K.E.J.; Ahlberg, E.; Knee, C.S. Proton conductivity in Sm$_2$Sn$_2$O$_7$ pyrochlores. *Solid State Ion.* **2010**, *181*, 1577–1585. [CrossRef]

241. Omata, T.; Otsuka-Yao-Matsuo, S. Electrical Properties of Proton-Conducting Ca-Doped La$_2$Zr$_2$O$_7$ with a Pyrochlore-Type Structure. *J. Electrochem. Soc.* **2001**, *148*, E252–E261. [CrossRef]

242. Shimura, T.; Komori, M.; Iwahara, H. Ionic conduction in pyrochlore-type oxides containing rare earth elements at high temperature. *Solid State Ion.* **1996**, *86–88*, 685–689. [CrossRef]

243. Ma, W.; Gong, S.; Xu, H.; Cao, X. On improving the phase stability and thermal expansion coefficients of lanthanum cerium oxide solid solutions. *Scr. Mater.* **2006**, *54*, 1505–1508. [CrossRef]

244. Besikiotis, V.; Ricote, S.; Jensen, M.H.; Norby, T.; Haugsrud, R. Conductivity and hydration trends in disordered fluorite and pyrochlore oxides: A study on lanthanum cerate–zirconate based compounds. *Solid State Ion.* **2012**, *229*, 26–32. [CrossRef]

245. Kalland, L.E.; Norberg, S.T.; Kyrklund, J.; Hull, S.; Eriksson, S.G.; Norby, T.; Mohn, C.E.; Knee, C.S. C-type related order in the defective fluorites La$_2$Ce$_2$O$_7$ and Nd$_2$Ce$_2$O$_7$ studied by neutron scattering and ab initio MD simulations. *Phys. Chem. Chem. Phys.* **2016**, *18*, 24070–24080. [CrossRef] [PubMed]

246. Zhang, F.X.X.; Tracy, C.L.L.; Lang, M.; Ewing, R.C.C. Stability of fluorite-type La$_2$Ce$_2$O$_7$ under extreme conditions. *J. Alloys Compd.* **2016**, *674*, 168–173. [CrossRef]

247. Wang, J.; Bai, S.; Zhang, H.; Zhang, C. The structure, thermal expansion coefficient and sintering behavior of Nd^{3+}-doped $La_2Zr_2O_7$ for thermal barrier coatings. *J. Alloys Compd.* **2009**, *476*, 89–91. [CrossRef]

248. Lehmann, H.; Pitzer, D.; Pracht, G.; Vassen, R.; Stöver, D. Thermal conductivity and thermal expansion coefficients of the lanthanum rare-earth-element zirconate system. *J. Am. Ceram. Soc.* **2003**, *86*, 1338–1344. [CrossRef]

249. Haugsrud, R. Defects and transport properties in Ln_6WO_{12} (Ln = La, Nd, Gd, Er). *Solid State Ion.* **2007**, *178*, 555–560. [CrossRef]

250. Zayas-Rey, M.J.; dos Santos-Gómez, L.; Marrero-López, D.; León-Reina, L.; Canales-Vázquez, J.; Aranda, M.A.G.; Losilla, E.R. Structural and Conducting Features of Niobium-Doped Lanthanum Tungstate, $La_{27}(W_{1-x}Nb_x)_5O_{55.55-\delta}$. *Chem. Mater.* **2013**, *25*, 448–456. [CrossRef]

251. Magrasó, A.; Haugsrud, R. Effects of the La/W ratio and doping on the structure, defect structure, stability and functional properties of proton-conducting lanthanum tungstate $La_{28-x}W_{4+x}O_{54+\delta}$. A review. *J. Mater. Chem. A* **2014**, *2*, 12630–12641. [CrossRef]

252. Magrasó, A.; Hervoches, C.H.; Ahmed, I.; Hull, S.; Nordström, J.; Skilbred, A.W.B.; Haugsrud, R. In situ high temperature powder neutron diffraction study of undoped and Ca-doped $La_{28-x}W_{4+x}O_{54+3x/2}$ (x = 0.85). *J. Mater. Chem. A* **2013**, *1*, 3774–3782. [CrossRef]

253. Hancke, R.; Magrasó, A.; Norby, T.; Haugsrud, R. Hydration of lanthanum tungstate (La/W=5.6 and 5.3) studied by TG and simultaneous TG–DSC. *Solid State Ion.* **2013**, *231*, 25–29. [CrossRef]

254. Quarez, E.; Kravchyk, K.V.; Joubert, O. Compatibility of proton conducting La_6WO_{12} electrolyte with standard cathode materials. *Solid State Ion.* **2012**, *216*, 19–24. [CrossRef]

255. Seeger, J.; Ivanova, M.E.; Meulenberg, W.A.; Sebold, D.; Stöver, D.; Scherb, T.; Schumacher, G.; Escolástico, S.; Solís, C.; Serra, J.M. Synthesis and characterization of nonsubstituted and substituted proton-conducting $La_{6-x}WO_{12-y}$. *Inorg. Chem.* **2013**, *52*, 10375–10386. [CrossRef] [PubMed]

256. Zayas-Rey, M.J.; dos Santos-Gómez, L.; Cabeza, A.; Marrero-López, D.; Losilla, E.R. Proton conductors based on alkaline-earth substituted $La_{28-x}W_{4+x}O_{54+3x/2}$. *Dalton Trans.* **2014**, *43*, 6490–6499. [CrossRef] [PubMed]

257. Peng, R.; Wu, T.; Liu, W.; Liu, X.; Meng, G. Cathode processes and materials for solid oxide fuel cells with proton conductors as electrolytes. *J. Mater. Chem.* **2010**, *20*, 6218–6225. [CrossRef]

258. Merkle, R.; Poetzsch, D.; Maier, J. Oxygen Reduction Reaction at Cathodes on Proton Conducting Oxide Electrolytes: Contribution from Three Phase Boundary Compared to Bulk Path. *ECS Trans.* **2015**, *66*, 95–102. [CrossRef]

259. Téllez Lozano, H.; Druce, J.; Cooper, S.J.; Kilner, J.A. Double perovskite cathodes for proton-conducting ceramic fuel cells: Are they triple mixed ionic electronic conductors? *Sci. Technol. Adv. Mater.* **2017**, *18*, 977–986. [CrossRef] [PubMed]

260. Strandbakke, R.; Cherepanov, V.A.; Zuev, A.Y.; Tsvetkov, D.S.; Argirusis, C.; Sourkouni, G.; Prünte, S.; Norby, T. Gd- and Pr-based double perovskite cobaltites as oxygen electrodes for proton ceramic fuel cells and electrolyser cells. *Solid State Ion.* **2015**, *278*, 120–132. [CrossRef]

261. Zohourian, R.; Merkle, R.; Maier, J. Proton uptake into the protonic cathode material $BaCo_{0.4}Fe_{0.4}Zr_{0.2}O_{3-\delta}$ and comparison to protonic electrolyte materials. *Solid State Ion.* **2017**, *299*, 64–69. [CrossRef]

262. Bernuy-Lopez, C.; Rioja-Monllor, L.; Nakamura, T.; Ricote, S.; O'Hayre, R.; Amezawa, K.; Einarsrud, M.A.; Grande, T. Effect of Cation Ordering on the Performance and Chemical Stability of Layered Double Perovskite Cathodes. *Materials* **2018**, *11*, 196. [CrossRef] [PubMed]

263. Zhao, L.; He, B.; Ling, Y.; Xun, Z.; Peng, R.; Meng, G.; Liu, X. Cobalt-free oxide $Ba_{0.5}Sr_{0.5}Fe_{0.8}Cu_{0.2}O_{3-\delta}$ for proton-conducting solid oxide fuel cell cathode. *Int. J. Hydrog. Energy* **2010**, *35*, 3769–3774. [CrossRef]

264. Grimaud, A.; Mauvy, F.; Bassat, J.M.; Fourcade, S.; Rocheron, L.; Marrony, M.; Grenier, J.C. Hydration Properties and Rate Determining Steps of the Oxygen Reduction Reaction of Perovskite-Related Oxides as H^+-SOFC Cathodes. *J. Electrochem. Soc.* **2012**, *159*, B683–B694. [CrossRef]

265. Dailly, J.; Fourcade, S.; Largeteau, A.; Mauvy, F.; Grenier, J.C.; Marrony, M. Perovskite and A_2MO_4-type oxides as new cathode materials for protonic solid oxide fuel cells. *Electrochim. Acta* **2010**, *55*, 5847–5853. [CrossRef]

266. Shang, M.; Tong, J.; O'Hayre, R. A promising cathode for intermediate temperature protonic ceramic fuel cells: $BaCo_{0.4}Fe_{0.4}Zr_{0.2}O_{3-\delta}$. *RSC Adv.* **2013**, *3*, 15769–15775. [CrossRef]

267. Tao, Z.; Bi, L.; Zhu, Z.; Liu, W. Novel cobalt-free cathode materials $BaCe_xFe_{1-x}O_{3-\delta}$ for proton-conducting solid oxide fuel cells. *J. Power Sources* **2009**, *194*, 801–804. [CrossRef]

268. Poetzsch, D.; Merkle, R.; Maier, J. Proton conductivity in mixed-conducting BSFZ perovskite from thermogravimetric relaxation. *Phys. Chem. Chem. Phys.* **2014**, *16*, 16446–16453. [CrossRef] [PubMed]

269. Poetzsch, D.; Merkle, R.; Maier, J. Proton uptake in the H$^+$-SOFC cathode material Ba$_{0.5}$Sr$_{0.5}$Fe$_{0.8}$Zn$_{0.2}$O$_{3-\delta}$: Transition from hydration to hydrogenation with increasing oxygen partial pressure. *Faraday Discuss.* **2015**, *182*, 129–143. [CrossRef] [PubMed]

270. Mukundan, R.; Davies, P.K.; Worrell, W.L. Electrochemical Characterization of Mixed Conducting Ba(Ce$_{0.8-y}$Pr$_y$Gd$_{0.2}$)O$_{2.9}$ Cathodes. *J. Electrochem. Soc.* **2001**, *148*, A82–A86. [CrossRef]

271. Yang, L.; Liu, Z.; Wang, S.; Choi, Y.; Zuo, C.; Liu, M. A mixed proton, oxygen ion, and electron conducting cathode for SOFCs based on oxide proton conductors. *J. Power Sources* **2010**, *195*, 471–474. [CrossRef]

272. Tao, Z.; Bi, L.; Yan, L.; Sun, W.; Zhu, Z.; Peng, R.; Liu, W. A novel single phase cathode material for a proton-conducting SOFC. *Electrochem. Commun.* **2009**, *11*, 688–690. [CrossRef]

273. Wu, T.; Zhao, Y.; Peng, R.; Xia, C. Nano-sized Sm$_{0.5}$Sr$_{0.5}$CoO$_{3-\delta}$ as the cathode for solid oxide fuel cells with proton-conducting electrolytes of BaCe$_{0.8}$Sm$_{0.2}$O$_{2.9}$. *Electrochim. Acta* **2009**, *54*, 4888–4892. [CrossRef]

274. Upasen, S.; Batocchi, P.; Mauvy, F.; Slodczyk, A.; Colomban, P. Chemical and structural stability of La$_{0.6}$Sr$_{0.4}$Co$_{0.2}$Fe$_{0.8}$O$_{3-\delta}$ ceramic vs. medium/high water vapor pressure. *Ceram. Int.* **2015**, *41*, 14137–14147. [CrossRef]

275. Pu, T.; Tan, W.; Shi, H.; Na, Y.; Lu, J.; Zhu, B. Steam/CO$_2$ electrolysis in symmetric solid oxide electrolysis cell with barium cerate-carbonate composite electrolyte. *Electrochim. Acta* **2016**, *190*, 193–198. [CrossRef]

276. Lin, B.; Dong, Y.; Yan, R.; Zhang, S.; Hu, M.; Zhou, Y.; Meng, G. In situ screen-printed BaZr$_{0.1}$Ce$_{0.7}$Y$_{0.2}$O$_{3-\delta}$ electrolyte-based protonic ceramic membrane fuel cells with layered SmBaCo$_2$O$_{5+x}$ cathode. *J. Power Sources* **2009**, *186*, 446–449. [CrossRef]

277. Kim, J.; Sengodan, S.; Kwon, G.; Ding, D.; Shin, J.; Liu, M.; Kim, G. Triple-Conducting Layered Perovskites as Cathode Materials for Proton-Conducting Solid Oxide Fuel Cells. *ChemSusChem* **2014**, *7*, 2811–2815. [CrossRef] [PubMed]

278. Lin, B.; Zhang, S.; Bi, L.; Ding, H.; Liu, X.; Gao, J.; Meng, G. Prontonic ceramic membrane fuel cells with layered GdBaCo$_2$O$_{5+x}$ cathode prepared by gel-casting and suspension spray. *J. Power Sources* **2008**, *177*, 330–333. [CrossRef]

279. Brieuc, F.; Dezanneau, G.; Hayoun, M.; Dammak, H. Proton diffusion mechanisms in the double perovskite cathode material GdBaCo$_2$O$_{5.5}$: A molecular dynamics study. *Solid State Ion.* **2017**, *309*, 187–191. [CrossRef]

280. Zhao, L.; He, B.; Lin, B.; Ding, H.; Wang, S.; Ling, Y.; Peng, R.; Meng, G.; Liu, X. High performance of proton-conducting solid oxide fuel cell with a layered PrBaCo$_2$O$_{5+\delta}$ cathode. *J. Power Sources* **2009**, *194*, 835–837. [CrossRef]

281. Ding, H.; Xue, X.; Liu, X.; Meng, G. A novel layered perovskite cathode for proton conducting solid oxide fuel cells. *J. Power Sources* **2010**, *195*, 775–778. [CrossRef]

282. Nian, Q.; Zhao, L.; He, B.; Lin, B.; Peng, R.; Meng, G.; Liu, X. Layered SmBaCuCoO$_{5+\delta}$ and SmBaCuFeO$_{5+\delta}$ perovskite oxides as cathode materials for proton-conducting SOFCs. *J. Alloys Compd.* **2010**, *492*, 291–294. [CrossRef]

283. Zhao, L.; He, B.; Nian, Q.; Xun, Z.; Peng, R.; Meng, G.; Liu, X. In situ drop-coated BaZr$_{0.1}$Ce$_{0.7}$Y$_{0.2}$O$_{3-\delta}$ electrolyte-based proton-conductor solid oxide fuel cells with a novel layered PrBaCuFeO$_{5+\delta}$ cathode. *J. Power Sources* **2009**, *194*, 291–294. [CrossRef]

284. Taillades, G.; Dailly, J.; Taillades-Jacquin, M.; Mauvy, F.; Essouhmi, A.; Marrony, M.; Lalanne, C.; Fourcade, S.; Jones, D.J.; Grenier, J.C.; et al. Intermediate temperature anode-supported fuel cell based on BaCe$_{0.9}$Y$_{0.1}$O$_3$ electrolyte with novel Pr$_2$NiO$_4$ cathode. *Fuel Cells* **2010**, *10*, 166–173. [CrossRef]

285. Nasani, N.; Ramasamy, D.; Mikhalev, S.; Kovalevsky, A.V.; Fagg, D.P. Fabrication and electrochemical performance of a stable, anode supported thin BaCe$_{0.4}$Zr$_{0.4}$Y$_{0.2}$O$_{3-\delta}$ electrolyte Protonic Ceramic Fuel Cell. *J. Power Sources* **2015**, *278*, 582–589. [CrossRef]

286. Upasen, S.; Batocchi, P.; Mauvy, F.; Slodczyk, A.; Colomban, P. Protonation and structural/chemical stability of Ln$_2$NiO$_{4+\delta}$ ceramics vs. H$_2$O/CO$_2$: High temperature/water pressure ageing tests. *J. Alloys Compd.* **2014**, *622*, 1074–1085. [CrossRef]

287. Zhao, H.; Mauvy, F.; Lalanne, C.; Bassat, J.M.; Fourcade, S.; Grenier, J.C. New cathode materials for ITSOFC: Phase stability, oxygen exchange and cathode properties of La$_{2-x}$NiO$_{4+\delta}$. *Solid State Ion.* **2008**, *179*, 2000–2005. [CrossRef]

288. Wang, J.; Zhou, J.; Wang, T.; Chen, G.; Wu, K.; Cheng, Y. Decreasing the polarization resistance of LaSrCoO$_4$ cathode by Fe substitution for Ba(Zr$_{0.1}$Ce$_{0.7}$Y$_{0.2}$)O$_3$ based protonic ceramic fuel cells. *J. Alloys Compd.* **2016**, *689*, 581–586. [CrossRef]

289. Acuña, W.; Tellez, J.F.; Macías, M.A.; Roussel, P.; Ricote, S.; Gauthier, G.H. Synthesis and characterization of BaGa$_2$O$_4$ and Ba$_3$Co$_2$O$_6$(CO$_3$)$_{0.6}$ compounds in the search of alternative materials for Proton Ceramic Fuel Cell (PCFC). *Solid State Sci.* **2017**, *71*, 61–68. [CrossRef]

290. Danilov, N.A.; Tarutin, A.P.; Lyagaeva, J.G.; Pikalova, E.Y.; Murashkina, A.A.; Medvedev, D.A.; Patrakeev, M.V.; Demin, A.K. Affinity of YBaCo$_4$O$_{7+\delta}$-based layered cobaltites with protonic conductors of cerate-zirconate family. *Ceram. Int.* **2017**, *43*, 15418–15423. [CrossRef]

291. Kinyanjui, F.G.; Norberg, S.T.; Knee, C.S.; Eriksson, S.G. Proton conduction in oxygen deficient Ba$_3$In$_{1.4}$Y$_{0.3}$M$_{0.3}$ZrO$_8$ (M = Ga^{3+} or Gd^{3+}) perovskites. *J. Alloys Compd.* **2014**, *605*, 56–62. [CrossRef]

292. Yahia, H.B.; Mauvy, F.; Grenier, J.C. Ca$_{3-x}$La$_x$Co$_4$O$_{9+\delta}$ (x = 0, 0.3): New cobaltite materials as cathodes for proton conducting solid oxide fuel cell. *J. Solid State Chem.* **2010**, *183*, 527–531. [CrossRef]

293. Macias, M.A.; Sandoval, M.V.; Martinez, N.G.; Vázquez-Cuadriello, S.; Suescun, L.; Roussel, P.; Świerczek, K.; Gauthier, G.H. Synthesis and preliminary study of La$_4$BaCu$_5$O$_{13+\delta}$ and La$_{6.4}$Sr$_{1.6}$Cu$_8$O$_{20\pm\delta}$ ordered perovskites as SOFC/PCFC electrode materials. *Solid State Ion.* **2016**, *288*, 68–75. [CrossRef]

294. Lin, B.; Ding, H.; Dong, Y.; Wang, S.; Zhang, X.; Fang, D.; Meng, G. Intermediate-to-low temperature protonic ceramic membrane fuel cells with Ba$_{0.5}$Sr$_{0.5}$Co$_{0.8}$Fe$_{0.2}$O$_{3-\delta}$-BaZr$_{0.1}$Ce$_{0.7}$Y$_{0.2}$O$_{3-\delta}$ composite cathode. *J. Power Sources* **2009**, *186*, 58–61. [CrossRef]

295. Yang, L.; Wang, S.; Lou, X.; Liu, M. Electrical conductivity and electrochemical performance of cobalt-doped BaZr$_{0.1}$Ce$_{0.7}$Y$_{0.2}$O$_{3-\delta}$ cathode. *Int. J. Hydrog. Energy* **2011**, *36*, 2266–2270. [CrossRef]

296. Sun, W.; Yan, L.; Lin, B.; Zhang, S.; Liu, W. High performance proton-conducting solid oxide fuel cells with a stable Sm$_{0.5}$Sr$_{0.5}$Co$_{3-\delta}$–Ce$_{0.8}$Sm$_{0.2}$O$_{2-\delta}$ composite cathode. *J. Power Sources* **2010**, *195*, 3155–3158. [CrossRef]

297. Sun, W.; Zhu, Z.; Jiang, Y.; Shi, Z.; Yan, L.; Liu, W. Optimization of BaZr$_{0.1}$Ce$_{0.7}$Y$_{0.2}$O$_{3-\delta}$-based proton-conducting solid oxide fuel cells with a cobalt-free proton-blocking La$_{0.7}$Sr$_{0.3}$FeO$_{3-\delta}$–Ce$_{0.8}$Sm$_{0.2}$O$_{2-\delta}$ composite cathode. *Int. J. Hydrog. Energy* **2011**, *36*, 9956–9966. [CrossRef]

298. Fabbri, E.; Licoccia, S.; Traversa, E.; Wachsman, E.D. Composite cathodes for proton conducting electrolytes. *Fuel Cells* **2009**, *9*, 128–138. [CrossRef]

299. Yang, C.; Xu, Q. A functionally graded cathode for proton-conducting solid oxide fuel cells. *J. Power Sources* **2012**, *212*, 186–191. [CrossRef]

300. Fabbri, E.; Bi, L.; Pergolesi, D.; Traversa, E. High-performance composite cathodes with tailored mixed conductivity for intermediate temperature solid oxide fuel cells using proton conducting electrolytes. *Energy Environ. Sci.* **2011**, *4*, 4984–4993. [CrossRef]

301. Vert, V.B.; Solís, C.; Serra, J.M. Electrochemical properties of PSFC-BCYb composites as cathodes for proton conducting solid oxide fuel cells. *Fuel Cells* **2011**, *11*, 81–90. [CrossRef]

302. Yang, C.; Zhang, X.; Zhao, H.; Shen, Y.; Du, Z.; Zhang, C. Electrochemical properties of BaZr$_{0.1}$Ce$_{0.7}$Y$_{0.1}$Yb$_{0.1}$O$_{3-\delta}$-Nd$_{1.95}$NiO$_{4+\delta}$ composite cathode for protonic ceramic fuel cells. *Int. J. Hydrog. Energy* **2015**, *40*, 2800–2807. [CrossRef]

303. Dailly, J.; Taillades, G.; Ancelin, M.; Pers, P.; Marrony, M. High performing BaCe$_{0.8}$Zr$_{0.1}$Y$_{0.1}$O$_{3-\delta}$-Sm$_{0.5}$Sr$_{0.5}$CoO$_{3-\delta}$ based protonic ceramic fuel cell. *J. Power Sources* **2017**, *361*, 221–226. [CrossRef]

304. Li, G.; Zhang, Y.; Ling, Y.; He, B.; Xu, J.; Zhao, L. Probing novel triple phase conducting composite cathode for high performance protonic ceramic fuel cells. *Int. J. Hydrog. Energy* **2016**, *41*, 5074–5083. [CrossRef]

305. Bausá, N.; Solís, C.; Strandbakke, R.; Serra, J.M. Development of composite steam electrodes for electrolyzers based on barium zirconate. *Solid State Ion.* **2017**, *306*, 62–68. [CrossRef]

306. Li, H.; Chen, X.; Chen, S.; Wu, Y.; Xie, K. Composite manganate oxygen electrode enhanced with iron oxide nanocatalyst for high temperature steam electrolysis in a proton-conducting solid oxide electrolyzer. *Int. J. Hydrog. Energy* **2015**, *40*, 7920–7931. [CrossRef]

307. Ding, H.; Sullivan, N.P.; Ricote, S. Double perovskite Ba$_2$FeMoO$_{6-\delta}$ as fuel electrode for protonic-ceramic membranes. *Solid State Ion.* **2017**, *306*, 97–103. [CrossRef]

308. Robinson, S.; Manerbino, A.; Coors, W.G. Galvanic hydrogen pumping in the protonic ceramic perovskite BaCe$_{0.2}$Zr$_{0.7}$Y$_{0.1}$O$_{3-\delta}$. *J. Membr. Sci.* **2013**, *446*, 99–105. [CrossRef]

309. Kyriakou, V.; Garagounis, I.; Vourros, A.; Vasileiou, E.; Manerbino, A.; Coors, W.G.; Stoukides, M. Methane steam reforming at low temperatures in a $BaZr_{0.7}Ce_{0.2}Y_{0.1}O_{2.9}$ proton conducting membrane reactor. *Appl. Catal. B Environ.* **2016**, *186*, 1–9. [CrossRef]

310. Shen, C.T.; Lee, Y.H.; Xie, K.; Yen, C.P.; Jhuang, J.W.; Lee, K.R.; Lee, S.W.; Tseng, C.J. Correlation between microstructure and catalytic and mechanical properties during redox cycling for Ni-BCY and Ni-BCZY composites. *Ceram. Int.* **2017**, *43*, S671–S674. [CrossRef]

311. Nasani, N.; Ramasamy, D.; Antunes, I.; Perez, J.; Fagg, D.P. Electrochemical behaviour of Ni-BZO and Ni-BZY cermet anodes for Protonic Ceramic Fuel Cells (PCFCs)—A comparative study. *Electrochim. Acta* **2015**, *154*, 387–396. [CrossRef]

312. Nasani, N.; Ramasamy, D.; Brandão, A.D.; Yaremchenko, A.A.; Fagg, D.P. The impact of porosity, pH_2 and pH_2O on the polarisation resistance of $Ni–BaZr_{0.85}Y_{0.15}O_{3-\delta}$ cermet anodes for Protonic Ceramic Fuel Cells (PCFCs). *Int. J. Hydrog. Energy* **2014**, *39*, 21231–21241. [CrossRef]

313. Pikalova, E.; Medvedev, D. Effect of anode gas mixture humidification on the electrochemical performance of the $BaCeO_3$-based protonic ceramic fuel cell. *Int. J. Hydrog. Energy* **2016**, *41*, 4016–4025. [CrossRef]

314. Park, Y.E.; Ji, H.I.; Kim, B.K.; Lee, J.H.; Lee, H.W.; Park, J.S. Pore structure improvement in cermet for anode-supported protonic ceramic fuel cells. *Ceram. Int.* **2013**, *39*, 2581–2587. [CrossRef]

315. Taillades, G.; Pers, P.; Mao, V.; Taillades, M. High performance anode-supported proton ceramic fuel cell elaborated by wet powder spraying. *Int. J. Hydrog. Energy* **2016**, *41*, 12330–12336. [CrossRef]

316. Li, G.; Jin, H.; Cui, Y.; Gui, L.; He, B.; Zhao, L. Application of a novel $(Pr_{0.9}La_{0.1})_2(Ni_{0.74}Cu_{0.21}Nb_{0.05})O_{4+\delta}$-infiltrated $BaZr_{0.1}Ce_{0.7}Y_{0.2}O_{3-\delta}$ cathode for high performance protonic ceramic fuel cells. *J. Power Sources* **2017**, *341*, 192–198. [CrossRef]

317. Ricote, S.; Bonanos, N.; Lenrick, F.; Wallenberg, R. $LaCoO_3$: Promising cathode material for protonic ceramic fuel cells based on $BaCe_{0.2}Zr_{0.7}Y_{0.1}O_{3-delta}$ electrolyte. *J. Power Sources* **2012**, *218*, 313–319. [CrossRef]

318. Babiniec, S.M.; Ricote, S.; Sullivan, N.P. Characterization of ionic transport through $BaCe_{0.2}Zr_{0.7}Y_{0.1}O_{3-\delta}$ membranes in galvanic and electrolytic operation. *Int. J. Hydrog. Energy* **2015**, *40*, 9278–9286. [CrossRef]

319. Strandbakke, R.; Vøllestad, E.; Robinson, S.A.; Fontaine, M.L.; Norby, T. $Ba_{0.5}Gd_{0.8}La_{0.7}Co_2O_{6-\delta}$ Infiltrated in Porous $BaZr_{0.7}Ce_{0.2}Y_{0.1}O_3$ Backbones as Electrode Material for Proton Ceramic Electrolytes. *J. Electrochem. Soc.* **2017**, *164*, F196–F202. [CrossRef]

320. Song, S.H.; Yoon, S.E.; Choi, J.; Kim, B.K.; Park, J.S. A high-performance ceramic composite anode for protonic ceramic fuel cells based on lanthanum strontium vanadate. *Int. J. Hydrog. Energy* **2014**, *39*, 16534–16540. [CrossRef]

321. Lapina, A.; Chatzichristodoulou, C.; Holtappels, P.; Mogensen, M. Composite $Fe-BaCe_{0.2}Zr_{0.6}Y_{0.2}O_{2.9}$ Anodes for Proton Conductor Fuel Cells. *J. Electrochem. Soc.* **2014**, *161*, F833–F837. [CrossRef]

322. Miyazaki, K.; Okanishi, T.; Muroyama, H.; Matsui, T.; Eguchi, K. Development of Ni–Ba(Zr,Y)O_3 cermet anodes for direct ammonia-fueled solid oxide fuel cells. *J. Power Sources* **2017**, *365*, 148–154. [CrossRef]

323. Rioja-Monllor, L. In Situ Exsolution Synthesis of Composite Cathodes for Protonic Ceramic Fuel Cells. Ph.D. Thesis, Norwegian University of Science and Technology, Trondheim, Norway, 2018.

324. Lee, K.T.; Manthiram, A. Comparison of $Ln_{0.6}Sr_{0.4}CoO_{3-\delta}$ (Ln = La, Pr, Nd, Sm, and Gd) as Cathode Materials for Intermediate Temperature Solid Oxide Fuel Cells. *J. Electrochem. Soc.* **2006**, *153*, A794–A798. [CrossRef]

325. Taguchi, H.; Komatsu, T.; Chiba, R.; Nozawa, K.; Orui, H.; Arai, H. Characterization of $LaNi_xCo_yFe_{1-x-y}O_3$ as a cathode material for solid oxide fuel cells. *Solid State Ion.* **2011**, *182*, 127–132. [CrossRef]

326. Tietz, F.; Arul Raj, I.; Zahid, M.; Stöver, D. Electrical conductivity and thermal expansion of $La_{0.8}Sr_{0.2}(Mn,Fe,Co)O_{3-\delta}$ perovskites. *Solid State Ion.* **2006**, *177*, 1753–1756. [CrossRef]

327. Petric, A.; Huang, P.; Tietz, F. Evaluation of La–Sr–Co–Fe–O perovskites for solid oxide fuel cells and gas separation membranes. *Solid State Ion.* **2000**, *135*, 719–725. [CrossRef]

328. Tai, L.W.; Nasrallah, M.M.; Anderson, H.U.; Sparlin, D.M.; Sehlin, S.R. Structure and electrical properties of $La_{1-x}Sr_xCo_{1-y}Fe_yO_3$. Part 1. The system $La_{0.8}Sr_{0.2}Co_{1-y}Fe_yO_3$. *Solid State Ion.* **1995**, *76*, 259–271. [CrossRef]

329. Pelosato, R.; Cordaro, G.; Stucchi, D.; Cristiani, C.; Dotelli, G. Cobalt based layered perovskites as cathode material for intermediate temperature Solid Oxide Fuel Cells: A brief review. *J. Power Sources* **2015**, *298*, 46–67. [CrossRef]

330. Rath, M.K.; Lee, K.T. Investigation of aliovalent transition metal doped $La_{0.7}Ca_{0.3}Cr_{0.8}X_{0.2}O_{3-\delta}$ (X = Ti, Mn, Fe, Co, and Ni) as electrode materials for symmetric solid oxide fuel cells. *Ceram. Int.* **2015**, *41*, 10878–10890. [CrossRef]

331. Wei, B.; Lü, Z.; Jia, D.; Huang, X.; Zhang, Y.; Su, W. Thermal expansion and electrochemical properties of Ni-doped $GdBaCo_2O_{5+\delta}$ double-perovskite type oxides. *Int. J. Hydrog. Energy* **2010**, *35*, 3775–3782. [CrossRef]

332. Kharton, V.; Naumovich, E.; Kovalevsky, A.; Viskup, A.; Figueiredo, F.; Bashmakov, I.; Marques, F.M. Mixed electronic and ionic conductivity of LaCo(M)O$_3$ (M = Ga, Cr, Fe or Ni): IV. Effect of preparation method on oxygen transport in $LaCoO_{3-\delta}$. *Solid State Ion.* **2000**, *138*, 135–148. [CrossRef]

333. Radaelli, P.G.; Cheong, S.W. Structural phenomena associated with the spin-state transition in LaCoO$_3$. *Phys. Rev. B* **2002**, *66*, 094408. [CrossRef]

334. Zobel, C.; Kriener, M.; Bruns, D.; Baier, J.; Grüninger, M.; Lorenz, T.; Reutler, P.; Revcolevschi, A. Evidence for a low-spin to intermediate-spin state transition in (formula presented). *Phys. Rev. B Condens. Matter Mater. Phys.* **2002**, *66*, 1–4. [CrossRef]

335. Ullmann, H.; Trofimenko, N.; Tietz, F.; Stöver, D.; Ahmad-Khanlou, A. Correlation between thermal expansion and oxide ion transport in mixed conducting perovskite-type oxides for SOFC cathodes. *Solid State Ion.* **2000**, *138*, 79–90. [CrossRef]

336. Thommy, L.; Joubert, O.; Hamon, J.; Caldes, M.T. Impregnation versus exsolution: Using metal catalysts to improve electrocatalytic properties of LSCM-based anodes operating at 600 °C. *Int. J. Hydrog. Energy* **2016**, *41*, 14207–14216. [CrossRef]

337. Jiang, Z.; Xia, C.; Chen, F. Nano-structured composite cathodes for intermediate-temperature solid oxide fuel cells via an infiltration/impregnation technique. *Electrochim. Acta* **2010**, *55*, 3595–3605. [CrossRef]

338. Li, G.; He, B.; Ling, Y.; Xu, J.; Zhao, L. Highly active YSB infiltrated LSCF cathode for proton conducting solid oxide fuel cells. *Int. J. Hydrog. Energy* **2015**, *40*, 13576–13582. [CrossRef]

339. Tucker, M.C. Progress in metal-supported solid oxide fuel cells: A review. *J. Power Sources* **2010**, *195*, 4570–4582. [CrossRef]

340. Kim, J.-H.; Manthiram, A. $LnBaCo_2O_{5+\delta}$ Oxides as Cathodes for Intermediate-Temperature Solid Oxide Fuel Cells. *J. Electrochem. Soc.* **2008**, *155*, B385–B390. [CrossRef]

341. Riza, F.; Ftikos, C.; Tietz, F.; Fischer, W. Preparation and characterization of $Ln_{0.8}Sr_{0.2}Fe_{0.8}Co_{0.2}O_{3-\delta}$ (Ln = La, Pr, Nd, Sm, Eu, Gd). *J. Eur. Ceram. Soc.* **2001**, *21*, 1769–1773. [CrossRef]

342. Klyndyuk, A.I. Thermal and chemical expansion of $LnBaCuFeO_{5+\delta}$ (Ln = La, Pr, Gd) ferrocuprates and $LaBa_{0.75}Sr_{0.25}CuFeO_{5+\delta}$ solid solution. *Russ. J. Inorg. Chem.* **2007**, *52*, 1343–1349. [CrossRef]

343. Klyndyuk, A.I.; Chizhova, E.A. Properties of perovskite-like phases $LnBaCuFeO_{5+\delta}$ (Ln = La, Pr). *Glass Phys. Chem.* **2008**, *34*, 313–318. [CrossRef]

344. Jin, F.; Xu, H.; Long, W.; Shen, Y.; He, T. Characterization and evaluation of double perovskites $LnBaCoFeO_{5+\delta}$ (Ln = Pr and Nd) as intermediate-temperature solid oxide fuel cell cathodes. *J. Power Sources* **2013**, *243*, 10–18. [CrossRef]

345. Che, X.; Shen, Y.; Li, H.; He, T. Assessment of $LnBaCo_{1.6}Ni_{0.4}O_{5+\delta}$ (Ln = Pr, Nd, and Sm) double-perovskites as cathodes for intermediate-temperature solid-oxide fuel cells. *J. Power Sources* **2013**, *222*, 288–293. [CrossRef]

346. Mori, M.; Hiei, Y.; Sammes, N.M.; Tompsett, G.A. Thermal-Expansion Behaviors and Mechanisms for Ca- or Sr-Doped Lanthanum Manganite Perovskites under Oxidizing Atmospheres. *J. Electrochem. Soc.* **2000**, *147*, 1295–1302. [CrossRef]

347. Shao, Z.; Yang, W.; Cong, Y.; Dong, H.; Tong, J.; Xiong, G. Investigation of the permeation behavior and stability of a $Ba_{0.5}Sr_{0.5}Co_{0.8}Fe_{0.2}O_{3-\delta}$ oxygen membrane. *J. Membr. Sci.* **2000**, *172*, 177–188. [CrossRef]

348. Shao, Z.; Dong, H.; Xiong, G.; Cong, Y.; Yang, W. Performance of a mixed-conducting ceramic membrane reactor with high oxygen permeability for methane conversion. *J. Membr. Sci.* **2001**, *183*, 181–192. [CrossRef]

349. Shao, Z.; Haile, S.M. A high-performance cathode for the next generation of solid-oxide fuel cells. *Nature* **2004**, *431*, 170–173. [CrossRef] [PubMed]

350. Lin, Y.; Ran, R.; Zheng, Y.; Shao, Z.; Jin, W.; Xu, N.; Ahn, J. Evaluation of $Ba_{0.5}Sr_{0.5}Co_{0.8}Fe_{0.2}O_{3-\delta}$ as a potential cathode for an anode-supported proton-conducting solid-oxide fuel cell. *J. Power Sources* **2008**, *180*, 15–22. [CrossRef]

351. Patra, H.; Rout, S.K.; Pratihar, S.K.; Bhattacharya, S. Thermal, electrical and electrochemical characteristics of $Ba_{1-x}Sr_xCo_{0.8}Fe_{0.2}O_{3-\delta}$ cathode material for intermediate temperature solid oxide fuel cells. *Int. J. Hydrog. Energy* **2011**, *36*, 11904–11913. [CrossRef]

352. Wei, B.; Lü, Z.; Huang, X.; Miao, J.; Sha, X.; Xin, X.; Su, W. Crystal structure, thermal expansion and electrical conductivity of perovskite oxides $Ba_xSr_{1-x}Co_{0.8}Fe_{0.2}O_{3-\delta}$ ($0.3 \leq x \leq 0.7$). *J. Eur. Ceram. Soc.* **2006**, *26*, 2827–2832. [CrossRef]

353. McIntosh, S.; Vente, J.F.; Haije, W.G.; Blank, D.H.A.; Bouwmeester, H.J.M. Oxygen Stoichiometry and Chemical Expansion of $Ba_{0.5}Sr_{0.5}Co_{0.8}Fe_{0.2}O_{3-\delta}$ Measured by in Situ Neutron Diffraction. *Chem. Mater.* **2006**, *18*, 2187–2193. [CrossRef]

354. Kriegel, R.; Kircheisen, R.; Töpfer, J. Oxygen stoichiometry and expansion behavior of $Ba_{0.5}Sr_{0.5}Co_{0.8}Fe_{0.2}O_{3-\delta}$. *Solid State Ion.* **2010**, *181*, 64–70. [CrossRef]

355. Zhu, Q.; Jin, T.; Wang, Y. Thermal expansion behavior and chemical compatibility of $Ba_xSr_{1-x}Co_{1-y}Fe_yO_{3-\delta}$ with 8YSZ and 20GDC. *Solid State Ion.* **2006**, *177*, 1199–1204. [CrossRef]

356. Wang, H.; Tablet, C.; Feldhoff, A.; Caro, J. Investigation of phase structure, sintering, and permeability of perovskite-type $Ba_{0.5}Sr_{0.5}Co_{0.8}Fe_{0.2}O_{3-\delta}$ membranes. *J. Membr. Sci.* **2005**, *262*, 20–26. [CrossRef]

357. Hwang, H.J.; Moon, J.W.; Lee, S.; Lee, E.A. Electrochemical performance of LSCF-based composite cathodes for intermediate temperature SOFCs. *J. Power Sources* **2005**, *145*, 243–248. [CrossRef]

358. Richter, J.; Holtappels, P.; Graule, T.; Nakamura, T.; Gauckler, L.J. Materials design for perovskite SOFC cathodes. *Monatshefte für Chemie* **2009**, *140*, 985–999. [CrossRef]

359. Tai, L.W.; Nasrallah, M.M.; Anderson, H.U.; Sparlin, D.M.; Sehlin, S.R. Structure and electrical properties of $La_{1-x}Sr_xCo_{1-y}Fe_yO_3$. Part 2. The system $La_{1-x}Sr_xCo_{0.2}Fe_{0.8}O_3$. *Solid State Ion.* **1995**, *76*, 273–283. [CrossRef]

360. Xu, Q.; Huang, D.; Zhang, F.; Chen, W.; Chen, M.; Liu, H. Structure, electrical conducting and thermal expansion properties of $La_{0.6}Sr_{0.4}Co_{0.8}Fe_{0.2}O_{3-\delta}$–$Ce_{0.8}Sm_{0.2}O_{2-\delta}$ composite cathodes. *J. Alloys Compd.* **2008**, *454*, 460–465. [CrossRef]

361. Fan, B.; Yan, J.; Yan, X. The ionic conductivity, thermal expansion behavior, and chemical compatibility of $La_{0.54}Sr_{0.44}Co_{0.2}Fe_{0.8}O_{3-\delta}$ as SOFC cathode material. *Solid State Sci.* **2011**, *13*, 1835–1839. [CrossRef]

362. Kim, S.; Kim, S.H.; Lee, K.S.; Yu, J.H.; Seong, Y.H.; Han, I.S. Mechanical properties of LSCF ($La_{0.6}Sr_{0.4}Co_{0.2}Fe_{0.8}O_{3-\delta}$)–GDC ($Ce_{0.9}Gd_{0.1}O_{2-\delta}$) for oxygen transport membranes. *Ceram. Int.* **2017**, *43*, 1916–1921. [CrossRef]

363. Kostogloudis, G.C.; Ftikos, C. Properties of A-site-deficient $La_{0.6}Sr_{0.4}Co_{0.2}Fe_{0.8}O_{3-\delta}$-based perovskite oxides. *Solid State Ion.* **1999**, *126*, 143–151. [CrossRef]

364. Kharton, V.V.; Yaremchenko, A.A.; Patrakeev, M.V.; Naumovich, E.N.; Marques, F.M.B. Thermal and chemical induced expansion of $La_{0.3}Sr_{0.7}(Fe,Ga)O_{3-\delta}$ ceramics. *J. Eur. Ceram. Soc.* **2003**, *23*, 1417–1426. [CrossRef]

365. Wang, S.; Katsuki, M.; Dokiya, M.; Hashimoto, T. High temperature properties of $La_{0.6}Sr_{0.4}Co_{0.8}Fe_{0.2}O_{3-\delta}$ phase structure and electrical conductivity. *Solid State Ion.* **2003**, *159*, 71–78. [CrossRef]

366. Duan, C.; Hook, D.; Chen, Y.; Tong, J.; O'Hayre, R. Zr and Y co-doped perovskite as a stable, high performance cathode for solid oxide fuel cells operating below 500 °C. *Energy Environ. Sci.* **2017**, *10*, 176–182. [CrossRef]

367. Wang, F.; Yan, D.; Zhang, W.; Chi, B.; Pu, J.; Jian, L. $LaCo_{0.6}Ni_{0.4}O_{3-\delta}$ as cathode contact material for intermediate temperature solid oxide fuel cells. *Int. J. Hydrog. Energy* **2013**, *38*, 646–651. [CrossRef]

368. Kharton, V.; Viskup, A.P.; Bochkoc, D.M.; Naumovich, E.N.; Reut, O.P. Mixed electronic and ionic conductivity of $LaCo(M)O_3$ (M = Ga, Cr, Fe or Ni) III. Diffusion of oxygen through $LaCo_{1-x-y}Fe_xNi_yO_3$ ceramics. *Solid State Ion.* **1998**, *110*, 61–68. [CrossRef]

369. Chiba, R.; Yoshimura, F.; Sakurai, Y. An investigation of $LaNi_{1-x}Fe_xO_3$ as a cathode material for solid oxide fuel cells. *Solid State Ion.* **1999**, *124*, 281–288. [CrossRef]

370. Grande, T.; Tolchard, J.R.; Selbach, S.M. Anisotropic Thermal and Chemical Expansion in Sr-Substituted $LaMnO_{3+\delta}$: Implications for Chemical Strain Relaxation. *Chem. Mater.* **2012**, *24*, 338–345. [CrossRef]

371. Huang, Y.; Dass, R.; Xing, Z.; Goodenough, J. Double perovskites as anode materials for solid-oxide fuel cells. *Science* **2006**, *312*, 254–257. [CrossRef] [PubMed]

372. Wei, T.; Zhang, Q.; Huang, Y.H.; Goodenough, J.B. Cobalt-based double-perovskite symmetrical electrodes with low thermal expansion for solid oxidefuel cells. *J. Mater. Chem.* **2012**, *22*, 225–231. [CrossRef]

373. Cherepanov, V.A.; Aksenova, T.V.; Gavrilova, L.Y.; Mikhaleva, K.N. Structure, nonstoichiometry and thermal expansion of the $NdBa(Co,Fe)_2O_{5+\delta}$ layered perovskite. *Solid State Ion.* **2011**, *188*, 53–57. [CrossRef]

374. West, M.; Manthiram, A. Layered LnBa$_{1-x}$Sr$_x$CoCuO$_{5+\delta}$ (Ln = Nd and Gd) perovskite cathodes for intermediate temperature solid oxide fuel cells. *Int. J. Hydrog. Energy* **2013**, *38*, 3364–3372. [CrossRef]

375. Kim, J.H.; Irvine, J.T.S. Characterization of layered perovskite oxides NdBa$_{1-x}$Sr$_x$Co$_2$O$_{5+\delta}$ (x = 0 and 0.5) as cathode materials for IT-SOFC. *Int. J. Hydrog. Energy* **2012**, *37*, 5920–5929. [CrossRef]

376. Li, X.; Jiang, X.; Xu, H.; Xu, Q.; Jiang, L.; Shi, Y.; Zhang, Q. Scandium-doped PrBaCo$_{2-x}$Sc$_x$O$_{6-\delta}$ oxides as cathode material for intermediate-temperature solid oxide fuel cells. *Int. J. Hydrog. Energy* **2013**, *38*, 12035–12042. [CrossRef]

377. Kim, J.; Choi, S.; Park, S.; Kim, C.; Shin, J.; Kim, G. Effect of Mn on the electrochemical properties of a layered perovskite NdBa$_{0.5}$Sr$_{0.5}$Co$_{2-x}$Mn$_x$O$_{5+\delta}$ (x = 0, 0.25, and 0.5) for intermediate-temperature solid oxide fuel cells. *Electrochim. Acta* **2013**, *112*, 712–718. [CrossRef]

378. Wang, Y.; Zhao, X.; Lü, S.; Meng, X.; Zhang, Y.; Yu, B.; Li, X.; Sui, Y.; Yang, J.; Fu, C.; et al. Synthesis and characterization of SmSrCo$_{2-x}$Mn$_x$O$_{5+\delta}$ (x = 0.0, 0.2, 0.4, 0.6, 0.8, 1.0) cathode materials for intermediate-temperature solid-oxide fuel cells. *Ceram. Int.* **2014**, *40*, 11343–11350. [CrossRef]

379. Phillipps, M.B.; Sammes, N.M.; Yamamoto, O. Gd$_{1-x}$A$_x$Co$_{1-y}$Mn$_y$O$_3$ (A = Sr, Ca) as a cathode for the SOFC. *Solid State Ion.* **1999**, *123*, 131–138. [CrossRef]

380. Liu, L.; Guo, R.; Wang, S.; Yang, Y.; Yin, D. Synthesis and characterization of PrBa$_{0.5}$Sr$_{0.5}$Co$_{2-x}$Ni$_x$O$_{5+\delta}$ (x = 0.1, 0.2 and 0.3) cathodes for intermediate temperature SOFCs. *Ceram. Int.* **2014**, *40*, 16393–16398. [CrossRef]

381. Jo, S.H.; Muralidharan, P.; Kim, D.K. Enhancement of electrochemical performance and thermal compatibility of GdBaCo$_{2/3}$Fe$_{2/3}$Cu$_{2/3}$O$_{5+\delta}$ cathode on Ce$_{1.9}$Gd$_{0.1}$O$_{1.95}$ electrolyte for IT-SOFCs. *Electrochem. Commun.* **2009**, *11*, 2085–2088. [CrossRef]

382. Zhang, Y.; Yu, B.; Lü, S.; Meng, X.; Zhao, X.; Ji, Y.; Wang, Y.; Fu, C.; Liu, X.; Li, X.; et al. Effect of Cu doping on YBaCo$_2$O$_{5+\delta}$ as cathode for intermediate-temperature solid oxide fuel cells. *Electrochim. Acta* **2014**, *134*, 107–115. [CrossRef]

383. Jin, F.; Shen, Y.; Wang, R.; He, T. Double-perovskite PrBaCo$_{2/3}$Fe$_{2/3}$Cu$_{2/3}$O$_{5+\delta}$ as cathode material for intermediate-temperature solid-oxide fuel cells. *J. Power Sources* **2013**, *234*, 244–251. [CrossRef]

384. Jiang, L.; Wei, T.; Zeng, R.; Zhang, W.X.; Huang, Y.H. Thermal and electrochemical properties of PrBa$_{0.5}$Sr$_{0.5}$Co$_{2-x}$Fe$_x$O$_{5+\delta}$ (x = 0.5, 1.0, 1.5) cathode materials for solid-oxide fuel cells. *J. Power Sources* **2013**, *232*, 279–285. [CrossRef]

385. Zhao, L.; Shen, J.; He, B.; Chen, F.; Xia, C. Synthesis, characterization and evaluation of PrBaCo$_{2-x}$Fe$_x$O$_{5+\delta}$ as cathodes for intermediate-temperature solid oxide fuel cells. *Int. J. Hydrog. Energy* **2011**, *36*, 3658–3665. [CrossRef]

386. Ni, D.W.; Charlas, B.; Kwok, K.; Molla, T.T.; Hendriksen, P.V.; Frandsen, H.L. Influence of temperature and atmosphere on the strength and elastic modulus of solid oxide fuel cell anode supports. *J. Power Sources* **2016**, *311*, 1–12. [CrossRef]

387. Kim, Y.N.; Kim, J.H.; Manthiram, A. Effect of Fe substitution on the structure and properties of LnBaCo$_{2-x}$Fe$_x$O$_{5+\delta}$ (Ln = Nd and Gd) cathodes. *J. Power Sources* **2010**, *195*, 6411–6419. [CrossRef]

388. Volkova, N.E.; Gavrilova, L.Y.; Cherepanov, V.A.; Aksenova, T.V.; Kolotygin, V.A.; Kharton, V.V. Synthesis, crystal structure and properties of SmBaCo$_{2-x}$Fe$_x$O$_{5+\delta}$. *J. Solid State Chem.* **2013**, *204*, 219–223. [CrossRef]

389. Xue, J.; Shen, Y.; He, T. Double-perovskites YBaCo$_{2-x}$Fe$_x$O$_{5+\delta}$ cathodes for intermediate-temperature solid oxide fuel cells. *J. Power Sources* **2011**, *196*, 3729–3735. [CrossRef]

390. Zhou, Q.; Wei, W.C.J.; Guo, Y.; Jia, D. LaSrMnCoO$_{5+\delta}$ as cathode for intermediate-temperature solid oxide fuel cells. *Electrochem. Commun.* **2012**, *19*, 36–38. [CrossRef]

391. Aksenova, T.V.; Gavrilova, L.Y.; Yaremchenko, A.A.; Cherepanov, V.A.; Kharton, V.V. Oxygen nonstoichiometry, thermal expansion and high-temperature electrical properties of layered NdBaCo$_2$O$_{5+\delta}$ and SmBaCo$_2$O$_{5+\delta}$. *Mater. Res. Bull.* **2010**, *45*, 1288–1292. [CrossRef]

392. Zhao, H.; Zheng, Y.; Yang, C.; Shen, Y.; Du, Z.; Świerczek, K. Electrochemical performance of Pr$_{1-x}$Y$_x$BaCo$_2$O$_{5+\delta}$ layered perovskites as cathode materials for intermediate-temperature solid oxide fuel cells. *Int. J. Hydrog. Energy* **2013**, *38*, 16365–16372. [CrossRef]

393. Zhao, L.; Nian, Q.; He, B.; Lin, B.; Ding, H.; Wang, S.; Peng, R.; Meng, G.; Liu, X. Novel layered perovskite oxide PrBaCuCoO$_{5+\delta}$ as a potential cathode for intermediate-temperature solid oxide fuel cells. *J. Power Sources* **2010**, *195*, 453–456. [CrossRef]

394. Mogni, L.; Prado, F.; Jiménez, C.; Caneiro, A. Oxygen order-disorder phase transition in layered GdBaCo$_2$O$_{5+\delta}$ perovskite: Thermodynamic and transport properties. *Solid State Ion.* **2013**, *240*, 19–28. [CrossRef]

395. Kuroda, C.; Zheng, K.; Świerczek, K. Characterization of novel GdBa$_{0.5}$Sr$_{0.5}$Co$_{2-x}$Fe$_x$O$_{5+\delta}$ perovskites for application in IT-SOFC cells. *Int. J. Hydrog. Energy* **2013**, *38*, 1027–1038. [CrossRef]

396. Tarancón, A.; Burriel, M.; Santiso, J.; Skinner, S.J.; Kilner, J.A. Advances in layered oxide cathodes for intermediate temperature solid oxide fuel cells. *J. Mater. Chem.* **2010**, *20*, 3799–3813. [CrossRef]

397. Hu, Y.; Bouffanais, Y.; Almar, L.; Morata, A.; Tarancon, A.; Dezanneau, G. La$_{2-x}$Sr$_x$CoO$_{4-\delta}$ (x = 0.9, 1.0, 1.1) Ruddlesden-Popper-type layered cobaltites as cathode materials for IT-SOFC application. *Int. J. Hydrog. Energy* **2013**, *38*, 3064–3072. [CrossRef]

398. Prado, F.; Mogni, L.; Cuello, G.J.; Caneiro, A. Neutron powder diffraction study at high temperature of the Ruddlesden-Popper phase Sr$_3$Fe$_2$O$_{6+\delta}$. *Solid State Ion.* **2007**, *178*, 77–82. [CrossRef]

399. Flura, A.; Dru, S.; Nicollet, C.; Vibhu, V.; Fourcade, S.; Lebraud, E.; Rougier, A.; Bassat, J.M.; Grenier, J.C. Chemical and structural changes in Ln$_2$NiO$_{4+\delta}$ (Ln = La, Pr or Nd) lanthanide nickelates as a function of oxygen partial pressure at high temperature. *J. Solid State Chem.* **2015**, *228*, 189–198. [CrossRef]

400. Amow, G.; Skinner, S.J. Recent developments in Ruddlesden-Popper nickelate systems for solid oxide fuel cell cathodes. *J. Solid State Electrochem.* **2006**, *10*, 538–546. [CrossRef]

401. Amow, G.; Davidson, I.J.; Skinner, S.J. A comparative study of the Ruddlesden-Popper series, La$_{n+1}$Ni$_n$O$_{3n+1}$ (n = 1, 2 and 3), for solid-oxide fuel-cell cathode applications. *Solid State Ion.* **2006**, *177*, 1205–1210. [CrossRef]

402. Skinner, S.J.; Kilner, J.A. Oxygen diffusion and surface exchange in La$_{2-x}$Sr$_x$NiO$_{4+\delta}$. *Solid State Ion.* **2000**, *135*, 709–712. [CrossRef]

403. Boehm, E.; Bassat, J.M.; Steil, M.C.; Dordor, P.; Mauvy, F.; Grenier, J.C. Oxygen transport properties of La$_2$Ni$_{1-x}$Cu$_x$O$_{4+\delta}$ mixed conducting oxides. *Solid State Sci.* **2003**, *5*, 973–981. [CrossRef]

404. Vibhu, V.; Rougier, A.; Nicollet, C.; Flura, A.; Grenier, J.C.; Bassat, J.M. La$_{2-x}$Pr$_x$NiO$_{4+\delta}$ as suitable cathodes for metal supported SOFCs. *Solid State Ion.* **2015**, *278*, 32–37. [CrossRef]

405. Kharton, V.V.; Kovalevsky, A.V.; Avdeev, M.; Tsipis, E.V.; Patrakeev, M.V.; Yaremchenko, A.A.; Naumovich, E.N.; Frade, J.R. Chemically induced expansion of La$_2$NiO$_{4+\delta}$-based materials. *Chem. Mater.* **2007**, *19*, 2027–2033. [CrossRef]

406. Boehm, E.; Bassat, J.M.; Dordor, P.; Mauvy, F.; Grenier, J.C.; Stevens, P. Oxygen diffusion and transport properties in non-stoichiometric Ln$_{2-x}$NiO$_{4+\delta}$ oxides. *Solid State Ion.* **2005**, *176*, 2717–2725. [CrossRef]

407. Lee, S.; Lee, K.; Jang, Y.H.; Bae, J. Fabrication of solid oxide fuel cells (SOFCs) by solvent-controlled co-tape casting technique. *Int. J. Hydrog. Energy* **2017**, *42*, 1648–1660. [CrossRef]

408. Dailly, J.; Ancelin, M.; Marrony, M. Long term testing of BCZY-based protonic ceramic fuel cell PCFC: Micro-generation profile and reversible production of hydrogen and electricity. *Solid State Ion.* **2017**, *306*, 69–75. [CrossRef]

409. Mahata, T.; Nair, S.R.; Lenka, R.K.; Sinha, P.K. Fabrication of Ni-YSZ anode supported tubular SOFC through iso-pressing and co-firing route. *Int. J. Hydrog. Energy* **2012**, *37*, 3874–3882. [CrossRef]

410. Ramousse, S.; Menon, M.; Brodersen, K.; Knudsen, J.; Rahbek, U.; Larsen, P.H. Manufacturing of anode-supported SOFC's: Processing parameters and their influence. *ECS Trans.* **2007**, *7*, 317–327. [CrossRef]

411. Majumdar, S.; Claar, T.; Flandermeyer, B. Stress and Fracture Behavior of Monolithic Fuel Cell Tapes. *J. Am. Ceram. Soc.* **1986**, *69*, 628–633. [CrossRef]

412. Prakash, B.S.; Kumar, S.S.; Aruna, S.T. Properties and development of Ni/YSZ as an anode material in solid oxide fuel cell: A review. *Renew. Sustain. Energy Rev.* **2014**, *36*, 149–179. [CrossRef]

413. Nasani, N.; Wang, Z.J.; Willinger, M.G.; Yaremchenko, A.A.; Fagg, D.P. In-situ redox cycling behaviour of NiBaZr$_{0.85}$Y$_{0.15}$O$_{3-\delta}$ cermet anodes for Protonic Ceramic Fuel Cells. *Int. J. Hydrog. Energy* **2014**, *39*, 19780–19788. [CrossRef]

414. Onishi, T.; Han, D.; Noda, Y.; Hatada, N.; Majima, M.; Uda, T. Evaluation of performance and durability of Ni-BZY cermet electrodes with BZY electrolyte. *Solid State Ion.* **2018**, *317*, 127–135. [CrossRef]

415. Sažinas, R.; Einarsrud, M.A.; Grande, T. Toughening of Y-doped BaZrO$_3$ proton conducting electrolytes by hydration. *J. Mater. Chem. A* **2017**, *5*, 5846–5857. [CrossRef]

416. Zhang, Y.; Xia, C. A durability model for solid oxide fuel cell electrodes in thermal cycle processes. *J. Power Sources* **2010**, *195*, 6611–6618. [CrossRef]

417. Lee, K.T.; Vito, N.J.; Wachsman, E.D. Comprehensive quantification of Ni-$Gd_{0.1}Ce_{0.9}O_{1.95}$ anode functional layer microstructures by three-dimensional reconstruction using a FIB/SEM dual beam system. *J. Power Sources* **2013**, *228*, 220–228. [CrossRef]

418. Atkinson, A.; Barnett, S.; Gorte, R.J.; Irvine, J.T.S.; McEvoy, A.J.; Mogensen, M.; Singhal, S.C.; Vohs, J. Advanced anodes for high-temperature fuel cells. *Nat. Mater.* **2004**, *3*, 17–27. [CrossRef] [PubMed]

419. Zhu, W.Z.; Deevi, S.C. A review on the status of anode materials for solid oxide fuel cells. *Mater. Sci. Eng. A* **2003**, *362*, 228–239. [CrossRef]

420. Lee, S.; Park, I.; Lee, H.; Shin, D. Continuously gradient anode functional layer for BCZY based proton-conducting fuel cells. *Int. J. Hydrog. Energy* **2014**, *39*, 14342–14348. [CrossRef]

421. Anandakumar, G.; Li, N.; Verma, A.; Singh, P.; Kim, J.H. Thermal stress and probability of failure analyses of functionally graded solid oxide fuel cells. *J. Power Sources* **2010**, *195*, 6659–6670. [CrossRef]

422. Dippon, M.; Babiniec, S.M.; Ding, H.; Ricote, S.; Sullivan, N.P. Exploring electronic conduction through $BaCe_xZr_{0.9-x}Y_{0.1}O_{3-d}$ proton-conducting ceramics. *Solid State Ion.* **2016**, *286*. [CrossRef]

423. Biswas, S.; Nithyanantham, T.; Nambiappan Thangavel, S.; Bandopadhyay, S. High-temperature mechanical properties of reduced NiO–8YSZ anode-supported bi-layer SOFC structures in ambient air and reducing environments. *Ceram. Int.* **2013**, *39*, 3103–3111. [CrossRef]

424. Patki, N.S.; Manerbino, A.; Way, J.D.; Ricote, S. Galvanic hydrogen pumping performance of copper electrodes fabricated by electroless plating on a $BaZr_{0.9-x}Ce_xY_{0.1}O_{3-\delta}$ proton-conducting ceramic membrane. *Solid State Ion.* **2018**, *317*, 256–262. [CrossRef]

425. Stange, M.; Stefan, E.; Denonville, C.; Larring, Y.; Rørvik, P.M.; Haugsrud, R. Development of novel metal-supported proton ceramic electrolyser cell with thin film BZY15–Ni electrode and BZY15 electrolyte. *Int. J. Hydrog. Energy* **2017**, *42*, 13454–13462. [CrossRef]

MDPI

St. Alban-Anlage 66

4052 Basel

Switzerland

Tel. +41 61 683 77 34

Fax +41 61 302 89 18

www.mdpi.com

Crystals Editorial Office

E-mail: crystals@mdpi.com

www.mdpi.com/journal/crystals

www.ingramcontent.com/pod-product-compliance
Lightning Source LLC
Chambersburg PA
CBHW051856210326
41597CB00033B/5918